Enterprise Service Computing:

From Concept to Deployment

Robin G. Qiu, The Pennsylvania State University, USA

Acquisition Editor:	Kristin Klinger
Senior Managing Editor:	Jennifer Neidig
Managing Editor:	Sara Reed
Assistant Managing Editor:	Sharon Berger
Development Editor:	Kristin Roth
Copy Editor:	Shanelle Ramelb
Typesetter:	Marko Primorac
Cover Design:	Lisa Tosheff
Printed at:	Integrated Book Technology

Published in the United States of America by
Idea Group Publishing (an imprint of Idea Group Inc.)
701 E. Chocolate Avenue
Hershey PA 17033
Tel: 717-533-8845
Fax: 717-533-8661
E-mail: cust@idea-group.com
Web site: http://www.idea-group.com

and in the United Kingdom by
Idea Group Publishing (an imprint of Idea Group Inc.)
3 Henrietta Street
Covent Garden
London WC2E 8LU
Tel: 44 20 7240 0856
Fax: 44 20 7379 3313
Web site: http://www.eurospan.co.uk

Library of Congress Cataloging-in-Publication Data

Enterprise service computing : from concept to deployment / Robin G. Qui, editor.
p. cm.
Summary: "This book focuses on providing readers a comprehensive understanding of the development cycle of enterprise service computing. Covered topics range from concept development, system design, modeling, and development technologies, to final deployment. Both theoretical research results and practical applications are provided"--Provided by publisher.
Includes bibliographical references and index.
ISBN 1-59904-180-4 (hardcover) -- ISBN 1-59904-181-2 (softcover) -- ISBN 1-59904-182-0 (ebook)
1. Service industries--Information technology. 2. Service industries--Technological innovations. 3. Information technology I. Qui, Robin G., 1964-
HD9980.5.E58 2006
658.4'038--dc22
2006027706

British Cataloguing in Publication Data
A Cataloguing in Publication record for this book is available from the British Library.

Enterprise Service Computing:
From Concept to Deployment

Table of Contents

Preface .... **vii**

Section I:
Business Aspects of Enterprise Service Computing

Chapter I
Information Technology as a Service **1**
Robin G. Qiu, The Pennsylvania State University, USA

Chapter II
Aligning Business Processes with Enterprise Service Computing Infrastructure **25**
Wei Zhao, University of Alabama at Birmingham, USA
Jun-Jang Jeng, IBM T.J. Watson Research, USA
Lianjun An, IBM T.J. Watson Research, USA
Fei Cao, University of Alabama at Birmingham, USA
Barret R. Bryant, University of Alabama at Birmingham, USA
Rainer Hauser, IBM Zurich Research, Switzerland
Tao Tao, IBM T.J. Watson Research, USA

Chapter III
Service Portfolio Measurement (SPM): Assessing Finanacial Performance of Service-Oriented Information Systems **58**
Jan vom Brocke, European Reseach Center for Information Systems (ERCIS), University of Münster, Germany

Section II: Enterprise Service Computing: Requirements

Chapter IV
Requirements Engineering for Integrating the Enterprise 92
Raghvinder S. Sangwan, Penn State Great Valley, USA

Chapter V
Mobile Workforce Management in a Service-Oriented Enterprise: Capturing Concepts and Requirements in a Multi-Agent Infrastructure 105
Dickson K.W. Chiu, Dickson Computer Systems, Hong Kong
S. C. Cheung, Hong Kong University of Science and Technology, Hong Kong
Ho-fung Leung, The Chinese University of Hong Kong, Hong Kong

Section III: Enterprise Service Computing: Modeling

Chapter VI
Designing Enterprise Applications Using Model-Driven Service-Oriented Architectures 132
Marten van Sinderen, University of Twente, The Netherlands
João Paulo Andrade Almeida, Telematica Instituut, The Netherlands
Luís Ferreira Pires, University of Twente, The Netherlands
Dick Quartel, University of Twente, The Netherlands

Chapter VII
A Composite Application Model for Building Enterprise Information Systems in a Connected World 156
Jean-Jacques Dubray, SAP Labs, USA

Chapter VIII
Three-Point Service-Oriented Design and Modeling Methodology for Web Services Composition 176
Xiang Gao, Supercom Canada, Canada
Jen-Yao Chung, IBM T.J. Watson Research, USA

Section IV: Enterprise Service Computing: Technologies

Chapter IX
Data Replication Strategies in Wide-Area Distributed Systems..............211
Sushant Goel, University of Melbourne, Australia
Rajkumar Buyya, University of Melbourne, Australia

Chapter X
Web Services vs. ebXML: An Evaluation of Web Services and ebXML for E-Business Applications 242
Yuhong Yan, Canada National Research Council, Canada
Matthias Klein, University of New Brunswick, Canada

Chapter XI
Leveraging Pervasive and Ubiquitous Service Computing...................... 261
Zhijun Zhang, University of Phoenix, USA

Section V: Enterprise Service Computing: Formal Modeling

Chapter XII
A Petri Net-Based Specification Model Towards Verifiable Service Computing 285
Jia Zhang, Northern Illinois University, USA
Carl K. Chang, Iowa State University, USA
Seong W. Kim, Samsung Advanced Institute of Technology, Korea

Chapter XIII
Service Computing for Design and Reconfiguration of Integrated E-Supply Chains.. 322
Mariagrazia Dotoli, Politecnico di Bari, Italy
Maria Pia Fanti, Politecnico di Bari, Italy
Carlo Meloni, Politecnico di Bari, Italy
Mengchu Zhou, New Jersey Institute of Technology, USA

Section VI: Enterprise Service Computing: Best Practices and Deployment

Chapter XIV
Best Practice in Leveraging E-Business Technologies to Achieve Business Agility 356
Ehap H. Sabri, University of Texas at Dallas, USA

Chapter XV
Concepts and Operations of Two Research Projects on Web Services and Context at Zayed University 388
Zakaria Maamar, Zayed University, UAE

About the Authors 408

Index 418

Preface

The developed economy is shifting from being manufacturing based to services based. Different from the traditional manufacturing business, the services business is more complicated and dynamic, and end-user driven rather than product driven. To stay competitive, an enterprise thus has to rethink its business strategies and revamp its operational and organizational structures by taking advantage of its unique engineering expertise and application experience. With the fast industrialization of information technology, services sectors have expanded their territories substantially from traditional commercial transportation, logistics and distribution, health-care delivery, and issuance to financial engineering, e-commerce, e-retailing, e-entertainment (and "e-everything" if possible), supply chains, knowledge transformation and delivery, and services consulting.

Today's market reality is that the consumer or customer demands more innovative and flexible goods and services with high quality and shorter lead times. For a competitive enterprise, unique and satisfactory services differentiate the enterprise from its competitors; on the other hand, a highly satisfactory services delivery indeed drives more product sales. To meet the needs of the service-led economy, as a matter of fact, enterprises are gradually embracing defining and selling anything as a customer value service for competitive advantage (Rosmarin, 2006).

Research and education in services-enterprise engineering to cultivate and empower the ecosystem driven by services, technology, and management have lagged behind when compared to many other areas. Nevertheless, the services and services-enterprise engineering scope have evolved and expanded enormously as the world economy accelerates the pace of globalization. To ensure the delivery of superior services outcomes to end users, it is indispensable to align IT and business goals for efficiency and cost effectiveness (IBM, 2004; Karmarkar, 2004). However, little is really known about how IT as a service can be systematically applied for the delivery of a componentized business (Rangaswamy & Pal, 2005).

Only recently have there been some international initiatives to promote research and education in this emerging field in a comprehensive manner (Cherbakov, Galambos, Harishankar, Kalyana, & Rackham, 2005; IBM, 2004). The U.S. National Science Foundation (NSF, 2006) officially launched the Service Enterprise Engineering (SEE) program on October 1, 2002, aimed at promoting research on the design, planning, and control of operations and processes in commercial and institutional service enterprises. In parallel with many other worldwide initiatives, the Services Science Global (SSG, 2006), a nonprofit research consortium, initiated the Institute of Electrical and Electronics Engineering (IEEE) International Annual Conference on Service Operations and Logistics, and Informatics (SOLI) to provide worldwide scholars and professionals with a timely and effective platform to exchange their new findings, ideas, and experiences in research on services sciences, management, and engineering. The conference is the first one of its kind to provide such a collective forum in a comprehensive manner for the dissemination of services research, and to promote services sciences, management, and engineering research worldwide, aimed at facilitating the growth of this newly emerging interdisciplinary field from both research and educational perspectives. The first IEEE SOLI international conference was successfully held in Beijing from August 10 to 12, 2005, which drew about 200 attendees from over 30 countries; about 200 quality papers from over 300 submissions are published in the conference proceedings (SSG, 2005).

As information has always been one of the most competitive factors for today's business operations, IT as a service plays a critical role in deploying adaptive enterprise service computing. The delivery of the right information to the right user in a timely manner in an enterprise ensures that management makes informed decisions, productions produce high-quality products on schedule, and customers are provided satisfactory services. As we know, the componentization of the business is the mainstream of the day for constructing best-of-breed services components for delivering superior services to end users, while the successful operations of a componentized business require seamless enterprise integration (Cherbakov et al., 2005). However, enterprise integration in practice is still time consuming, costly, and inconsistent, which results in the fact that many integrated enterprise systems lack the flexibility of adapting to market fluctuations and the capability of evolving as technologies advance and human capitals migrate.

During the last decade or so, networking and computing technologies have made substantial advances, which makes it possible to establish effective communications among people, disparate control and management systems, and heterogeneous information silos across enterprises. To increase the degree of enterprise business-

(system) process automation and accordingly improve enterprise productivity, enterprises integration has gone through several waves, such as *ad hoc* point-to-point integration, data integration, application integration, and work-flow management. Enterprise service computing emerges as a new enterprise integration technology and implementation model that has significant potential in addressing enterprise systems integration challenges, such as agility, performance, scalability, security, heterogeneity, and adaptability.

Recently, a lot of research and development advances have been achieved in adaptive enterprise service computing, integration, and management, for instance, the emergence of business-process management, service computing, service-oriented architecture technology, and grid computing. The results of the advancement have stimulated the creation of a new class of mission-critical infrastructures, a new category of integration methods and software tools, and a new group of business platforms for cost-effectively exploiting and managing business processes. Inspired by the fast advancement in the emerging areas, this book covers many areas of interest to professionals, managers, graduate-school students, professors, and researchers working in the general field of adaptive enterprise service computing. However, as many chapters have broad discussions of services science, management, and engineering, we encourage you to read this book to have some fundamental understanding of this emerging and promising interdisciplinary field even though you might not exactly be in the field of enterprise service computing.

Book Organization

Enterprise service computing emerges as a new wave of IT in support of the best practices for enterprises. As the phrase "only change is certain" is the reality, the best practices of business operations change from time to time. To ensure that IT-driven businesses can keep their competitive edge, the underlying IT systems should be able to quickly and optimally align the changing business goals, which essentially requires a suite of new technologies to support the implementation of the IT systems from concept to deployment. Similar to a typical software-engineering approach, the book is essentially organized into six sections covering the life cycle of enterprise service computing's development. These six sections are as follows.

- **Section I: Business Aspects of Enterprise Service Computing** — The business view of adaptive IT-supporting systems
- **Section II: Enterprise Service Computing: Requirements** — Provides answers to the following questions: What is the specification? What are the essential needs for delivering the business from the IT perspective?

- **Section III: Enterprise Service Computing: Modeling** — Design methods and methodologies to ensure that the requirements will be met during the implementation
- **Section IV: Enterprise Service Computing: Technologies** — The right technologies for implementing the verified models of the needed IT system
- **Section V: Enterprise Service Computing: Formal Modeling** — The necessary verification and validation of complex systems, given the fact that adaptive enterprise service computing typically exists across the heterogeneous and geographically dispersed value net
- **Section VI: Enterprise Service Computing: Best Practices and Deployment** — The deployed solutions aligning the best practices of enterprises

Although the reader is encouraged to read from start to finish, it is absolutely appropriate to hop around from section to section. As the edited book is more or less research oriented, each chapter presents its own view of technology and business. To be self-contained, each contributed chapter in sequence does not need the prerequisite of having read previous ones. The given sequence of chapters in the book is just one possible way to help readers cluster their systematic thinking as they read through the book, or to simply help them locate their interests easily. The following paragraphs give a brief description of each contributed chapter.

Business Aspects of Enterprise Service Computing

IT as a service has many aspects, such as the points of view of the end user, business, operations, management, and technology. As the complexities, uncertainties, and changes reconfigure the business world from time to time, an enterprise should be able to quickly adapt to the change. Given that today's business is highly driven by IT systems, to appropriately design and develop an IT system to enable adaptive enterprise service computing, a comprehensive while visionary business aspect of the IT system lays the sound and solid foundation of the IT systems deployed across an enterprise. This section gives introductory discussions of enterprise service computing from certain business perspectives.

Chapter I, "Information Technology as a Service" by Robin Qiu, discusses the concept of IT as a service in great detail. To ensure the prompt and cost-effective delivery of innovative and satisfactory goods and services to customers, enterprises nowadays have to rethink their operational and organizational structures by overcoming a variety of distance, social, and cultural barriers. Qiu explains that IT plays a critical role in helping enterprises transform their business in an optimal manner. As challenges appear in many aspects from business strategy, marketing, modeling, innovations, design, and engineering to operations and management, the successful deployment of adaptive enterprise service computing aligning business goals will be the key to gaining the future competitive edge.

Chapter II, "Aligning Business Processes with Enterprise Service Computing Infrastructure" by Wei Zhao, Jun-Jang Jeng, Lianjun An, Fei Cao, Barret Bryant, Rainer Hauser, and Tao Tao, argues that for the IT-driven business operations of many enterprises today, there exists a natural gap and disconnection between the decision and evaluation at the business level, and the execution and metrics at the IT level. As a solution to bridging the gap, they provide (a) a model transformation framework that effectively transforms business-level decisions into IT-level executable representations based on the concept of SOA, (b) a framework based on a service-level agreement that can monitor and synthesize IT-level performance and metrics to optimally provide the end users with satisfactory services, and (c) certain techniques that enable the dynamic adaptation of IT infrastructure as needed.

Chapter III, "Service-Portfolio Measurement (SPM): Assessing Financial Performance of Service-Oriented Information Systems" by Jan vom Brocke, focuses on the decision making of the service portfolio enabled in a service-oriented enterprise, aimed at aligning the decision making and the company's financial situation to ensure better long-term economic consequences. For a selected service portfolio, Brocke provides a framework for facilitating the assessment of the various financial consequences, which thus can help management justify the strategic plan of developing and deploying a service-oriented information system to support the service-oriented business operations.

Enterprise Service Computing: Requirements

Successfully soliciting the comprehensive requirements of an IT system is the foundation of the design and development of enterprise service computing. Although there are numerous approaches to engineering requirements for IT systems in general, the following two chapters use examples in two different IT operational environments to demonstrate how the requirements can be best captured to ultimately support service-oriented businesses.

In **Chapter IV**, "Requirements Engineering for Integrating the Enterprise," Raghvinder Sangwan reasons that understanding the need and level of cooperation and collaboration among the different segments of an enterprise, its suppliers, and customers is indispensable to the success of integrating enterprises' disparate IT systems. Given the complexity and heterogeneous nature of IT systems deployed in enterprises, capturing the requirements in a comprehensive manner relies on solid formulism and modeling. Specifically, Sangwan reviews the research to date on model-driven requirements engineering and examines a case study on integrating health-care providers to gain insight into the challenges and issues.

In **Chapter V**, "Mobile Workforce Management in a Service-Oriented Enterprise: Capturing Concepts and Requirements in a Multiagent Infrastructure," Dickson Chiu, S. Cheung, and Ho-fung Leung look at the requirements engineering in support of a mobile workforce by considering the mobility needs in a service-oriented computing infrastructure. They indicate that the demand of the deployment of mobile workforce management (MWM) across multiple platforms is increasing substantially. They demonstrate that a multiagent information system (MAIS) infrastructure provides a suitable paradigm to capture the concepts and requirements of an MWM as well as a phased development and deployment. Through a case study, they show an approach to formulating a scalable, flexible, and intelligent MAIS with agent clusters.

Enterprise Service Computing: Modeling

Once the requirements are successfully solicited from the end users, appropriate design methods and methodologies should be adopted. Different models typically provide different points of view for the future implementation of the IT system, for instance, architectural, communication, functional, object-component, failure, security, and deployment models. Using different modeling techniques essentially ensures that the requirements will be met during the implementation.

Chapter VI, "Designing Enterprise Applications Using Model-Driven Service-Oriented Architectures" by Marten van Sinderen, João Paulo Andrade Almeida, Luís Ferreira Pires, and Dick Quartel, presents a model-driven service-oriented approach to the design of enterprise integration applications. As the model-driven architecture (MDA) is well recognized as an adequate approach to managing system and software complexity in distributed objected-oriented design, they argue that the combination of MDA and the modeling of service-oriented computing will be an applicable approach to facilitating the development and deployment of enterprise integrated applications to support the dynamic changes of the business processes of an enterprise.

In **Chapter VII**, "A Composite Application Model for Building Enterprise Information Systems in a Connected World," Jean-Jacques Dubray emphasizes that the Web as a ubiquitous, distributed computing platform has changed dramatically the IT systems in support of business operations today. As a specialization and composition of activities empowers each economic agent to use and contribute the best of its abilities in the business world, IT-connected systems should behave in the same manner to best support business operations. The service-oriented composite application model should be appropriately applied to the development of IT-connected systems. Dubray argues that the enabled real-time and federated information systems through enterprise service computing will allow units of work to be executed more cooperatively to optimally fulfill the determined business goals across enterprises.

More specifically, in **Chapter VIII**, Xiang Gao and Jen-Yao Chung elucidate their three-point service-oriented conceptual design and modeling methodology for Web-services composition based on the design principle of SOA, aimed at providing a design approach to warranting the semantic consistency in support of business operations. They identify issues on the semantic consistency of Web-services composition at the conceptual level, while using the standard schemes of an order-handling system to define and demonstrate the semantic consistency of Web-services composition. Moreover, a formal service model is also applied to formally define the semantics of services interactions and formal history semantic conformances within Web-services interactions and composition at the conceptual system level.

Enterprise Service Computing: Technologies

The right technologies for implementing the verified models of the needed IT system should be determined as they lead to the future scalability, maintainability, and adaptability concerns. Given the complexity and heterogeneity in the nature of enterprise service computing, different implementing technologies (e.g., networking, programming, data storage, and security) should be applied to different parts of the implementation. The pick of run-time technologies for leveraging the interoperability and adaptability is more critical for enterprise service computing as the deployed IT systems focus on the ongoing change of business needs across enterprises.

In **Chapter IX**, "Data-Replication Strategies in Wide-Area Distributed Systems," Sushant Goel and Rajkumar Buyya indicate the importance of fast and effective access to data in support of efficient business operations. As replication is a widely accepted phenomenon in distributed computing environments, replication protocols are the mechanisms to guarantee the proper synchronization between replicated data sources. As applications vary with business settings, different replication protocols may be suitable for different applications. In the chapter, the authors present a survey of replication algorithms for different distributed storage and content-management systems. The survey covers a variety of systems, from distributed database-management systems, service-oriented data grids, and peer-to-peer (P2P) systems to storage-area networks. In particular, the replication algorithms of contemporary architectures, data grids, and P2P systems are provided with great detail.

Chapter X, "Web Services vs. ebXML: An Evaluation of Web Services and ebXML for E-Business Applications" by Yuhong Yan and Matthias Klein, explains relevant aspects of the Web-services and ebXML technologies and compares their capabilities from an e-business point of view. By exploring the similarity and difference between Web services and ebXML, they argue the two technologies have many things in common and should complement each other. The Web-service technology provides an excellent solution to integrating heterogeneous systems over the network. When combined with business-process management initiatives, it is a perfect fit in adaptive enterprise service computing as an interoperable run-time technology.

In **Chapter XI**, "Leveraging Pervasive and Ubiquitous Service Computing," Zhijun Zhang reviews the different wireless networking technologies and latest mobile devices in the market of the day and discusses how the recent advances in pervasive computing can help enterprises better bridge the gap between their employees or customers and information. Zhang proposes a pervasive and ubiquitous service-computing-based service-oriented architecture to maximally leverage the provided mobile capability in adaptive enterprise service computing so that service-oriented businesses can be optimally supported and the right information service can be made available at the point of need.

Enterprise Service Computing: Formal Modeling

The necessary verification and validation of complex systems is very important given that adaptive enterprise service computing typically exists across the heterogeneous and geographically dispersed value net. As the dynamics and behavioral properties of IT supports come directly from design and development, some future needs might be skipped during the design or certain undesirable features (e.g., blocking, deadlocks, security implications) may be unexceptionally embedded or introduced. A formal approach to modeling the dynamics of the future deployed system can assure that the desirable trajectories of the IT system better align the business objectives currently and in the future.

The advances of Web-services technology gradually make the deployment of adaptive enterprise service computing easier as engineering enterprise Web applications in support of business operations can be rapidly realized by composing Web-services components. However, the services results of the composed network components (e.g., Web-services components) are hardly guaranteed to be satisfactory. Thus, it is necessary to have a formal approach to validate and reason about the properties of an enterprise system composed of Web-service components. In **Chapter XII**, "A Petri-Net-Based Specification Model Toward Verifiable Service Computing," Jia Zhang, Carl K. Chang, and Seong W. Kim introduce the Web-services net (WS-Net) for realizing the verification and validation of IT systems composed from Web-services components. WS-Net essentially is an executable architectural description language. By incorporating the semantics of colored petri nets with the style and understandability of the object-oriented concept and Web-services concept, WS-Net is able to facilitate the simulation, verification, and automated composition of Web services in enterprise service computing.

In **Chapter XIII**, "Service Computing for Design and Reconfiguration of Integrated E-Supply-Chains," Mariagrazia Dotoli, Maria Pia Fanti, Carlo Meloni, and Mengchu Zhou apply a formal approach to the design and reconfiguration of an integrated e-supply-chain (IESC) network. A three-level decision-support system (DSS) is proposed. The performance of all the potential IESC candidates is evaluated and the best ones are selected at the first level. Then at the second level, a multicriteria integer-linear optimization technique is used to configure the needed IESC network. Finally, the configured network is evaluated and validated at the third level to ensure that the identified services are enabled across the network.

Enterprise Service Computing: Best Practices and Deployment

The deployed solutions aligning the best practices of enterprises are the ultimate goal of adaptive enterprise service computing. On one hand, the research and development of IT (e.g., enterprise service computing) is a service. On the other hand, when IT helps enterprises streamline their business processes to deliver quality and competitive goods and services, it essentially functions as a knowledge service. To maximize the return of IT investment, the implementation of IT applications must also follow the best practices in service computing.

Chapter XIV, "Best Practice in Leveraging E-Business Technologies to Achieve Business Agility" by Ehap Sabri, presents the fact that the implementation of e-business solutions to enable best practice leads to achieving business cost reduction and agility. As an example, Sabri discusses the strategic and operational impact of e-business solutions on supply chains and explains the performance benefits and implementation challenges enterprises should expect when the best practices will be adopted. Sabri provides the best-practice framework in leveraging e-business applications to support process improvements aimed at eliminating non-value-adding activities and enabling real-time visibility and velocity for the supply chain.

Finally, in **Chapter XV**, "Concepts and Operations of Two Research Projects on Web Services and Context at Zayed University," Zakaria Maamar shows the deployment results of two research projects applying context in Web services. The success of enterprise service computing relies on the successful deployment of the developed IT system. The IT system should also deliver the desirable services to the end users to best support the business operations. In the first project, called ConCWS, the deployment focuses on using context during Web-services composition. In the second project, called ConPWS, the deployment then focuses on using context during Web-services personalization. In the two projects, Zakaria also demonstrates how agent technology can actively participate in enterprise service computing. Through the successful deployment of the two projects, Zakaria essentially shows how enterprise service computing can ultimately deliver the right information services to meet the needs of end users.

Disclaimer

No product or service mentioned in this book is endorsed by its maker or provider, nor are any claims of the capabilities of the product or service discussed or mentioned. Products and company names mentioned may be the trademarks or registered trademarks of their respective owners.

References

Cherbakov, L., Galambos, G., Harishankar, R., Kalyana, S., & Rackham, G. (2005). Impact of service orientation at the business level. *IBM Systems Journal, 44*(4), 653-668.

IBM. (2004). *Service science: A new academic discipline?* IBM. Retrieved February 4, 2006, from http://www.research.ibm.com/ssme

Karmarkar, U. (2004). Will you survive the services revolution? *Harvard Business Review, 82*(6), 100-107.

National Science Foundation (NSF). (2006). *Service enterprise engineering*. Retrieved February 4, 2006, from http://www.nsf.gov/funding/pgm_summ.jsp?pims_id=13343&org=NSF

Rangaswamy, A., & Pal, N. (2005). Service innovation and new service business models: Harnessing e-technology for value co-creation (eBRC white paper). *2005 Workshop on Service Innovation and New Service Business Models.* Retrieved September 10, 2005, from *http://www.smeal.psu.edu/ebrc/*

Rosmarin, R. (2006). *Sun's serviceman.* Retrieved February 6, 2006, from http://www.forbes.com/2006/01/13/sun-microsystems-berg_cx_rr_0113sunqa_print.html

Services Science Global (SSG). (2005). *The Proceedings of 2005 IEEE International Conference on Services Operations and Logistics, and Informatics* (Online program abstract). Retrieved September 10, 2005, from http://www.ssglobal.org/2005

Services Science Global (SSG). (2006). *Services Science Global: A non-profit research consortium.* Retrieved February 6, 2006, from http://www.ssglobal.org

Acknowledgments

The editor would like to acknowledge the help of all involved in the collation and review process of the book, without whose support the project could not have been satisfactorily completed.

Most of the authors of chapters served as referees for chapters written by other authors. Thanks go to all those who provided constructive and comprehensive reviews. Support of the Division of Engineering and Information Science at Penn State Great Valley is acknowledged for providing administrative services for the project from beginning to end.

Special thanks also go to all the staff at Idea Group Inc., whose contributions throughout the whole process from inception of the initial idea to final publication have been invaluable. In particular, I am very grateful for Kristin Roth, who continuously prodded via e-mail for keeping the project on schedule, and for Mehdi Khosrow-Pour, whose enthusiasm motivated me to initially accept his invitation for taking on this project.

In closing, I wish to thank all of the authors for their insights and excellent contributions to this book. I also want to thank all of the people who assisted me in the reviewing process. Finally, I want to thank my wife, Lihua Zhang, and children, Lawrence and Jason, for their love and support throughout this project.

Robin G. Qiu, PhD

Philadelphia, Pennsylvania, USA

June 2006

Section I

Business Aspects of Enterprise Service Computing

Chapter I

Information Technology as a Service

Robin G. Qiu, The Pennsylvania State University, USA

Abstract

In this information era, both business and living communities are truly IT driven and service oriented. As the globalization of the world economy accelerates with the fast advance of networking and computing technologies, IT plays a more and more critical role in assuring real-time collaborations for delivering needs across the world. Nowadays, world-class enterprises are eagerly embracing service-led business models aimed at creating highly profitable service-oriented businesses. They take advantage of their own years of experience and unique marketing, engineering, and application expertise and shift gears toward creating superior outcomes to best meet their customers' needs in order to stay competitive. IT has been considered as one of the high-value services areas. In this chapter, the discussion will focus on IT as a service. We present IT development, research, and outsourcing as a knowledge service; on the other hand, we argue that IT as a service helps enterprises align their business operations, workforce, and technologies to maximize their profits by continuously improving their performance. Numerous research and development aspects of service-enterprise engineering from a business perspective will be briefly explored, and then computing methodologies and technologies to enable adaptive enterprise service computing in support of service-enterprise engineering will be simply studied and analyzed. Finally, future development and research avenues in this emerging interdisciplinary field will also be highlighted.

Introduction

With the significant advances in networks, telecommunication, and computing technologies, people, organizations, systems, and heterogeneous information sources now can be linked together more efficiently and cost effectively than ever before. The quick advances of IT in general significantly transform not only science and engineering research, but also expectations of how people live, learn, and work as we witnessed during the last decade or so. Life at home, work, and leisure gets easier, better, and more enjoyable.

In the business world, because of rich information linkages, the right data and information in the right context can be delivered to the right user (e.g., people, machines, devices, software components, etc.) in the right place and at the right time, resulting in the substantial increase of the degree of business-process automation, the continual increment of production productivity and services quality, the reduction of service lead time, and the improvement of end users' satisfaction. As a variety of devices, hardware, and software become network aware, almost everything is capable of being handled over a network. Many tasks can be done on site or remotely, and in the same manner, so are a variety of services provided or even self-performed over the Internet. At the end of day, end users or consumers do not care about how and where the product was made, by whom, and how it was delivered; what the end users or consumers essentially care about is that their needs are met in a satisfactory manner.

In manufacturing, the deployment of integrated information systems is accelerating (Qiu, 2004). A typical IT-driven manufacturing business can be created by deploying enterprise-wide information systems managing the life cycle of both the business and its electronic aspects; that is, an order is taken over the Internet, and the products are made and delivered as promised. For instance, customers submit their orders via Internet browsers directly through a sales-force automation center, which automatically triggers the generation of the appropriate material releases and production requirements. It also informs all the other relevant planning systems, such as those for advance production schedule, finance, supply chain, logistics, and customer-relationship management, of the new order entry. The scheduler then assigns or configures an on-site or remote production line through the production control in the most efficient way possible, taking into account raw material, procurement, and production capacity. A shop-floor production-execution schedule is then generated, in which problems are anticipated and appropriate adjustments are made accordingly in a corresponding manufacturing execution system. In the designated facility, the scheduled work is accomplished automatically through a computer-controlled production line in an efficient and cost-effective manner. As soon as the work is completed, the ordered product gets automatically warehoused and/or distributed. Ultimately, the customers should be provided the least cost and best quality goods, as well as the most satisfactory services (Qiu, Wysk, & Xu, 2003).

No matter what is made and how services are delivered, in reality, high living standards with a better quality of life are what we are pursuing as human beings. When the communities we are living in are deeply studied, we understand that our communities are truly IT driven and service oriented in the information era. Here are a few daily noticeable, inescapable, and more contemporary service examples that could be on demand at any time and place (Dong & Qiu, 2004; Qiu, 2005).

- A passenger traveling in a rural and unfamiliar area suddenly has to go to a hospital due to sickness, so local hospital information is immediately required at the point of need. The passenger and his or her companions wish to get local hospital information through their cellular phones. Generally speaking, when travelers are in an unfamiliar region for tourism or business, handy and accurate information on routes and traffic, weather, restaurants, hotels, hospitals, and attractions and entertainment in the destination region will be very helpful.
- A truck fully loaded with hazardous chemical materials is overturned on a suburban highway. Since the chemicals could be poisonous, people on site need critical knowledge (i.e., intelligent assistance) to quickly perform life-saving and other critical tasks after one calls 911 (in the United States). However, people on site most likely cannot perform the tasks effectively due to limited knowledge and resources. Situations could be worse if tasks are not done appropriately, which could lead to an irreversible and horrible result. Intelligent assistant services are necessary at the point of need. Obviously, the situation demands a quick response from the governmental IT-driven emergency response systems.
- Transportation plays a critical role in warranting the quality service and effectiveness of a supply chain. When a truck is fully loaded with certain goods, certain attention might be required from the driver from time to time, for instance, to the air, temperature, and/or humidity requirements. Only when the requirements are met on the road can the goods in transit be maintained with good quality. Otherwise, the provided transportation service will not be satisfactory. Due to the existence of a variety of goods, it is impossible for

drivers to master all the knowledge on how the goods can be best monitored and accordingly protected on the highway as many uncertain events might occur during the transportation of the goods. On-demand services to assure warranty are the key for an enterprise to lead competitors. As manufacturing and services become global, more challenges are added into this traditional service.

- The growing elderly population draws much attention throughout the world, resulting in issues on the shortage of laborers and more importantly the lack of effective health-care delivery. Studies show that elderly patients (65 or older) are twice as likely to be harmed by a medication error because they often receive complex drug regimens and suffer from more serious ailments that make them particularly vulnerable to harmful drug mistakes. Outpatient prescription-drug-related injuries are common in elderly patients, but many could be prevented. For instance, about 58% adverse drug events could be prevented if continuity-of-care record plans and related health-care information systems are adopted for providing prompt assistant services; over 20% of drug-related injuries could be prevented if the given medication instructions are provided at the point of need so the instructions are adhered to by the patients.

Apparently, the real-time flow of information and quick delivery of relevant information and knowledge at the point of need from an information-service provider or system is essential for providing quality services to meet the on-demand needs descried in all the above scenarios.

In a broader view, the service-oriented society is clearly evidenced by the largest labor migration in history around the world. According to Rangaswamy and Pal (2005), total-solution services (enabled performances for the customer's benefit rather than a physical good) constitute the prime marketplace battlefield of the day. For instance, in the U.S. economy, over 70% of the 2005 gross domestic product was generated from services businesses. Similar numbers dominate many other developed economies worldwide. Even in some developing countries like China and India, 35% of labor is service oriented, and the number continues to climb every year. The world-developed economy is clearly heading to one that is IT driven, technology based, and services led.

Because of the fast development of IT, the globalization of the world economy is accelerating. Under the umbrella of global virtual enterprises through collaborative partnerships, enterprises can provide best-of-breed goods and services at a more competitive price while meeting the changing needs of today's on-demand business environment. As competition in the globalizing economy unceasingly intensifies, it becomes essential for enterprises to rethink their operational and organizational structures to meet the consumers' fluctuating demand for innovation, flexibility, and shorter lead times for their provided goods and services. For example, for farsighted manufacturers in the developed economy, as their product technologies might quickly lose their competitiveness, they recognize that only their services components would distinguish themselves from their competitors. Therefore, enterprises are keen on building highly profitable service-oriented businesses by taking advantage of their own unique engineering and application expertise, aimed at shifting gears toward creating superior outcomes to best meet their customers' needs in order to stay competitive (Rangaswamy & Pal, 2005).

The value of provided goods or delivered services lies in their ability to satisfy an end user's need, which is not simply and strictly seen in the physical attributes of the provided product or the technical characteristics of the delivered service. For today's competitive enterprise, a superior outcome provided to its customer inevitably constitutes services contributing to the entire customer solution through the well-established and highly collaborated value net, including the support of solution engineering, the sale of a physical product, product sustaining, personnel training, and/or a knowledge-transformation service in a satisfactory manner. This new and emerging field is truly interdisciplinary in nature and explores new frontiers of research, attempting to build a true science and engineering base and establish the foundation for understanding future competitiveness (IBM, 2004; Rangaswamy & Pal, 2005).

As discussed above, IT plays a critical role in facilitating today's geographically dispersed manufacturing and services delivery. It is IT that enables real-time information flow. When enterprise-wide information systems are fully integrated throughout the whole customer order fulfillment process, all of the order information, lot travelers, material consumptions, customer services, and accounting ledgers can be continuously updated. Therefore, top management can keep abreast of how efficiently the enterprises are running and thus make the optimal decisions possible with real-time information on sales, finances, resources, and capacity utilizations. As a result, enterprises are staying due to better competitive advantages as customers are totally satisfied with their on-demand needs. Total satisfaction typically drives further sales.

As the economy shifts from being manufacturing based to information-services based, better understanding of services marketing, innovation, design, engineering, operations, and management will be essential. Evidently, IT is a service from an end user's perspective. By aggregating the concepts and research from the latest literature, this chapter presents the author's point of view on this emerging interdisciplinary field, focusing on how enterprise service computing might be evolving from the current research and development. In the remainder of the chapter, numerous research and development aspects of service-enterprise engineering from the business perspective will be first discussed, and then some technologies to enable adaptive enterprise service computing in support of future service-enterprise engineering will be briefly introduced and analyzed.

Aspects of Services-Enterprise Engineering

With the push of the ongoing industrialization of information technologies, enterprises must aggregate products and services into total solutions by implementing an integrated and complete value net over all of their geographically dispersed collaborative partners to deliver services-led solutions in order to stay competitive (Figure 1). The essential goal of applying total solutions to value networks is to enable the discovery, design, deployment, execution, operation, monitoring, optimization, analysis, transformation, and creation of coordinated business processes across the value chain: a collaborating ecosystem. Ultimately, the profit across the whole value network can be maximized as it becomes the top business objective in today's global business environment (Karmarkar, 2004).

The shift from a manufacturing base to a services base makes enterprises rethink their business strategies and revamp their operational and organizational structures to meet the customers' fluctuating demands for services delivered in a satisfactory fashion. Enterprises across the board in general are eager to seek new business opportunities by streamlining their business processes; building complex, integrated, and more efficient IT-driven systems; and embracing the worldwide Internet-based marketplace. It is well recognized that business-process automation, outsourcing, customization, offshore sourcing, business-process transformation, and self-services became another business wave in today's evolving global services-led economy.

Services sectors nowadays cover commercial transportation, logistics and distribution, health-care delivery, financial engineering, e-commerce, retailing, hospitality and entertainment, issuance, supply chains, knowledge transformation and delivery, and consulting. In the developed countries, services enterprises are the new industrial base of the economy. For instance, almost four out of five jobs are currently offered by these services sectors in the United States. On the other hand, in the developing countries, more traditional labor-intensive manufacturing and services are still the core businesses. However, the developing countries actively participate in the global

Figure 1. Service-oriented business value net

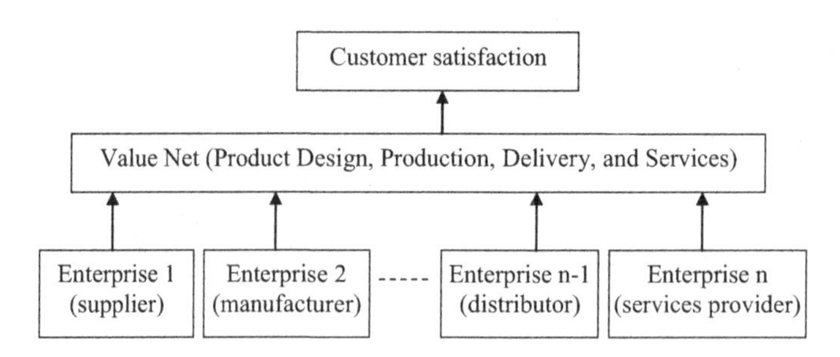

economy, mainly providing the physical attributes of goods and services through the manufacturing and delivering of labor-intensive services and business functions. The global economy in the 21st century requiring a business ecosystem for better effectiveness, efficiency, and manageability thus involves all the collaborated business-function organisms with their operational settings. The global ecosystem indeed includes participants from both the developed and developing countries.

Although this new wave seems to be repeating the trends that afflicted U.S. manufacturing in the 1970s, it gets more complicated in demanding higher efficiency and better cost effectiveness across the geographically dispersed value chains. Moreover, compared to industry's knowledge of mature manufacturing business practices, services-enterprise engineering is still substantially uncharted territory (Rangaswamy & Pal, 2005). According to IBM (2004):

Services sciences, Management and Engineering [as an emerging interdisciplinary field] hopes to bring together ongoing work in computer science, operations research, industrial engineering, business strategy, management sciences, social and cognitive sciences, and legal sciences to develop the skills [and knowledge] required in a services-led economy.

Only recently have there been some international initiatives to promote research and education in this emerging field in a comprehensive manner.

Research and education in services-enterprise engineering to cultivate and empower the ecosystem driven by services, technology, and management have been lagging behind when compared to many other areas as many scholars from leading universities and professionals from industrial bellwethers have only recently started to pay more attention to services-enterprise-engineering research. More importantly, the services and services-enterprise-engineering scope have evolved and expanded enormously as the world economy accelerates the pace of globalization due to the tremendous advances in IT (including computing, networks, software, and management science). Consequently, little is really known about how services sciences, management, and engineering can be systematically applied for the delivery of a services-led value chain from end to end.

As mentioned earlier, today's services concept evolves beyond the traditional nonagricultural and/or nonmanufacturing performance for the consumer's benefit. For example, many new and emerging high-value areas, such as IT outsourcing, post-sales training, and on-demand innovations consulting (including any work helping customers improve their products, business processes, goods and services delivery, and supportive IT systems), are well recognized as services, drawing substantial attention from many industrial bellwethers (Fitzgerald, 2005; Rosmarin, 2006). On one hand, unique and satisfactory services differentiate an enterprise from its competitors; on the other hand, highly satisfactory services delivery frequently drives more product sales. As the shift from a manufacturing base to a services base becomes inescapable for both the developed countries and the developing countries, enterprises are gradually embracing defining and selling anything as a service (Rosmarin).

To ensure the prompt and cost-effective delivery of innovative and satisfactory services for customers throughout the geographically dispersed value net, enterprises nowadays have to rethink their operational and organizational structures by overcoming a variety of social and cultural barriers. Challenges appear in many aspects from business strategy, marketing, modeling, innovations, design, and engineering to operations and management. Since the general services topic is too broad and vague, the following discussions mainly focus on the needs for research and education in IT services, aimed at providing some fundamental understanding for business executives, managers, knowledge workers, and professionals in the relevant business sectors and research fields.

Business Strategy

As an enterprise is diving into building a highly profitable service-oriented business by taking advantage of its own unique engineering expertise and services knowledge, aimed at shifting gears toward creating superior outcomes to best meet customer needs, an adequate business-service strategy will be vital for the enterprise's growth in the long run. As discussed earlier, it is the mainstream for enterprises to collaborate with their worldwide partners to deliver best-of-breed services to their customers.

Despite the recognition of the importance of service-enterprise-engineering research, the shift to focus on services in the information era has created a research gap due to the overwhelming complexity of interdisciplinary issues across service-business modeling, information technologies, and workforce management. Filling the gap is essential. According to Rust (2004, p. 211):

We can move the field forward not only by understanding and serving the customer but by designing efficient systems of service delivery; training and motivating service providers; using new service technologies; and understanding how service affects the marketplace, the economy, and government policy.

The development of a business strategy meeting the long-term growth of a services enterprise should ensure that the defined business road map organically integrates corporate strategy and culture with organizational structure and functional strategy, and allows managing the interface of strategy and technology in a flexible and cost-effective fashion.

In general, the development of business strategy for enterprises adaptable to a new business environment requires great understanding of incorporations of solutions to addressing at least the following challenges in the services-led economy (Cherbakov, Galambos, Harishankar, Kalyana, & Rackham, 2005; Wright, Filatotchev, Hoskisson, & Peng, 2005).

- **Maximizing the total value across the value chain:** The outcome of the value chain nowadays is clearly manifested by customer satisfaction, which is mainly dependent upon the capability of providing on-demand, customizable, and innovative services across the value net.
- **The international transferability of staying competitive:** Enterprises reconstruct themselves by taking advantage of globalization in improving their profit margins, resulting in the fact that subcontracting and specialization prevail. Radically relying on efficient and cost-effective collaborations, a services provider essentially becomes a global ecosystem in which international transferability plays a critical role. International transferability could cover a variety of aspects from human capital, worldwide trade and finance, social structures, and natural resources to cultures and customs.
- **Organizational learning as competitive advantage:** The globalization of the services workforce creates new and complex issues due to the differences in cultures, time, and skills.
- **Coping with the complexities, uncertainties, and changes:** Change is the only certain thing today and tomorrow. As the complexities, uncertainties, and changes are reconfiguring the business world, an enterprise should be able to quickly adapt to the change.
- **Aligning business goals and technologies to execute world-class best practices:** Business componentization cultivates value nets embracing best-of-breed components throughout collaborative partnerships. The value nets essentially are social-technical systems and operate in a network characterized by more dynamic interactions, real-time information flows, and integrated IT systems. Apparently, aligning business goals and IT is indispensable to the successful execution of applied world-class best practices in services enterprises.

Services Marketing

"[Today's] business reality is that goods are commodities; the service sells the product," says Roland Rust (2004, p. 211), a leading professor and the David Bruce Smith chair in marketing at the Robert H. Smith School of Business at the University of Maryland. It is not a secret that quality services essentially lead to high customer satisfaction. Satisfaction characterized as a superior outcome then further drives customer decisions. The services-led total solution measured by performance for the customer's final benefit rather than the functionality of physical goods become the prime competition in the global services-led marketplaces.

There are many new business opportunities in numerous newly expanded areas under the new concept of services, for example, e-commence, e-services, auctions, and IT consulting (Menor, Tatikonda, & Sampson, 2002). Although these emerging services have gained much popularity with consumers, many new issues solicit more exploration for a better understanding of marketing to ensure that business goals can be met in the long run. Rust and Lemon (2001) discuss that Internet-based e-services can better serve consumers and exceed their expectations through real-time interactive, customizable, and personalized services. Effective e-service strategy and marketing play a significant role in growing the overall value of the services provider. A set of research questions in many customer-centric areas is proposed, aimed at leading to a stronger understanding of e-service and consumer behavior. Cao, Gruca, and Klenz (2003) model the relationships between e-retailer pricing, price satisfaction, and customer satisfaction so a more competitive business can be operated.

According to Rangaswamy and Pal (2005), service marketing as a fundamental service-value driver is much less understood compared to product marketing. Typically, a service outcome is freshly "manufactured" or "remanufactured" at the customer's site when it is delivered; it depends heavily on a well-defined and consistent process applied by trained personnel time after time, and leads to winning future competition through future innovations. It is hardly an easy transition from traditional business or consumer-product marketing techniques.

Services Design and Engineering

There have been many publications in the literature illustrating a variety of approaches to services design and engineering across industries. Although some of them present their scientific methodologies to realize the targeted goals specified by customers, the majority of them simply show their empirical and heuristic methods to deliver services design and engineering processes. In the emerging high-value IT-services area, there is a great need for methodologies for the design and engineering of long-term, high-quality, and sustaining services in order to meet the defined business strategy of a services provider.

Zhang and Prybutok (2005) study the design and engineering factors impacting service quality in the e-commerce services sector. Introducing new products and services indeed would certainly help new revenue generation. However, retaining customers' high satisfaction and alluring them for purchasing further products and services are highly dependent upon other numerous critical factors, for instance, system reliability, ease of use, localization and cultural affinity, personalization, and security. As the levels of price satisfaction might not be increased simply by lowering prices, competing on price hence is not a viable long-term strategy for online retailers. Cao et al. (2003) model the relationships between e-retailer pricing, price satisfaction, and customer satisfaction through analyzing the whole services process. They find that the design and engineering of a satisfactory ordering process generates higher overall ratings for fulfillment satisfaction, which better retains loyal customers and accordingly helps a services provider to stay competitive.

As discussed earlier, services sectors cover both traditional services (e.g., commercial transportation, logistics and distribution, health-care delivery, retailing, hospitality and entertainment, issuance, and product after-sale services) and contemporary services (e.g., supply chains, knowledge transformation and delivery, financial engineering, e-commerce, and consulting). The competitiveness of today's services substantially depends on efficient and effective services delivery networks that are constructed using talent and comprehensive knowledge with a combination of business, management, and technology. The services processes should be flexibly engineered by effectively bridging the science of modeling and algorithms on one hand, and business processes, people skills, and cultures on the other hand.

Services Modeling and Innovation Framework

Services innovations are the key to stay a step or two ahead of competitors. James Spohrer, director of IBM Services Research, has an insightful view of the need for the investigation of service innovations and modeling. He states:

Increasingly over the past ten years, the new frontier of service research and teaching has shifted more and more towards business-to-business process transformation models. Process reengineering, IT productivity paradox, and other case studies highlight the need to constantly redesign work to improve productivity through multiple types of innovation in demand, business value, process, and organization.

A well-defined services model and innovation framework will effectively guide services enterprises to best design, develop, and execute their well-defined strategic plan for long-term growth.

New service-delivery models are essentially derived by working closely with customers to cocreate innovative and unique solutions best meeting customers' inevitably changing needs. According to Rangaswamy and Pal (2005), a competitive service business model for an enterprise should be clearly described using a service innovation framework (Figure 2):

Figure 2. Services innovations framework and modeling

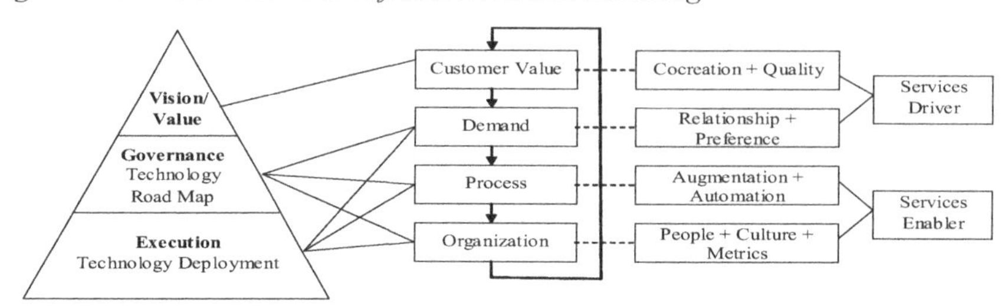

The framework can guide the creation of customer value and demand, and the processes and organizations that deliver services successfully—all of it catalyzed by emerging technologies.

Although detailed panel views of customer value, demand, process, and organization have been given in the white paper by Rangaswamy and Pal (2005), there is still the lack of a systematic approach to address how such a model and innovation framework can be enabled in practice. Given the tremendous complexity and variance from service to service, vertical service-domain knowledge of modeling and frameworks should be first investigated. Only when a better understanding of a variety of services domains is accomplished can an integrated and comprehensive methodology to address the services model and innovation framework across industries be explored and acquired.

Services Operations and Management

Operations research and management with focus on business-internal efficiency has made significant progress and developed a huge body of knowledge during the last 65 years or so. The relevant research and algorithm development has been mainly conducted in the areas of optimization, statistics, stochastic processes, and queuing theory. Current applications cover areas from vehicle routing and staffing, supply-chain modeling and optimization, transportation modeling, revenue management, risk management, services-industry resource planning and scheduling to airline optimization and forecasting. In general, operations research has unceasingly improved living standards as it has been widely applied in practice for the improvement of production management and applications productivity.

Operations research and management originated from practice and has been growing as a more quantitative, mathematical, and technical field. Larson (2005) argues that practice makes perfect operations research. As new problems are identified and framed, formulated, and solved by applying operations-research approaches, tremendous impact will be provided and accordingly a new theory might be created. Sociotechnical services systems show a more practical nature and are extremely complex, and they are typically modeled and formulated using qualitative approaches. An understanding of such a complex problem involves deep and thoughtful discussion and analysis using common sense, basic principles, and modeling. Through new initiatives, the operations-research body of knowledge can be perfectly applied to these practical problems. Services operations and management are essentially operations research and management applied to services settings.

As discussed earlier, on one hand, the research and development of IT is a service. On the other hand, when IT helps enterprises streamline their business processes to deliver quality and competitive goods and services, it essentially functions as a knowledge service. However, efficient IT-service delivery to meet the needs of adaptive enterprises requires talent and comprehensive knowledge with a combination of business, management, and IT. Therefore, service-based operations research and management is in demand as it matches the emerging realization of the importance of the customer and a more customer-oriented view of operations. Services operations and management fits well with the growing economic trend of globalization, which requires operations research in services practice.

According to Bell (2005), operations research applied to services has much to offer that could improve the lives of everyone. He presents seven useful operations research frameworks that can be effectively used in addressing practical and complex problems like services delivery networks. Moreover, services operations are closely synchronized with the business operations of other collaborative partners as well as customers, aimed at cocreating value for customers in a satisfactory manner while meeting the business objectives across the value net. Given the industrialization of services and the economy of globalization, reorganizing, realigning, redesigning, and restructuring enterprises' strategies, processes, IT systems, and people for the challenges ahead are essential for ensuring that services providers are agile and adaptive, and stay competitive (Karmarkar, 2004).

In summary, given the increasing complexity of building sociotechnical services systems for improving living standards by applying operation research and management science in practice, services operations and management should cover more initiatives for the rooted practical aspects of research, linking operational performance to business drivers, performance measurement and operations improvement, service design, service technology, human capital, the design of internal networks, and the management of service capacity (Johnston, 1999). The study should also take into consideration high performance, distributed computing, humans' and systems' behavioral and cognitive aspects (which emerges as the new look of the interface to systems engineering), and highly collaborative interaction natures.

Adaptive Enterprise Service Computing

Enterprises are eagerly embracing building highly profitable service-oriented businesses through properly aligning business and technology and cost effectively collaborating with their worldwide partners so that the best-of-breed services will be generated to meet the changing needs of customers. To be competitive in the long run, it is critical for enterprises to be adaptive given the extreme dynamics and complexity of conducting businesses in today's global economy. In an adaptive enterprise, people, processes, and technology should be organically integrated across the enterprise in an agile, flexible, and responsive fashion. As such, the enterprise can quickly turn changes and challenges into new opportunities in this on-demand business environment.

IT service is a high-value services area that plays a pivotal role in support of business operations, logistics, health-care delivery, and so forth. IT service in general requires people who are knowledgeable about the business, IT, and organization structures, as well as human behavior and cognition that go deep into successful services operations (IBM, 2004). For IT systems to better serve the service-oriented enterprise, service-oriented business components based on business-domain functions are necessary (Cherbakov et al., 2005). The question is what systematic approach and adequate computing technologies will be suitable for IT development leading to the success of building an adaptive enterprise.

Computing technologies (e.g., software development) unceasingly increase in their complexities and dependencies. Aiming to find a better approach to managing complexities and dependencies within a software system, the practice of software development has gone through several methods (e.g., conventional structural programming, object-oriented methods, interface-based models, and component-based constructs). The emergence of developing coarse-grained granularity constructs as a computing service allows components to be defined at a more abstract and business-semantic level. That is, a group of lower level and finer grained object functions, information, and implementation software objects and components can be choreographically composed as coarse-grained computing components, supporting and aligning business services.

The componentization of the business is the key to the construction of best-of-breed components for delivering superior services to the customers. Successful operations of a componentized business require seamless enterprise integration. Thus, service-oriented IT systems should be able to deal with more amounts of interaction among heterogeneous and interconnected components, and be more flexible and adaptive. Obviously, adaptive and semantic computing services representing business functions meet the needs of the service-oriented IT systems. When computing components manifest business services at the semantics level, an IT system is a component network, fundamentally illustrating a logic assembly of interconnecting computing components:

The need for flexibility across the value net requires that the component network be flexible; that is, the enterprise can "in-source" an outsourced component and vice versa; replace, on demand, a current partner with a different partner; change the terms of the contract between the two components, and so on. (Cherbakov et al., 2005)

A generic service-oriented IT computing architecture for the development of a component network is illustrated in Figure 3. The top two layers represent services operations from the business-process perspective while the bottom three layers show the value-adding services processes from the computing perspective. Apparently, how to optimally align enterprise-level business strategies with value-adding operations and activities is the key to the success of the deployment of an agile enterprise service-oriented IT system (Qiu, in press).

However, the exploitation, establishment, control, and management of dynamic, interenterprise, and cross-enterprise resource-sharing relationships and the realization of agility in a service-oriented IT system require new methodologies and technologies. The remaining discussions focus on the following four emerging synergic IT research and development areas aimed at providing some basic understanding of the emerging methodologies and technologies in support of the future deployment of IT services that enable adaptive enterprise service computing.

Figure 3. Service-oriented component-network architectural model

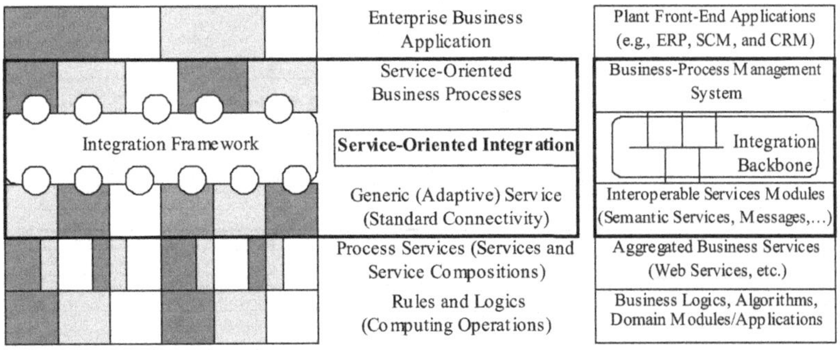

Integration Framework
Enterprise Business Application
Service-Oriented Business Processes
Service-Oriented Integration
Generic (Adaptive) Service (Standard Connectivity)
Process Services (Services and Service Compositions)
Rules and Logics (Computing Operations)

Plant Front-End Applications (e.g., ERP, SCM, and CRM)
Business-Process Management System
Integration Backbone
Interoperable Services Modules (Semantic Services, Messages,...)
Aggregated Business Services (Web Services, etc.)
Business Logics, Algorithms, Domain Modules/Applications

(a) The enterprise service computing architectural model *(b) An implementation*

- **Service-oriented architecture (SOA):** SOA is considered the design principle and mechanism for defining business services and computing models, thus effectively aligning business and IT.
- **Component-process model (CPM):** A component business-process model facilitates the construction of the business of an enterprise as an organized collection of business components (Cherbakov et al., 2005).
- **Business-process management (BPM):** BPM essentially provides mechanisms to transform the behaviors of disparate and heterogeneous systems into standard and interoperable business processes, aimed at effectively facilitating the conduct of IT-system integration at the semantics level (Smith & Fingar, 2003).
- **Web services:** Web services are simply a suite of software-development technologies based on Internet protocols, which provide the best interoperability between IT systems over the network.

Service-Oriented Architecture

According to Datz (2004), "SOA is higher level of [computing] application development (also referred to as coarse granularity) that, by focusing on business processes and using standard interfaces, helps mask the underlying complexity of the IT environment." Simply put, SOA is considered the design principle and mechanism for defining business services and computing models, thus effectively aligning business and IT (Figure 4; Newcomer & Lomow, 2005).

Based on the concept of SOA, a deployed service-oriented IT system can provide a common way of cost effectively and efficiently managing and executing distributed

Figure 4. Aligning business and information technology

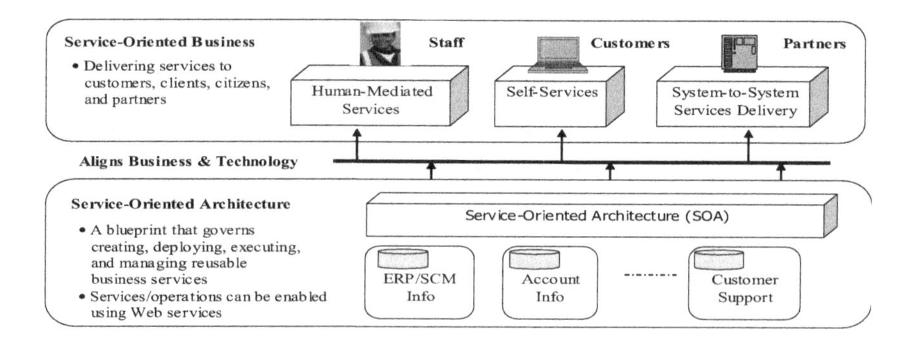

heterogeneous services across enterprises. To properly implement service-oriented IT systems complying with SOA, three major levels of abstraction throughout collaborated IT systems are necessary (Zimmermann, Krogdahl, & Gee, 2004).

- **Business processes:** A business process typically consists of a set of actions or activities executed with specifically defined long-term business goals. A business process usually requires multiple computing services. Service invocations frequently involve business components across the network. Examples of business processes are the initiation of a new employee, the selling of products or services, a project's status, and order-fulfillment information.
- **Services:** A service represents a logical group of low-level computing operations. For example, if customer profiling is defined as a service, then looking up customers from data sources by telephone number, listing customers by name and postal code on the Web, and updating data for new requests represent the associated operations.
- **Operations:** A computing operation represents a single logical unit of computation. In general, the execution of an operation will cause one or more data sets to be read, written, or modified. In a well-defined SOA implementation, operations have a specific, structured interface and return structured responses. An SOA operation can also be composed of other SOA operations for better structures and maintainability.

SOA as a design principle essentially is concerned with designing and developing integrated systems using heterogeneous network-addressable and standard interface-based computing services. Over the last few years, SOA and service computing technology have gained tremendous momentum with the introduction of Web services (a series of standard languages and tools for the implementation, registration, and invocation of services). Enterprise-wide integrated IT systems based on SOA ensure the interconnections among integrated applications in a loosely coupled, asynchronous, and interoperable fashion. It is believed that BPM (as transformation technologies) and SOA enable the best platform for integrating existing assets and future deployments (Bieberstein, Bose, Walker, & Lynch, 2005).

Component-Process Model

Given the increasing complexity and dynamics of the global business environment, the success of a business highly relies on its underlying IT-supportive systems to support the changing best practices. In adaptive enterprise service computing, the appropriate design of IT-driven business operations mainly depends on well-defined constructs of business processes, services, and operations. Hence, to make this promising SOA-based component-network architectural model able to be implemented, it is essential to have a well-defined, process-driven analytical and computing model that can help analysts and engineers understand and optimally construct the business model of an enterprise for IT implementation.

A business process typically consists of a series of services. As a business process acts in response to business events, the process should be dynamically supported by a group of services invoked in a legitimate sequence. To ascertain the dynamic and optimal behavior of a process, the group of underlying computing services should be selected, sequenced, and executed in a choreographed rather than predefined manner according to a set of business rules. A service is made of an ordered sequence of operations. CPM basically is a design and analytical method and platform to ensure that well-designed operation, service, and process abstractions can be characterized and constructed systematically for adaptive enterprise service computing (Cherbakov et al., 2005; Kano, Koide, Liu, & Ramachandran, 2005; Zimmermann et al., 2004).

CPM essentially provides a framework for organizing and grouping business functions as a collection of business components in a well-structured manner so that the components based on business processes can be modeled as logical business-service building blocks representing corresponding business functions. Figure 5 schematically illustrates a simplified components-process model for a service provider (Cherbakov et al., 2005). Just like many business-analysis diagrams, CPM can also be refined into a hierarchy. In other words, a process can be composed of a number of refined processes in a recursive fashion.

As CPM can accurately model business operations using well-defined services in SOA terms, CPM helps analyze a business and develop its componentized view

Figure 5. Component business-process schematic view

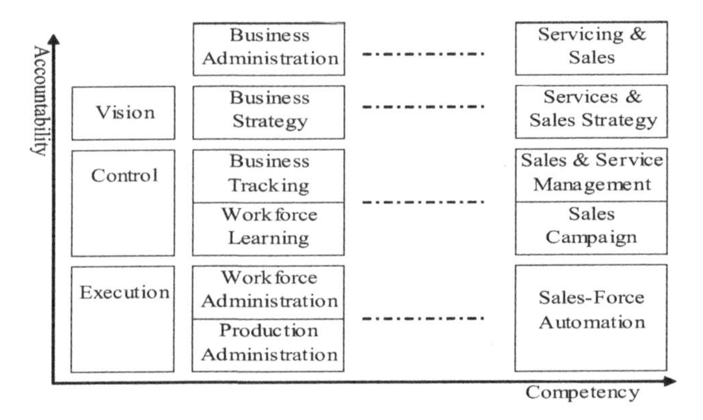

of the business. Furthermore, the developed model for the business will define components concentrating on the interfaces and service-level agreements between the services. As a result, each business component will be supported by a set of IT-enabled services, while meeting the requirements of the deployment of adaptive enterprise service computing. Most importantly, as the business evolves, CPM can help analyze the hot spot of the business operations. When business-performance transformation is required as the business settings change, CPM and the underlying IT systems can be quickly transformed to meet the needs of on-demand businesses (Cherbakov et al., 2005).

Business-Process Management

BPM emerges as a promising guiding principle and technology for integrating existing assets and future deployments. BPM is new in the sense that it describes existing disparate and heterogeneous systems as business-process services when conducting IT-system integration for better business agility rather than simply integrating those systems using EAIs (enterprise application integrations), APIs (application programming interfaces), Web-services orchestration, and the like. By providing mechanisms to transform the behaviors of disparate and heterogeneous systems into standard and interoperable business processes, BPM essentially aims at enabling a platform to effectively facilitate the conduct of IT-system integration at the semantics level (Smith & Fingar, 2003). Since an SOA computing service at the system level essentially is the business function provided by a group of components that are network addressable and interoperable, and might be dynamically discovered and used, BPM and SOA computing services can be organically while

Figure 6. BPM merging with SOA services

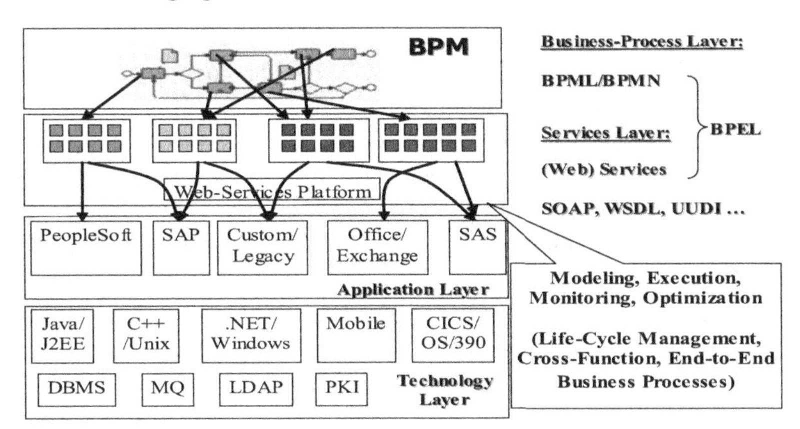

flexibly and choreographically integrated, which is schematically illustrated in Figure 6 (Newcomer & Lomow, 2005).

In essence, BPM takes a holistic approach to enterprise service computing from the business-process execution perspective, substantially leveraging the power of standardization, virtualization, and management. BPM initiatives include a suite of protocols and specifications, including the business process modeling language (BPML), business process modeling notation (BPMN), and business process execution language (BPEL). By treating the business-process executions as real-time data flows, BPM provides the capability of addressing a range of choreographic business challenges and improving business operations in nearly real time.

BPML is defined for modeling complex business processes. Using the BPML specification to describe the business model of an enterprise provides the abstract model of the enterprise. The abstracted model is programmatically structured and represented using extensible markup language (XML) syntax to express the defined executable business processes and supporting entities for the enterprise. BPMN provides the capability of defining and understanding internal and external business operations for the enterprise through a business-process diagram. Through visualization, it gives the enterprise the ability to communicate these modeling and development procedures in a standard manner. BPEL for Web services then defines a standard way of representing executable flow models, which essentially extends the reach of business-process models from analysis to implementation through leveraging the power of Web-service technologies.

The emergence of BPM introduces an innovative platform for conducting IT-system integration. BPM enables service-oriented IT systems over the network to be able to dynamically and promptly coordinate the behaviors of disparate and heterogeneous computing services across enterprises. It is through BPM that business agility is retained while the return of IT investment is maximized.

Web Services

Apart from traditional software technologies, Web technology in general is non-proprietary and platform independent. Using standard Internet protocols, a Web service is a self-contained, self-describing, and network computing component. A Web service can be conveniently deployed, published, located, and invoked across the network. As Web services can be assembled and reassembled as needed across the network, the needs of adaptive enterprise computing of a business can be cost effectively supported.

Web-services technology essentially consists of a stack of protocols and specifications for defining, creating, deploying, publishing, locating, and invoking black network components. The stack mainly includes the simple object access protocol (SOAP), XML and XML namespaces, Web service description language (WSDL), and universal description, discovery, and integration (UDDI).

A computing service deployed as a Web service has to strictly comply with the stack of protocols and specifications. SOAP is the underlying communication protocol between the service provider and consumer, and explicitly defines how the service provider and consumer interact and what the enabled computation results in. WSDL is the language for defining the computing service, and basically specifies the location of the computing service and the operations the service exposes. UDDI then provides the formal interface contract and the global base for the registration and discovery of the deployed computing service.

Web services are standard run-time technologies over the Internet, providing best-ever mechanisms for addressing heterogeneous computing issues. By converging SOA and Web technology, Web services represent the evolution of Web technology to support high performance, scalability, reliability, interoperability, and availability of distributed service-oriented IT systems across enterprises around the whole world.

Conclusion

This chapter aimed at providing a basic understanding of the IT-driven, service-led economy. By discussing the challenges of services marketing, innovations, design, engineering, operations, and management from an IT perspective, this chapter gave the author's point of view on how service-enterprise engineering should be evolving from the current research and development. For an enterprise to be adaptive and able to quickly turn changes and challenges into opportunities so that the needs of the on-demand business can be optimally met, the workforce, processes, and technologies have to be organically aligned and integrated across the enterprise in an agile, flexible, and responsive fashion.

The following four design and computing methodologies and technologies are currently proposed as the necessities of enabling adaptive enterprise service computing. SOA is the design methodology to ensure the best aligning of the business and IT-driven system. CPM is a structured view of a business, which helps analysts and designers to optimally construct the long-term architectural and functional models for IT implementation. BPM is a rigorous method to embody the design and development of CPM, which essentially provides mechanisms to transform the behaviors of disparate and heterogeneous systems into standard and interoperable business processes so that the conduct of IT-system integration can be accomplished at the semantics level. BPEL and Web services are the run-time technologies suited for this need of BPM materialization.

Grid computing is emerging as a new and powerful computing technology to enable resource sharing for complex problem solving across businesses, institutions, research labs, and universities. Well-informed operational, tactical, and strategic decisions can only be made when nearly perfect and real-time visibility of the fulfillment of products and services can be provided in the on-demand e-business environments. It is envisioned that grid computing will join services science, management, and engineering in support of IT-driven system deployment for enabling real-time adaptive enterprise service computing in the near future.

References

Bell, P. (2005). Operations research for everyone (including poets). *OR/MS Today, 32*(4), 22-27.

Bieberstein, N., Bose, S., Walker, L., & Lynch, A. (2005). Impact of service-oriented architecture on enterprise systems, organizational structures, and individuals. *IBM Systems Journal, 44*(4), 691-708.

Cao, Y., Gruca, T., & Klemz, B. (2003). Internet pricing, price satisfaction, and customer satisfaction. *International Journal of Electronic Commerce, 8*(2), 31-50.

Cherbakov, L., Galambos, G., Harishankar, R., Kalyana, S., & Rackham, G. (2005). Impact of service orientation at the business level. *IBM Systems Journal, 44*(4), 653-668.

Datz, T. (2004). What you need to know about service-oriented architecture. *CIO Magazine*, 78-85.

Dong, M., & Qiu, R. (2004). An approach to the design and development of an intelligent highway point-of-need service system. *The 7th International IEEE Conference on Intelligent Transport Systems*, 673-678.

Fitzgerald, M. (2005). Research in development. *MIT Technology Review.* Retrieved February 4, 2006, from http://www.technologyreview.com/articles/05/05/issue/brief_ibm.asp

IBM. (2004). *Service science: A new academic discipline?* Retrieved February 4, 2006, from http://www.research.ibm.com/ssme

Johnston, R. (1999). Service operations management: Return to roots. *International Journal of Operations & Production Management, 19*(2), 104-124.

Kano, M., Koide, A., Liu, T., & Ramachandran, B. (2005). Analysis and simulation of business solutions in a service-oriented architecture. *IBM Systems Journal, 44*(4), 669-690.

Karmarkar, U. (2004). Will you survive the services revolution? *Harvard Business Review, 82*(6), 100-107.

Larson, R. (2005). Practice makes perfect O.R. *OR/MS Today, 32*(2), 6-7.

Menor, L., Tatikonda, M., & Sampson, S. (2002). New service development: Areas for exploitation and exploration. *Journal of Operations Management, 20*(2), 135-157.

Newcomer, E., & Lomow, G. (2005). *Understanding SOA with Web services.* Addison-Wesley Professional.

Qiu, R. (2004). Manufacturing grid: A next generation manufacturing model. *2004 IEEE International Conference on Systems, Man and Cybernetics*, 4667-4672.

Qiu, R. (2005). An Internet computing model for ensuring continuity of healthcare. *2005 IEEE International Conference on Systems, Man and Cybernetics*, 2813-2818.

Qiu, R. (in press). A service-oriented integration framework for semiconductor manufacturing systems. *International Journal of Manufacturing Technology and Management.*

Qiu, R., Wysk, R., & Xu, Q. (2003). An extended structured adaptive supervisory control model of shop floor controls for an e-manufacturing system. *International Journal of Production Research, 41*(8), 1605-1620.

Rangaswamy, A., & Pal, N. (2005). Service innovation and new service business models: Harnessing e-technology for value co-creation (eBRC white paper). *2005 Workshop on Service Innovation and New Service Business Models.* Retrieved September 10, 2005, from http://www.smeal.psu.edu/ebrc/

Rosmarin, R. (2006). *Sun's serviceman.* Retrieved February 6, 2006, from http://www.forbes.com/2006/01/13/sun-microsystems-berg_cx_rr_0113sunqa_print.html

Rust, R. (2004). A call for a wider range of service research. *Journal of Service Research, 6*, 211.

Rust, R., & Lemon, K. (2001). E-service and the consumer. *International Journal of Electronic Commerce, 5*(3), 85-101.

Services Science Global (SSG). (2006). *Services Science Global: A non-profit research consortium.* Retrieved February 6, 2006, from http://www.ssglobal.org

Smith, H., & Fingar, P. (2003). *Business process management: The third wave.* Tampa, FL: Meghan-Kiffer Press.

Wright, M., Filatotchev, I., Hoskisson, R., & Peng, M. (2005). Strategy research in emerging economies: Challenging the conventional wisdom. *Journal of Management Studies, 42*(1), 1-33.

Zhang, X., & Prybutok, V. (2005). A consumer perspective of e-service quality. *IEEE Transactions on Engineering Management, 52*(4), 461-477.

Zimmermann, O., Krogdahl, O., & Gee, C. (2004). *Elements of service-oriented analysis and design.* Retrieved February 5, 2006, from http://www-128.ibm.com/developerworks/library/ws-soad1/

Chapter II

Aligning Business Processes with Enterprise Service Computing Infrastructure

Wei Zhao, University of Alabama at Birmingham, USA

Jun-Jang Jeng, IBM T.J. Watson Research, USA

Lianjun An, IBM T.J. Watson Research, USA

Fei Cao, University of Alabama at Birmingham, USA

Barret R. Bryant, University of Alabama at Birmingham, USA

Rainer Hauser, IBM Zurich Research, Switzerland

Tao Tao, IBM T.J. Watson Research, USA

Abstract

Multisourced and federated business operations and IT services are the backbone of today's enterprise. However, in most companies, there exists a natural gap and disconnection between the decision and evaluation at the business level and the execution and metrics at the IT level. This disconnection can lead to end-user dissatisfaction, diminished profit, and missed business objectives. In this chapter, we study the problem of this disconnection and provide the following frameworks and techniques toward bridging the gap: (a) We provide a model-transformation frame-

work that effectively transforms business-level decisions documented as business-process models into IT-level executable representations based on service-oriented infrastructure, (b) a framework is described that is able to monitor and synthesize IT-level performance and metrics to meet service-level agreements between business management and end users, and (c) techniques and experiments are discussed that enable dynamic adaptation of IT infrastructure according to business decision changes.

Introduction

Multisourced and federated business operations and IT services are the backbone of today's enterprise. However, in most companies, there exists a natural gap and disconnection between the decision and evaluation at the business level and the execution and metrics at the IT level. This disconnection can lead to end-user dissatisfaction, diminished profit, and missed business objectives. In this chapter, we will discuss some frameworks and techniques to bridge the gap.

First of all, we define the scope of businesses that are of particular interest in this chapter. The content of this chapter is suitable for a particular kind of business that is called dynamic e-business (DeB; Keller, Kar, Ludwig, Dan, & Hellerstein, 2002), although traditional types of business might also benefit from this chapter with some adaptation. Dynamic e-business, also called the virtual enterprise (Hoffner, Field, Grefen, & Ludwig, 2001), consists of an interconnection of loosely coupled and dynamically bound services provided by possibly different service providers with long and short business relationships. Those services together offer an end-to-end service to customers.

There are three aspects of the disconnection: how the business decisions are executed by the IT professionals, how the IT services are evaluated and synthesized according to business needs, and how to effectively reflect changes from one side of the gap to the other.

1. On one hand, senior management and lines of business tend to prescribe their decision on business operations in the form of informal drawings and policy rules, while IT-level professionals execute these decisions, after a manual translation, in terms of IT-domain technologies such as objects, classes, procedure calls, databases, and so forth. We first describe a model-transformation architecture that effectively transforms business-level decisions documented

as business-process models in a DeB environment into an IT-level executable representation based on service-oriented infrastructure. This IT infrastructure utilizes the Web and the Internet as the underlying operation environment and the Web-service technology family to realize the execution.

2. On the other hand, the IT department usually gauges its IT services based on individual IT components such as database transaction rate, Web-server response time, and network bandwidth availability, while business managers measure their business supported by those IT components in terms of overall business services such as overall user experience and supply-chain management. Our work on system dynamics modeling (SDM; Sterman, 2002) offers a comprehensive framework for service-level agreement (SLA) management (Bitpipe, 2005). SDM establishes SLA contractual commitments at the business level, continuously monitors the IT services delivered, and synthesizes IT performance metrics against business commitment.
3. Change management is always a hard problem. How the IT system responds to the changes in the business environment is called the moving-target problem (Schach, 2005). The agility of an enterprise depends on the responsiveness of its IT infrastructure. On the other side of the mirror, how the enterprise tolerates the changes in the IT environment is called technology drift. In this chapter, we are only concerned with the first problem. We discuss techniques that enable the dynamic adaptation of IT infrastructure with realignment to the changed business decisions and reconciliation with the existing running infrastructure. These techniques are particularly useful for a business that is mission critical and has high availability requirements.

Business Processes and Their Life Cycle

Business-Process Life Cycle

Business processes capture automated solutions for the business operations of an enterprise. Adaptive business processes have a closed-loop life cycle (Nainani, 2004) as shown in Figure 1. A closed-loop life cycle includes the following phases.

1. **Modeling phase:** Business analysts create a graphical process model using a particular business-process management (BPM; BPMI, 2005) product.

Figure 1. Business-process life cycle

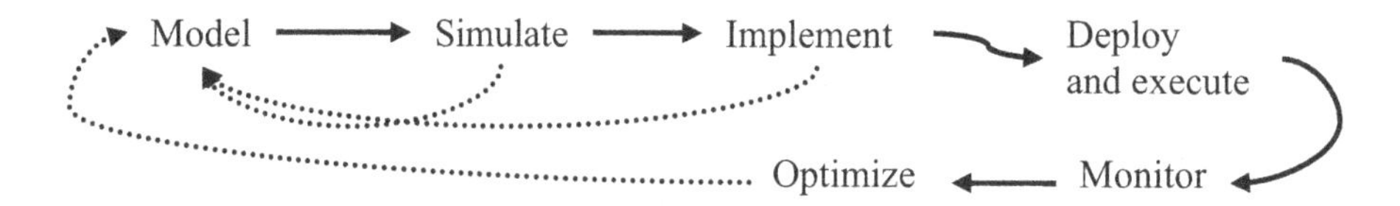

2. **Simulation phase:** The process model runs through some hypothetical scenarios to identify critical paths and bottlenecks. During the simulation phase, dynamic analysis is performed to refine the process.
3. **Implementation phase:** The finalized process model is transformed into an executable representation. Static analysis is performed during transformation to provide the modeler more information to refine the process model.
4. **Deployment and execution phase:** The process is deployed into a run-time environment for execution.
5. **Monitoring phase:** The running process is monitored continuously. Various key performance indicators (KPIs) are synthesized and reflected to the "dashboard" for viewing and controlling by the management personnel.
6. **Optimization phase:** Based on the performance metrics from the monitoring phase, the process model can be modified to achieve better performance.

Reverse engineering is necessary in both the implementation phase and the optimization phase as indicated by the dotted feedback lines in Figure 1. It is possible that software engineers will directly change the implementation. In order to make the model and implementation consistent, reverse engineering has to be performed to reflect implementation changes at the model level. Similarly, in the optimization phase, engineers could tune the implementation in order to achieve a higher performance requirement. This change again has to be reflected at the model level through reverse engineering.

Business-Process Model and the Abstract Process Graph

Diagramming and drawing are common and effective ways to communicate the understanding of business operations among management members. A visualized documentation of business processes resulting from the modeling phase (in the business-process life cycle) is usually a formalization of such drawings and is called the business-process model. A business-process model concerns the dynamic behavior of an enterprise. Other aspects of the enterprise such as information (e.g., products and documents), organization structure, policy rules, and security are not the primary concerns of process models but might be connected with the process model.

Zhao, Bhattacharya, Bryant, Cao, and Hauser (2005) have investigated the classification of available business-process modeling tools. Similar to traditional software design (Schach, 2005), although at a higher level of abstraction, there are two categories: the tools that are based on the data-driven principle, and the tools that are based on the operation-driven principle. To model business processes, the data-driven principle takes business artifacts (Nigam & Caswell, 2003) as the first-class entity. Artifact processing is a way to describe the operations of a business. The end-to-end processing of a specific artifact, from creation to completion and archiving, are captured in the artifact life cycle. The artifact life cycle is usually represented as a state machine (Kumaran & Nandi, 2003) that orchestrates a set of business activities and events that trigger state transitions in the artifact life cycle. The operation-driven principle takes business tasks as first-class entities and defines the operational order of a set of atomic business tasks and subprocesses that again consist of atomic business tasks and subprocesses. This principle underlies workflow modeling languages such as the unified modeling language (UML; *UML 2.0 Superstructure Final Adopted Specification*, 2003) and business process modeling notation (BPMN; *BPMN Specification*, 2004).

Different tools offer different modeling notations. However, all these models can be normalized into an abstract model called a process graph. A process graph can be treated as a finite automaton (FA) with three differences

1. The process graph contains concurrency that is absent in an FA.
2. In a process graph, each edge is assigned a unique identifier, while in an FA, the input symbols annotated on the edges can be repeated.
3. For each state in an FA, there is an outgoing edge for each input alphabet symbol. This rule does not need to be true in a process graph.

A process graph is defined formally as follows.

Definition 1: A process graph is a five-tuple $(Q, \Sigma, \delta, q_0, q_{end})$, where

1. Q is a finite set called the nodes,
2. Σ is a finite set called the edges,
3. the mapping rule $\delta: \Sigma \rightarrow (q_1, q_2)$ where $q_1 \in Q$ and $q_2 \in Q$ applies (i.e., for each edge $e \in \Sigma$, there is a pair of nodes (q_1, q_2) such that q_1 is the tail of the

edge and q_2 is the head of the edge. For example, $A \xrightarrow{a} B$ is an edge a with two nodes A and B; A is the tail of a and B is the head of a.),

4. $q_0 \in Q$ is the start node, and
5. $q_{end} \in Q$ is the final node.

Bridging the Gap in the Business-Process Life Cycle

The disconnection between business and IT introduced previously reveals a disconnection between phases in the business-process life cycle.

1. Primarily there are two ways to bring a business-process model from the modeling phase to an executable form in the implementation phase: manual transformation and automated transformation. In traditional business-process development, software engineers take the process model (or documentation of some other format) from the business analysts and manually reinterpret it into an executable language. As explained previously, manual transformation makes its way to the cultural gap between business-level documentations and IT representations. The model-transformation algorithm described in this chapter automatically transforms a business-process model into an executable representation. Since the business process execution language for Web services (BPEL4WS, or simply BPEL; *Business Process Execution Language for Web Services, Version 1.1*, 2003) has emerged as the standard executable business-process representation, we chose BPEL to be our target implementation language.
2. During the monitoring phase, the running system has to be continuously monitored and the low-level IT metrics have to be synthesized and translated into high-level measurements understandable by the business-level personnel. SDM is particularly useful in bridging the gap during the monitoring phase.
3. In the optimization phase, a business-process model can be statically changed and redeployed, or directly and dynamically updated. Dynamically updating the running business process is an effective way to manage changes in the optimization phase.

We will discuss model transformation, SDM, and dynamic updating later in the chapter.

Figure 2. An example process graph

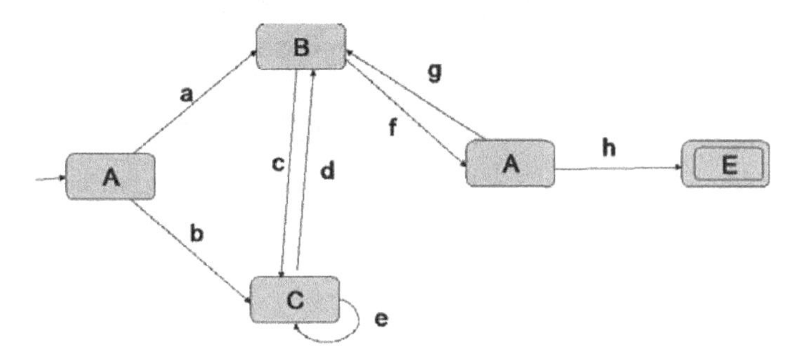

Model Transformation

Transformation Goal

The specific goals for transforming a business-process model into BPEL are (a) to generate the implementation code of an optimal size for any arbitrary process model, (b) to preserve the natural structure of business-process models in the generated code, and (c) to transform the concurrent processes. The preservation of process structure is necessary because from the process model we can only generate abstract BPEL code. Software engineers and integrators have to go through the generated code to perform tasks such as implementing needed services and performing bindings to the existing executable services.

The technical difficulty involved in this transformation comes from the fact that BPEL is a structured language, whereas the computation style of the business-process models is based on "go-to." It is difficult to transform the unstructured go-to control flows into structured statements when the process model is irreducible. Our algorithm is designed specifically for solving an irreducibility problem for which a satisfactory solution has not been found in existing research efforts and tool implementations. Some individual parts of the algorithm have been presented in our earlier papers (Zhao et al., 2005; Zhao, Bryant, Cao, Hauser, Bhattacharya, & Tao, in press; Zhao, Hauser, Bhattacharya, Bryant, & Cao, 2006). Pointers to a particular paper about a specific topic will be given.

We first look at what the irreducibility problem is. Figure 2 is an example process graph. A naïve transformation algorithm traverses the graph of Figure 2 and outputs two types of code at each node: a conditional statement such as "switch" or "if-else," and a loop statement such as "while" if the node is the entry of a natural loop. A natural loop is a loop that has a dominating entry, for example, the loop with nodes B and D where B is the dominating entry (Aho, Lam, Sethi, & Ullman, 2007). Based on this scheme, the pseudocode for Figure 2 is:

```
invoke A
switch
   case a, invoke B
        while f
              invoke D
              switch
                   case h, invoke E, exit
                   case g, invoke B
   case b, invoke C
        ......? .
```

However, the loop involved with B and C cannot be represented because it does not have a dominating entry; that is, two entries B and C do not dominate each other. This type of loop is called an irreducible loop (Aho et al., 2007). Zhao et al. (2006) have an in-depth analysis of the definitions and the problems with irreducible loops. A graph that contains irreducible loops is called an irreducible graph. Irreducibility is a classic problem in compiler theory. Traditionally, irreducibility is solved by using node-splitting techniques (Cocke & Miller, 1969) to translate an irreducible graph to a reducible graph. However, node-splitting techniques result in an equivalent but exponential reducible graph (Carter, Ferrante, & Thomborson, 2003), or at most one with a controllable size but worst-case exponential complexity (Janssen & Corporall, 1997).

Related Model-Transformation Methods

Different frameworks and methods have been proposed for transforming models to models or to code (Czarnecki & Helsen, 2003). However, none of these frameworks solve the irreducibility problem.

- The visitor pattern (Gamma, Helm, Johnson, & Vlissides, 1995) is a simple way of code generation. Code is output at each node when an object structure is traversed. The above naïve transformation algorithm can be implemented using the visitor pattern. In fact, the visitor pattern does not provide a more mechanical advantage over the traditional code-generation approaches of compiler theory, but rather a better separation of concerns. The visitor pattern separates the operation (e.g., code generation in this case) from the object structure that it is operated on. Shown in the above naïve transformation algorithm, generating code by simply traversing the object structure does not solve the irreducibility problem.

- Template-based model-to-code or model-to-model transformation is introduced in Cleaveland (2001). According to Czarnecki and Helsen (2003), "A template usually consists of the target text containing splices of meta-code to access information from the source and to perform code selection and iterative expansion." One typical example of the template-based approach is using XSLT (Cleaveland). The template-based approach is a major model-transformation approach underlying many model driven architecture (MDA) tools such as rational software architect (IBM, 2005a) and Codagen Architect (*Codagen*

Figure 3. A business-process model in IBM WBI Modeler

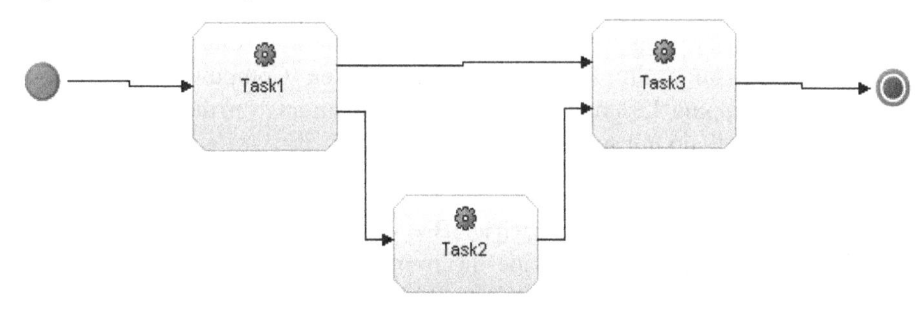

Figure 4. The corresponding BPEL code of Figure 3

```
<flow>
    <links>
        <link name ="link1"/>
        <link name ="link2"/>
        <link name ="link3"/>
    </links>

    <invoke name="Task1">
        <source linkName="link1"/>
        <source linkName="link2"/>
    </invoke>

    <invoke name="Task2">
        <target linkName="link2"/>
        <source linkName="link3"/>
    </invoke>

    <invoke name="Task3">
        <target linkName="link1"/>
        <target linkName="link3"/>
    </invoke>
</flow>
```

Architect 3.0, 2006). However, the template-based approach does not provide a solution to unravel unstructured and arbitrary go-tos into structured statements.

- A relational approach is proposed by Akehurst and Kent (2002) to define and implement transformations between metamodels. The idea is to treat the source and the target elements as a mathematical relation. A language is used to specify the relations. The goal is to generate transformation tools from the specification. Nevertheless, the task in our problem domain is not a definition of simple mappings from the entities in the source to the entities of the target, but a mapping from the structure of the entities in the source model to the structure of the entities in the target model. More importantly, the source and the target are structuring entities based on different principles.
- For the same reason as above, the graph-transformation (Andries et al., 1999; Agrawal, Karsai, & Shi, 2003) and structure-driven (Compuware, 2005) approaches, mapping the graph pattern and model element from the source model to target model, do not solve our target problem.

Some commercially available tools such as IBM WebSphere® Business Integration (WBI) Modeler (IBM, 2005b) provide the implementation of automated transformation from business-process models to BPEL. However, the implementation is based on a direct transformation scheme by using links (Kong, 2005; Mantell, 2003). A link is a BPEL construct that represents synchronization between multiple concurrent threads. This translation is illustrated in Figures 3 and 4. Based on the semantics of links, the target of a link has synchronization dependency on the source of the link. Therefore, links cannot form a cycle in a single BPEL-process specification to avoid the deadlock situation. As a result, this translation scheme restricts arbitrary cycles from being drawn from downstream nodes to upstream nodes in the process model.

The Proposed Transformation

Our approach is based on the observation that a regular expression (RE) is a theoretical model of the structure of the structured programming languages. RE has structured control-flow constructs: concatenation, or, and star. Therefore, we first transform the process graph to an RE; RE is then translated into BPEL using syntax-directed translation. Instead of treating an RE as a definition of a regular language, we consider an RE sentence a program written in a regular expression language (REL). Therefore, the REL can be easily customized to support specific features of our transformation system such as concurrency.

Figure 5. An electronic purchase system

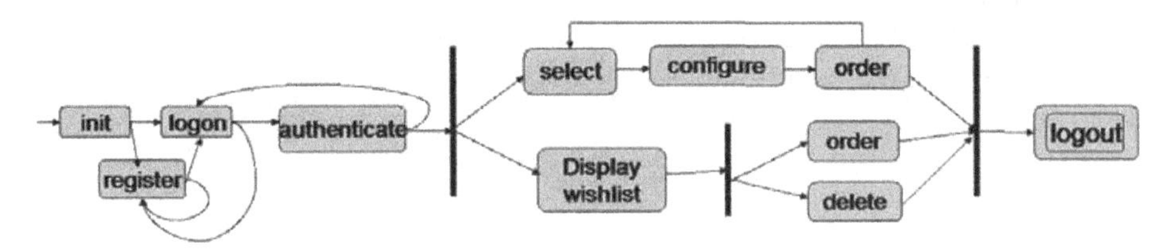

Figure 6. The abstracted process graph of Figure 5

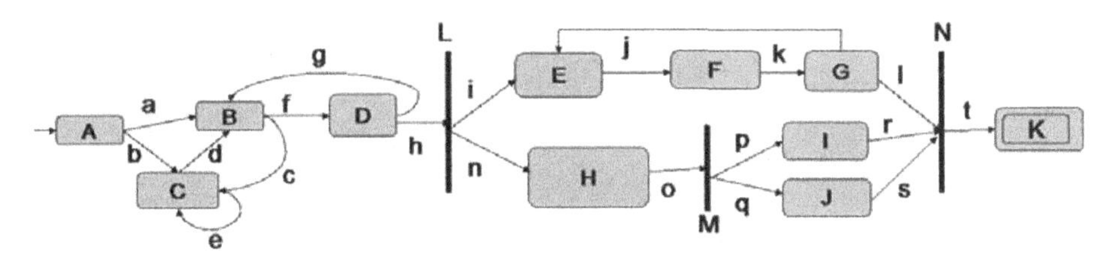

To illustrate our approach, we use an electronic purchase system as an example. The process of this example is shown in Figure 5. This process model is based on the operation-driven principle. The named rectangles denote tasks to be performed. The lines show the ordering of those tasks. The vertical bars are forks and joins. The description of the example is as follows.

From the initial page, an existing user has to log onto the system. If the log-on succeeds, the authentication service is invoked. Authentication can fail if the user provides the incorrect password, in which case the log-on is repeated. If the log-on detects the user does not exist, the user should be directed to the register function. From the initial page, new users should register. If they succeed, they can go to the log-on page. The registration page can be repeated in case of mistakes. After authentication, the user can simultaneously select a new product to purchase or display the wish list. The new-product purchase process consists of selecting a new item, configuring that item, and then placing an order for it. This process can be repeated for multiple items. After the user reviews the wish list, he or she can simultaneously place an order for some items and perform deletion on others. When the user finishes all the activity, he or she logs out of the system.

For the convenience of demonstrating the algorithm, the name of each task is abstracted as a single alphabetic letter. The abstracted process graph of Figure 5 is presented in Figure 6.

First, we demonstrate how to linearize a process model into an REL sentence. We use and extend the set-equation algorithm (Denning, Dennis, & Qualitz, 1978) for this purpose. The first step is to extract a set of equations. The set of equations for the graph in Figure 6 is as follows.

A=aB + bC (1)

B=cC + fD (2)

C=eC + dB (3)

D=gB + h**L** (4)

L={iE – nH} (5)

E=jF (6)

F=kG (7)

G=mE + l**N** (8)

H=o**M** (9)

M={pI – qJ} (10)

I=r**N** (11)

J=s**N** (12)

N=tK (13)

K=ε (since K is the final state) (14)

The + symbol means XOR. Each fork node starts a concurrent region denoted by the { and } symbols. Different threads are separated by the – symbol. The fork and join nodes are in bold face and underlined only for the purpose of indicating their specialty. It is not part of REL syntax. In these equations, the capital letters (nodes) can be treated as unknowns as in mathematical equations; the lower case letters (the edges) are the coefficients. To solve the equations, all the unknowns have to be substituted except the start node. The solution of the start node, node A in our example, would be the resulting RE.

There are four types of rules to solve the set of Equations 1 through 14.

1. Standard algebraic substitution: Substitute an unknown with its value. For example, A=aB, B=cC => A=acC.
2. Standard RE algebraic laws:
   a. **Commutative** +: R+S = S+R
   b. **Associative** +: R + (S+T) = (R+S) + T
   c. **Associative concatenation:** R (ST) = (RS) T

Figure 7. The dominator tree of Figure 6

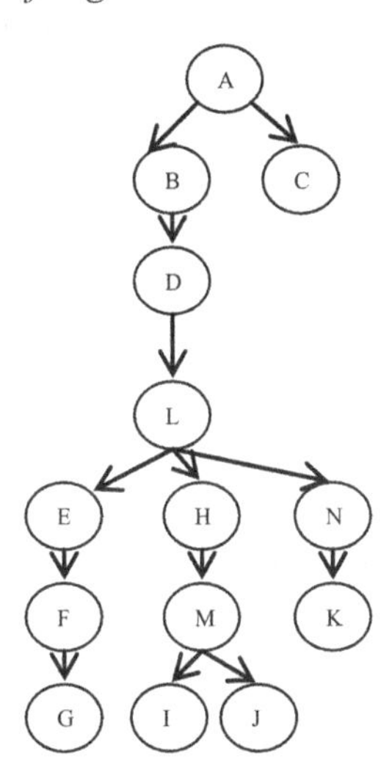

d. **Distributive +:** R(S+T) = RS+RT, (S+T)R = SR+TR

e. **Antidistributive +:** is the inverse of d.

3. Arden's rule for the removal of self-recursion: For example, C=eC+dB => C=e*dB. Please refer to Denning et al. (1978) for the proof of Arden's rule.
4. Merge of join node: If multiple threads coming out of a fork node have the same join node in the equation, this join node can be merged. For example, **A**={a**B** – b**B**} => **A**={a-b} **B**.

Given a set of equations, there exist multiple syntactically different but equivalent REs as the solution for the start state. Since the RE is an intermediate step through which we generate the target code, the complexity of the RE directly reflects the complexity of the target code. Three strategies for the application of equation-solving rules can help us to control the complexity of the generated REL. The detailed discussion and proof for Strategies 2 and 3 are presented in Zhao et al. (2006).

1. Regular states are substituted before concurrent states.
2. Arden's rule is applied only before the node is going to be substituted.
3. Among the regular states or concurrent states, the order of node substitution is from the bottom up in a dominator tree (Aho et al., 2007) of a graph. Node m dominates node n if and only if m is an ancestor of n in the dominator tree. The dominator tree of Figure 6 is shown in Figure 7.

The step-by-step guide of how to solve equations that contain only regular states based on the organization of a dominator tree using equation-solving rules 1, 2, and 3 can be found in Zhao et al. (2006). After all the regular states are substituted, the following set of equations result.

$$A=(a + b\, e^*d)\,(c\, e^*d + fg)^*fh\underline{\mathbf{L}} \tag{15}$$

$$\underline{\mathbf{L}}=\{i(j\ k\ m)^*jkl\underline{\mathbf{N}}\ -\ no\underline{\mathbf{M}}\ \} \tag{16}$$

$$\underline{\mathbf{M}}=\{pr\underline{\mathbf{N}} - qs\underline{\mathbf{N}}\} \tag{17}$$

$$\underline{\mathbf{N}}=t \tag{18}$$

Using rule 4, we get $\underline{\mathbf{M}}=\{pr\underline{\mathbf{N}} - qs\underline{\mathbf{N}}\} \Rightarrow \underline{\mathbf{M}}=\{pr - qs\}\underline{\mathbf{N}}.$ (19)

Substituting Equation 19 into Equation 16, we get the following set of equations.

$$A=(a + b\, e^*d)\,(c\, e^*d + fg)^*fh\underline{\mathbf{L}} \tag{20}$$

$$\underline{\mathbf{L}}=\{i(j\ k\ m)^*jkl\underline{\mathbf{N}}\ -\ no\{pr - qs\}\underline{\mathbf{N}}\ \} \tag{21}$$

$$\underline{\mathbf{N}}=t \tag{22}$$

Merge the node $\underline{\mathbf{N}}$ with Equation 21, and then substitute Equation 22 into Equation 21, and 21 into 20 to obtain the final equation:

$$A=(a + b\, e^*d)\,(c\, e^*d + fg)^*fh\ \{i(jkm)^*jkl\ -\ no\{pr - qs\}\}\ t. \tag{23}$$

The final REL sentence should go through the optimization step. There are two types of optimizations: (a) general RE optimizations and (b) loop look-ahead common subexpression elimination (or loop-exit optimization). General RE optimizations

Figure 8. The generated BPEL code for Figure 6

```
<process>
    <sequence>
        <invoke A/>
        <switch>
            <case condition= "a">
                <invoke B/>
            </case>
```

Figure 8. continued

```
            <case condition= "b" ...... />
        </switch>
        <assign thisExit= "false" />
        <while condition= "c or f and not thisExit">
            <switch>
                <case condition = "c" ..... />
                <case condition= "f">
                    <sequence>
                        <invoke D/>
                        <switch>
                            <case condition= "h">
                                <assign thisExit = "true"/>
                            </case>
                        </switch>
                        <switch>
                            <case condition= "g">
                                <invoke B/>
                            </case>

                        </switch>
                    </sequence>
                </case>
            </switch>
        </while>
        <flow>
            <sequence>
                ......
            </sequence>
            <sequence>
                <invoke H/>
                <flow>
                    ......
                </flow>
            </sequence>
        </flow>
        <invoke K/>
    </sequence>
</process>
```

include common subexpression elimination and loop normalizations. Interested readers may refer to Zhao et al. (2006) for a detailed discussion on the optimizations.

Loop-exit optimization puts the loop-exit indicator of brackets inside the body of the loop. After the loop-exit optimization, the REL sentence becomes:

A=(a + be*d) (ce*d + f [h] g)*h {i(jk [l] m)* l – no{pr – qs}} t. (24)

Through syntax-directed translation, Equation 24 can be translated into BPEL (a simplified version is presented in Figure 8). The loop-exit condition indicated by brackets in REL is represented by a variable *thisExit* in the *while* loop of BPEL. Also, as indicated by the BPEL code, this process has two concurrent regions, that is, two *flow* constructs, and one is nested inside the other.

However, based on this translation, some nodes and edges (i.e., paths) are repeated in the generated code whenever there is an irreducible loop. For example, the optimized REL sentence for Figure 6 contains repeated code "e*d" because there is an irreducible loop with nodes B and C. Strategies and optimizations can control the complexity of the generated code by minimizing the repetition but cannot eliminate the repetition completely. Therefore the upper bound (in the case of strongly connected components) of the generated code is exponential to the number of the nodes and edges in the input graph. The solution presented in Zhao et al. (2005) solves the problem by combining REL with another transformation scheme called the state machine controller (SMC) approach. The resulting combined method generates code of size $O(n)$, where n is the number of edges in the process model for an arbitrary process model. The detailed discussion on transforming and analyzing concurrent process models can be found in Zhao et al. (in press), where the intermediate language REL has been proved as a powerful mechanism through which some formal properties of concurrent process models such as properly nested concurrent regions, overlapping concurrent regions, and invalid concurrent regions can be detected, analyzed, and transformed or rejected.

In the next section, we will present our work on SDM and discuss how to synthesize IT-level metrics to meet SLAs concerning business-level users.

System Dynamics Modeling

Background on SDM

SDM can be used to study a variety of systems. Forrester (1961) initiated systems dynamics to model various industry problems. Sterman (2000) argued that systems dynamics can be used to study business dynamics. We have also been using system dynamics to study the demand conditioning process in supply chains (An & Ramachandran, 2005) and Web-service management (An & Jeng, 2005; An, Jeng, Ettl, & Chung, 2004). In this chapter, we are interested in discussing how to use it to model business processes to establish an autonomic control system for SLA management. Autonomic SLA management includes the establishment of a service-level agreement, continuous monitoring of running processes, the synthesis of performance metrics, and an adaptation of the business processes to optimize the overall service delivered. Some issues on how to dynamically adapt the running business processes are presented later. In this section, we show how to synthesize IT performance metrics to meet overall service agreement.

When applying SDM to business processes, two central aspects of a business process are addressed: dynamics and causality. SDM, commonly built based on system thinking and a thorough analysis of system behaviors, captures causal relationships among business tasks and the feedback-control loops for the target system. A model for system dynamics includes two major conceptual entities: stock and flow. A stock corresponds to an accumulative variable like stack or level. A flow represents dynamic changes in the content of the stock over time: inflow that increases the stock, and outflow that decreases the stock. To determine the flow rate, it is required to have system thinking about the target with a full dependency graph. Causal links from one variable to the other can be marked as positive or negative based on whether their values change in the same direction. Feedback loops are identified by checking whether all arrowed links form a loop. If the number of links with a negative sign on the loop is even, then the loop is called a positive feedback loop and causes self-reinforcement. If the number of links with a negative sign on the loop is odd, then the loop is called a negative feedback loop and causes self-correction (rebalancing).

An Example

We start by presenting an example business-process model of a supply chain shown in Figure 9 (drawn using the IBM WebSphere Business Integration Modeler). Then we will show how this process model is represented in an SDM and how the metrics can be synthesized. Major notations of this process model include activities (rectangular boxes), data repositories (cylinders), decision nodes (diamonds), merge nodes (triangles), and data (annotated lines) flowing among them. When this process model is normalized into the process graph, data repositories, decision nodes, and merge nodes can be ignored. The model starts with two parallel sequences of activities: "Order Parts" and "Receive Customer Order." In the second sequence, a branching logic is created to calculate the demand forecast and to put records in repositories "Product Demand" and "Part Demand." Both repositories are used in the activities "Make Product" and "Order Parts" of the first sequence. Repositories "Finished Product" and "Part Inventory" used in the second sequence are created and modified by the first sequence. The activity "Check Order and Product Status" is a synchronization point of the above two sequences. The rest of the model is self-explanatory.

Figure 10 shows the SDM (drawn using a software tool provided by Vensim, 2005) of the business-process model in Figure 9. There are four stocks (depicted as rectan-

Figure 9. The business-process model of a supply chain

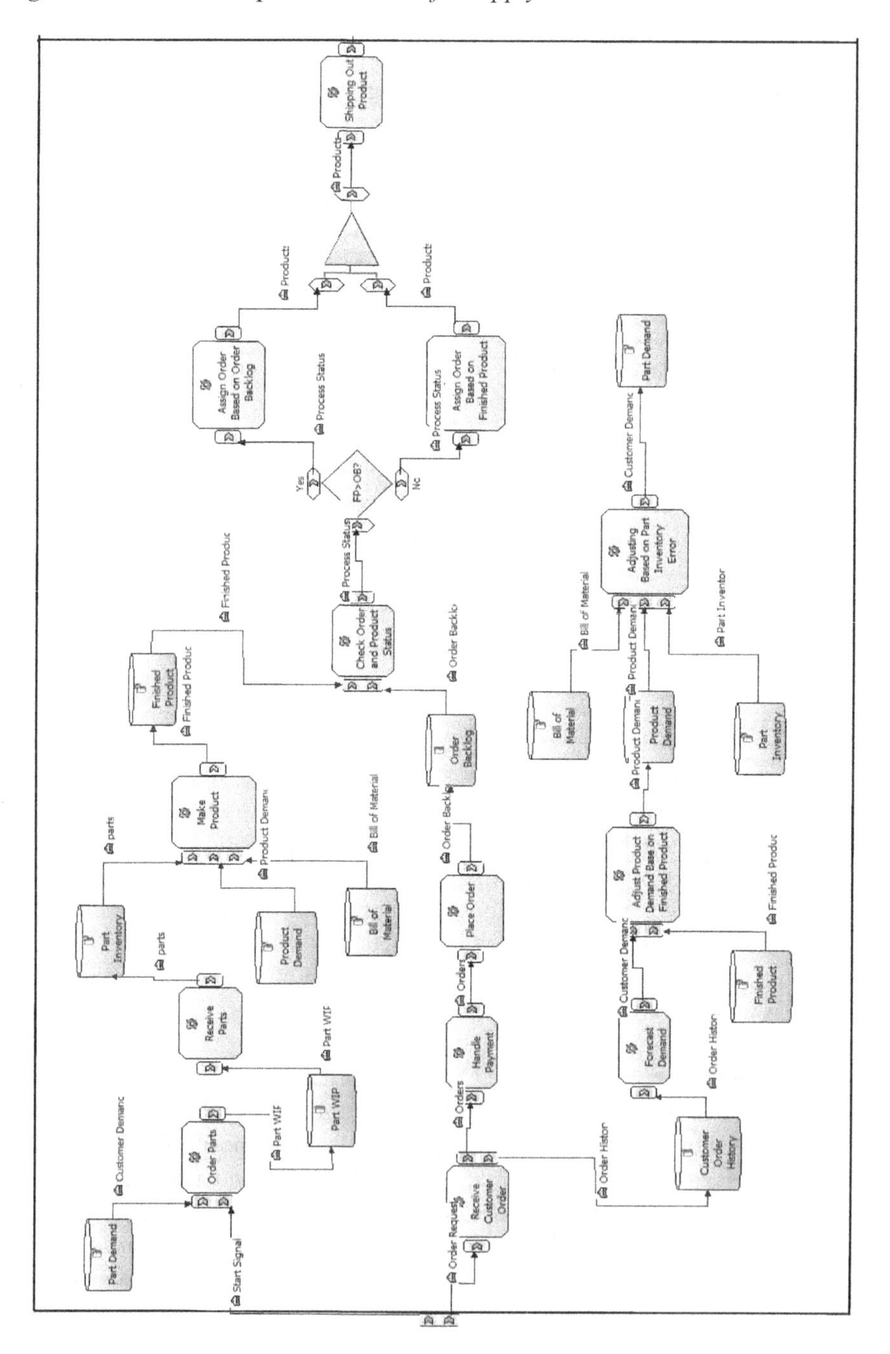

Figure 10. The SDM of the business-process model of Figure 9

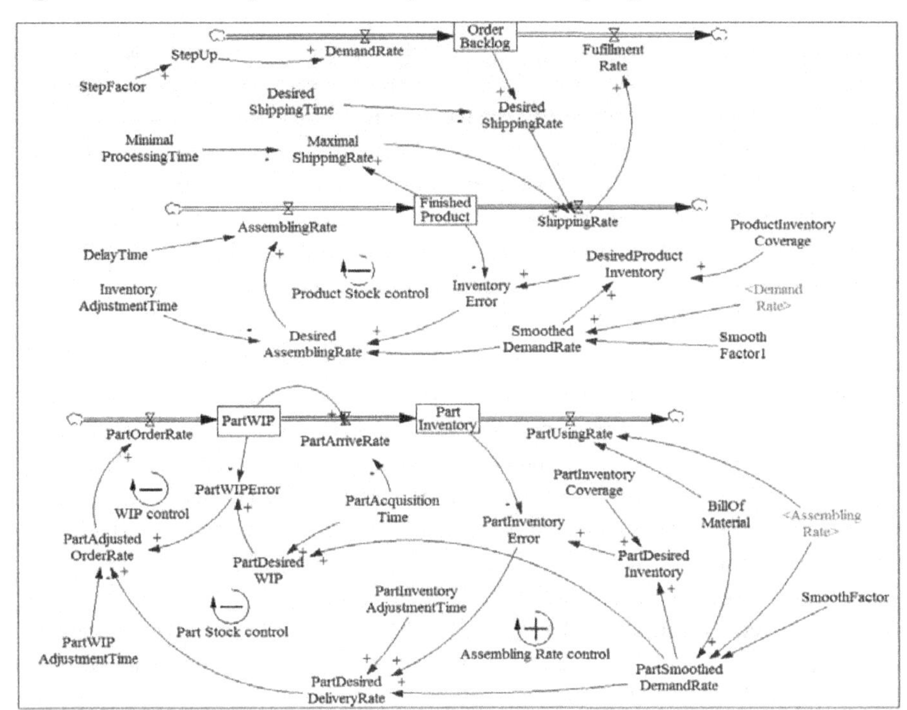

Figure 11. The causality tree for ShippingRate

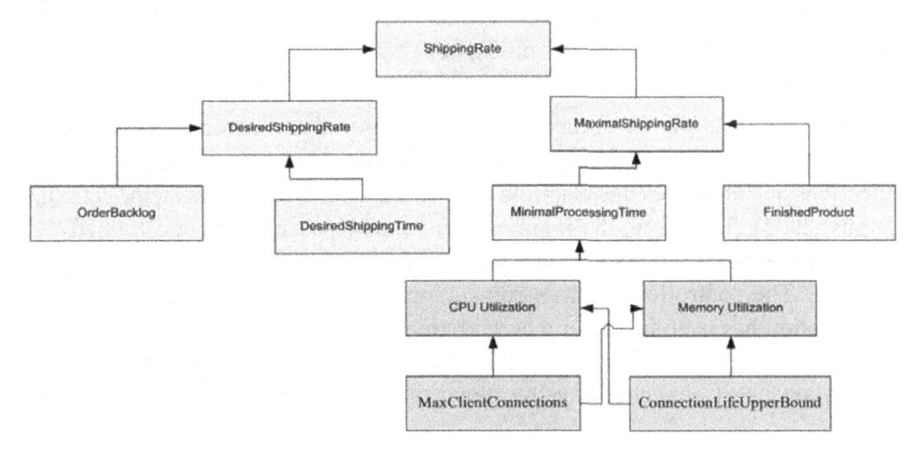

gular boxes) and each has its own in- and outflows. We omitted the part of handling payment for simplicity. The stock "Order Backlog" in Figure 10 corresponds to the data repository "Order Backlog" in Figure 9. The activity "Place Order" by the customer will have the effect of increasing the stock "Order Backlog" and changing the inflow "Demand Rate". Similarly, the activity "Shipping out Products" after the storage would have the effect of decreasing the stock and changing the outflow "Fulfillment Rate." The stock "Finished Product" is the result of the activity "Make Product" and will be reduced by the activity "Shipping out Products." The stock "Part WIP" is increased by "Order Parts" and is reduced by "Receive Parts", and the stock "Part Inventory" will be increased due to the effect of "Receive Parts" and reduced by the activity "Make Product."

It is less clear how the activities in a business-process model affect in- and outflow rates for each stock. In the business-process model, dependencies can be expressed graphically only through data streams. Any additional dependencies could be buried in the input criterion and output criterion of the data stream for each task. It is possible to use built-in expressions or plugged-in programming languages to express the input and output criteria. On the other hand, the dependencies in an SDM can be expressed graphically with the notion of polarity, wherein the positive polarity (denoted by a plus sign) represents reinforcement feedback loops and the negative polarity (denoted by a minus sign) represents those feedback loops that can reach equilibrium. The influence map can be created by connecting direction links. Based on the visual influence map, low-level metrics can be synthesized by capturing the causal relationships in the form of mathematical formulas. The key to bridging the gap between the business operations and the IT systems is to establish the dynamic causal relationships between IT-level and business-level performance metrics in the mathematical formulas. Figure 11 gives an example causality tree for "ShippingRate." The business-level variables are shown in yellow rectangular boxes; the IT-level variables are shown in green rectangular boxes. This causality tree corresponds to the business activities between "Check Order and Product Status" and "Shipping out Product" shown in Figure 9. We discuss this causality tree in a bottom-up order.

Let us assume the above inventory system is monitored and controlled by a main process wherein inventory processes can handle client requests through network connections one at a time (this example is adjusted from Diao, Hellerstein, Parekh, & Bigus, 2003). Therefore, the number of inventory processes is constrained by the maximum number of connections, say "MaxClientConnections," provided by the system. The controller monitors connections and manages their life cycles. If the connection has been idle over a certain amount of time, say "ConnectionLifeUpperBound," the connection is terminated or returned to the connection pool. A higher "MaxClientConnections" value allows the system to process more inventory requests and increases both CPU (central processing unit) and memory utilization. Decreasing the value of "ConnectionLifeUpperBound" potentially allows inventory processes to be more active, leading to higher CPU and memory utilization. The above description can be formulated as follows (Diao et al.):

$$CPU_{k+1} = A_{1,1} * CPU_k + A_{1,2} * MEM_k + B_{1,1} * MaxClientConnections_k + B_{1,2} * ConnectionLifeUpperBound_k$$

and

$$MEM_{k+1} = A_{2,1} * CPU_k + A_{2,2} * MEM_k + B_{2,1} * MaxClientConnections_k + B_{2,2} * ConnectionLifeUpperBound_k,$$

where CPU_k and MEM_k represent the values of CPU and memory utilization at the *k*th time interval. The metrics $MaxClientConnections_k$ and $ConnectionLifeUpperBound_k$ represent the values of "MaxClientConnections" and "ConnectionLifeUpperBound" at the *k*th time interval. The entries $A_{i,j}$ and $B_{i,j}$ represent modeling parameters at the IT level that can be obtained using statistical methods. In every time period, the controller has to decide how much of the available resources (CPU, memory) to allocate to each inventory process. How to realize this control strategy is beyond the scope of this chapter. Here, we only show how the business-level and IT-level metrics can be linked using the causality relationships of an SDM.

The metrics "MinimalProcessTime" at the business level depends on the metrics at the IT level such as CPU and memory utilization. Good "MinimalProcessTime" is ensured by reserving sufficient capability to handle workload spikes. A particular IT-system configuration by tuning parameters such as "MaxClientConnections" and "ConnectionLifeUpperBound" gives a particular "MinimalProcessTime."

"DesiredShippingTime" can be assigned a proper value based on the average values of real processing time. The "DesiredShippingRate" is determined by "OrderBacklog" and "DesiredShippingTime":

$$DesiredShippingRate = \frac{OrderBacklog}{DesiredShippingTime}.$$

The "MaximalShippingRate" is determined by "FinishedProduct" and "MinimalProcessingTime":

$$MaximalShippingRate = \frac{FinishedProduct}{MinimalProcessingTime}.$$

Finally, the "ShippingRate" is determined by "DesiredShippingRate" and "MaximalShippingRate":

$$ShippingRate = \min(DesiredShippingRate, MaximalShippingRate).$$

An SDM tends to represent a deterministic model and is used to study the overall behavior in a certain given timescale. Compared to the business-process model, the SDM aims to cover the detailed product-shipping activities by mimicking real-world events on a smaller scale. The assigned values in an SDM can be obtained from the simulation results of the business-process model or the average of the real data that are identified in the business artifacts of the business-process model.

We give a few more examples of how to synthesize low-level metrics to meet overall service requirements. The stock "Finished Product" is increased through the activity "Make Product" by using "Product Demand" that comes from "Forecast Demand" and "Adjust Product Based on Finished Product." The forecast model we have here is the exponential smoothing of historical data:

$$SmoothedDemandRate = Smooth(DemandRate, SmoothFactor),$$

where the function *smooth* is one of the built-in functions in the SDM tool. The adjustment from "Finished Product" introduces a negative feedback loop to rebalance the amount of products we should assemble:

$$DesiredProductInventory = smoothedDemandRate * ProductInventoryCoverage$$

$$DesiredAssemblingRate = SmoothedDemandRate + \frac{DesiredProductInventory - FinishedProduct}{InventoryAdjustmentTime}.$$

The real assembling rate could be delayed:

$$AssemblingRate = DelayFixed(DesiredAssemblingRate, DelayTime).$$

The part usage rate is transformed from "Assembling Rate" using "Bill of Material":

$$PartUsingRate[i] = \sum_{j} (BillOfMaterial[i, j] * AssemblingRate[j])$$

$$PartSmoothedDemandRate = Smooth(PartUsingRate, SmoothFactor).$$

The part demand, determining how much to order from suppliers, can be adjusted based on "Part Inventory" (another negative feedback loop):

$$PartDesiredInventory = PartSmoothedDemandRate * PartInventoryCoverage$$

$$PartDesiredDeliveryRate = PartSmoothedDemandRate + \frac{PartDesiredInventory - PartInventory}{PartInventoryAdjustmentTime}.$$

In an SDM, the dynamics of a business process is captured through stock flow diagrams with dependency graphs and mathematical formulations (systems of ordinary differential equations). A simulation of this system would expose its dynamic behavior in the time dimension. Because of the complexity of the system with nonlinearity and time delay, we may not be able to solve the system analytically. Based on available numerical methods for ordinary differential equations, like Euler's first-order finite difference and the Runge-Kutta second- and fourth-order finite-difference methods, the system can be solved numerically. Both Matlab (2005) and Vensim (2005) provide such equation solvers. Furthermore, the tool provided by Vensim helps modelers to determine visually what kind of action should be taken, such as changing system settings.

Dynamic Adaption

Background

We live in a dynamic and fast-changing environment. The agility of an enterprise not only depends on the sensibility of the management, but also on the responsiveness of its IT infrastructure in line with the changes in the business environment. This is the challenge to realize the so-called on-demand business (IBM, 2005c).

Primarily, there are two types of changes of business-level decisions that require the IT infrastructure to respond quickly: functional and nonfunctional. As the result of functional changes, the business-process model will change as well, for example, acquiring an additional arc, an additional node, or a modified attribute value. Any small adjustments to the business-process model potentially can change the proper-

Figure 12. An example flow graph from Aho et al. (2006)

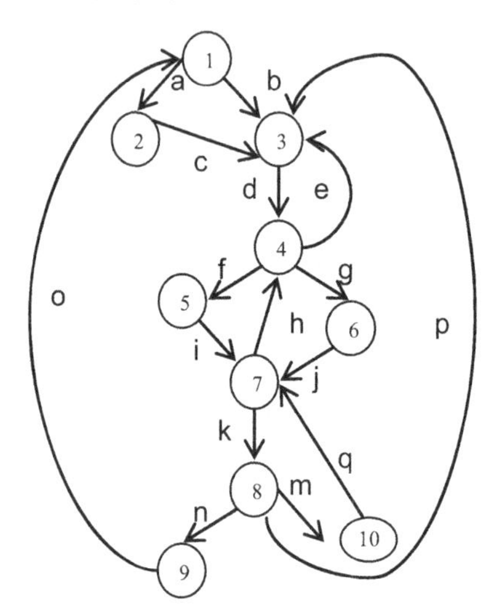

ties of the model as a whole. For instance, one more arc in the model could change a reducible process model to an irreducible one. Thereafter, the generated BPEL code should be updated accordingly. In general, it is not a good idea to regenerate the whole business model upon any changes, especially for large-scale process models. An algorithm is presented in the chapter that can scope and localize the changes so that only the subprocess in the scope will be regenerated.

A nonfunctional alteration of the business-process model does not change the properties of the process model, but reestablishes the linkage with the IT service that implements the same functionality while offering a better QoS (quality of service) value such as a higher response time and availability. We shall notice that our criterion of categorizing functional and nonfunctional adaptation is mechanical rather than semantic. To semantically verify whether the functionality of the process model is changed is not computable or at least fuzzy if we use the techniques to verify the process model against some formal functionality specification. Hence, a revision of the structure of a process model in order to achieve a higher overall QoS value, but with the same functionality, is categorized as a functional change according to our criterion.

Both functional and nonfunctional adaptation can be performed statically or dynamically. Static adaptation requires, in addition to a break in service availability, the regeneration of executable code from the process model, reloading, redeployment, and a relinking with remote IT services. In some mission-critical scenarios such as finance and military applications, there is a need for a continuous guarantee of service availability. Dynamic adaptation manipulates a business process in its IT-infrastructure run-time environment while maintaining the availability of services. Later, we discuss our experiment by using a technique that enables the dynamic adaptation of business processes running in the Microsoft .NET® (Microsoft, 2005) Web-service environment.

Localize the Changes

The main theory used to achieve change localization in a process model is the concept of a two-terminal region from control flow-graph analysis (Allen, 1970) in compiler theory. We start by introducing several relevant definitions.

Definition 2. Given a flow graph G with initial node N_0 and a node N of G, the interval with header N, denoted I(N), is defined as follows (Aho et al., 2007):

1. N is in I(N),
2. if all the predecessors of some node $M \neq N_0$ are in I(N), then M is in I(N), and
3. nothing else is in I(N).

The corresponding REL for this graph is ((a c + b) (d((f i + g j) (k [p n] mq)* h) * (p + e))* n o) *.

Based on Definition 2, we can obtain a set of intervals for Figure 12 as follows:

I(1) = {1, 2}

I(3) ={3}

I(4) ={4, 5, 6}

I(7) ={7, 8, 9, 10}.

One distinguished property of intervals is that all the intervals in one flow graph are disjoint.

A two-terminal region was originally defined as follows.

Definition 3. A two-terminal region is an interval with one exit node (Allen, 1970).

Figure 13. The correspondence of REL constructs and two-terminal regions

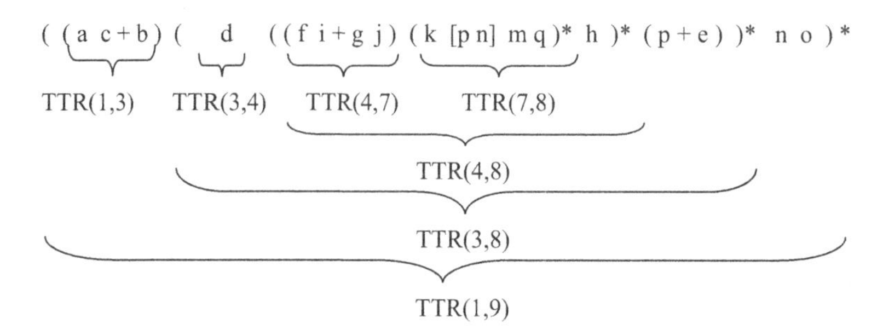

Intuitively, our goal of localizing the changes in a process model is to locate some regions within which changes do not affect other regions. Definition 3 does not satisfy our needs as a lot of regions are excluded from it; for example, I(1) and I(4) are not two-terminal regions. We thus adopted and modified the definition of two-terminal regions from Koehler, Hauser, Sendall, and Wahler (2005), in which two-terminal regions are used to partition the process model into subgraphs so that different subgraphs can be analyzed and transformed using different algorithms based on their specific properties. In fact, the definition of a two-terminal region in Koehler et al. results in different sets from that of Allen (1970); the definition we give in this chapter yields different sets from both Allen and Koehler et al. It is not our intention to compare the detailed differences in this chapter. Therefore, we straightly give our definition.

Definition 4. Node N dominates (or predominates) node M if every path from the initial node of the flow graph to M goes through N (Aho et al., 2007).

Definition 5. Node N postdominates node M if every path from M to the exit node goes through N (Allen, 1970).

Definition 6. A two-terminal region is a subgraph TTR(N,M) between N and M such that:

1. N predominates all nodes in the region, and
2. M postdominates all nodes in the region.

Based on Definition 6, we can obtain a set of two-terminal regions of Figure 12 as follows:

TTR(1, 3) ={1, 2, 3}

$TTR(3, 4) = \{3, 4\}$

$TTR(4, 7) = \{4, 5, 6, 7\}$

$TTR(7, 8) = \{7, 8, 10\}$

$TTR(4, 8) = TTR(4,7) \cup TTR(7,8)$

$TTR(3, 8) = TTR(3,4) \cup TTR(4,8)$

$TTR(1, 9) = TTR(1,3) \cup TTR(3, 8) \cup \{9\}.$

Different from intervals, two-terminal regions can be nested within each other or overlap with their entry or exit nodes. Now we are ready to define change localization.

Proposition 1. Changes can be localized into the closest enclosing two-terminal region if and only if the changes only affect the nodes (other than the entry or exit nodes) of this region.

Intuitively, a two-terminal region is a subgraph where everything coming from the outside of the region into the region goes through the entry node, and everything going from the inside of the region to the outside goes through the exit node. In REL (and hence in BPEL), a two-terminal region thus corresponds to a single construct. Shown in Figure 13, TTR(1, 3) corresponds to an "or" construct in REL and hence

Figure 14. The overview of the dynamic updating approach

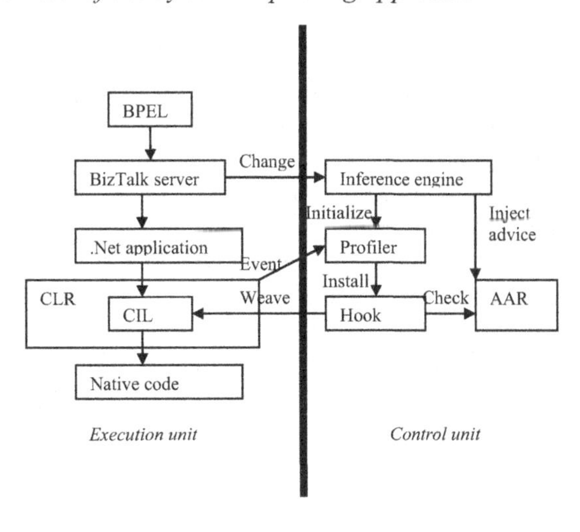

a "switch" construct in BPEL; TTR(7, 8) corresponds to a "star" construct in REL and hence a "while" construct in BPEL. Proposition 1 tells us to localize the changes onto one single construct. In implementation, a single construct is represented by a single syntax-tree node, therefore the adaptation is as easy as replacing a single syntax-tree node. To give an example, suppose the node 2 in Figure 12 has a self-loop denoted by letter x. Only TTR(1, 3) will be changed to (a x* c+b); the rest of the REL sentence remains the same.

Dynamic Updating

Dynamic updating requires the manipulation of the low-level run-time environment. Our first experiment was performed on the Microsoft .NET® Web-service environment. We only sketch out our infrastructure for this experiment in this section. Detailed discussion on this technique that was originally proposed for dynamic Web-service composition and provisioning can be found in Cao, Bryant, Liu, and Zhao (2005) and Cao, Bryant, Raje, et al. (2005).

Figure 14 provides an overview of the dynamic updating infrastructure. In the left pane of the execution unit, the generated BPEL is imported into the Microsoft BizTalk® Server (Microsoft, 2004) that is built on top of the .NET framework. .NET applications run over the common language runtime (CLR) environment where the .NET application is captured in the form of the common intermediate language (CIL; Gough, 2002). CIL code is just-in-time (JIT) compiled into native code and executed. Therefore, by manipulating CIL, a BPEL process can be adapted at run time.

The manipulation of CIL is illustrated in the right pane of the control unit. An inference engine detects changes from the model-editing environment and computes the location for new-code insertion. The manipulation of CIL at run time requires the interception of the managed execution. Instead of reimplementing the CLR, such as rewriting the open-source CLR Rotor (Stutz, Neward, & Shilling, 2003) to invasively add a listener for execution interception, compromising the portability of CLR, we use a pluggable, configurable CLR profiling interface (Microsoft, 2002) to achieve this goal, which can be enabled and disabled based on different needs. In contrast to the conventional publisher-listener model, which is often a client-server relationship, the profiler will be mapped into the same address space for the profiled application as an in-process server. The profiler can be initiated by the inference engine. The events generated from the CLR are the result of managed execution, including but not limited to garbage collection, class loading and unloading, CLR start-up and shutdown, and JIT compilation. The event of our interest is JIT compilation, for which we have implemented in-memory CIL manipulation for the event handler. The adapted CIL is then JIT compiled and executed, resulting in changed business-process behavior.

At the execution level, the functional and nonfunctional adaptations have the same mechanical effects and thus are not distinguished by the run-time adaptation system: Functional adaptation requires a replacement of a single construct that corresponds to a modified subprocess, and nonfunctional adaptation requires a replacement of a specific Web-service invocation statement. In some occasions, a construct need not be removed, but instead needs to be decorated and enriched to realize changes such as the addition of a new variable, security and access control, or a changed local business rule.

Because we only need two updating operations—replacement and decoration—we use the idea of aspect-oriented programming (AOP; Kiczales et al., 1997) to perform CIL-code manipulation. The reference engine can dynamically inject the predefined reusable adaptation advice in the compiled managed code form into the adaptation advice repository (AAR). The AAR is located in shared memory for fast access during in-memory CIL manipulation. The *hook* code, installed on demand by the profiler, will weave into CIL the applicable adaptation advice. There are three types of advice: before advice, after advice, and around advice. The before advice performs some preprocessing before the actual method execution, while the after advice performs some postprocessing immediately before the method execution returns. Both the before and after advice are for method decoration. The around advice is used for overriding original constructs or methods. Also included in the *hook* code are the instructions to check for whether an around advice is specified or not, and whether there is a jump instruction to redirect the execution to the exit point of the specific block if an around advice is specified.

Conclusion and Future Trends

The theme of this chapter is to bridge the gap between business operations and IT execution. We have explained this connection from three different perspectives: (a) how to transform business decisions into IT-level execution, (b) how to monitor IT execution and synthesize IT metrics to meet business-level commitments, and (c) how to quickly and dynamically update IT execution when business-level decisions change.

It is an emerging trend that companies are realizing the importance of business-process management, integration, and monitoring. SLA management also catches many enterprises' attention. Based on this demand, big players in consulting services and software tool construction have invested energy into tools and software for business-process management and SLA management, for example, IBM WebSphere Business Integration Modeler, the Microsoft BizTalk server, Oracle BPEL Process Manager (Oracle, 2005), and ARIS Design Platform (IDS-Scheer, 2005) in cooperation with SAP NetWeaver® (2006). It is important to establish communication between academic researchers and industry tool builders to stimulate new research topics and at the same time to disseminate research results into the real world.

There are still many open research problems such as novel model-transformation techniques, business-process-model analysis and optimization techniques, and round-trip (from modeling, execution, and monitoring to model modification) engineering. Part of our future work is to design a set of static model-analysis techniques based on the mechanical power of REL, similar to static analysis performed when a computer program is compiled. Those static analysis techniques could become an excellent resource for static model optimization.

References

Agrawal, A., Karsai, G., & Shi, F. (2003). *Graph transformations on domain-specific models* (Tech. Rep. No. ISIS-03-403). Vanderbilt University, Institute for Software Integrated Systems. Retrieved from http://www.isis.vanderbilt.edu/view.asp?GID=846&CAT=4

Aho, A. V., Lam, M. S., Sethi, R., & Ullman, J. D. (2007). *Compilers-principles, techniques, and tools* (2nd ed.). Addison-Wesley.

Akehurst, D. H., & Kent, S. (2002). A relational approach to defining transformations in a metamodel. *Proceedings of UML 2002*, (pp. 243-258).

Allen, F. (1970). Control flow analysis. *ACM SIGPLAN Notices*, *5*(7), 1-19.

An, L., & Jeng, J. J. (2005). Web service management using system dynamics. *Proceedings of the 2005 IEEE International Conference on Web Services (ICWS)*, (pp. 347-354).

An, L., Jeng, J. J., Ettl, M., & Chung, J. Y. (2004). A system dynamics framework for sense-and-response systems. *Proceedings of the IEEE International Conference on E-Commerce Technology for Dynamic E-Business (CEC-East'04)*, (pp. 6-13).

An, L., & Ramachandran, B. (2005). System dynamics model to understand demand conditioning dynamics in supply chain. *Proceedings of the 23rd International Conference of the System Dynamics Society*. Retrieved from http://www.systemdynamics.org/conf2005/proceed/papers/AN140.pdf

Andries, M., Engels, G., Habel, A., Hoffmann, B., Kreowski, H.-J., Kuske, S., et al. (1999). Graph transformation for specification and programming. *Science of Computer Programming, 34*(1), 1-54.

Bitpipe. (2005). Retrieved from http://www.bitpipe.com/tlist/SLA-Management-Services.html

BPMI. (2005). Retrieved from http://www.bpmi.org/

BPMN specification. (2004). Retrieved from http://www.bpmn.org/Documents/BPMN%20V1-0%20May%203%202004.pdf

Business process execution language for Web services, version 1.1. (2003). Retrieved from ftp://www6.software.ibm.com/software/developer/library/ws-bpel.pdf

Cao, F., Bryant, B. R., Liu, S.-H., & Zhao, W. (2005). A non-invasive approach to dynamic Web service provisioning. *Proceedings of the 2005 IEEE International Conference on Web Services*, (pp. 229-236).

Cao, F., Bryant, B. R., Raje, R. R., Olson, A. M., Auguston, M., Zhao, W., et al. (2005). A non-invasive approach to assertive and autonomous dynamic component composition in service-oriented paradigm. *Journal of Universal Computer Science, 11*(10), 1645-1675.

Carter, L., Ferrante, J., & Thomborson, C. (2003). Folklore confirmed: Reducible flow graphs are exponentially larger. *Proceedings of the 30th ACM SIGPLAN-SIGACT Symposium on Principles of Programming Languages*, (pp. 106-114).

Cleaveland, J. C. (2001). *Program generators with XML and JAVA.* Prentice Hall.

Cocke, J., & Miller, E. R. (1969). Some analysis techniques for optimizing computer programs. *Proceedings of the 2nd Hawaii International Conference on Systems Sciences (HICSS)*, (pp. 143-146).

Codagen Architect 3.0. (2006). Retrieved from http://www.codagen.com/products/architect/default.htm

Compuware. (2005). *OptimalJ.* Retrieved from http://www.compuware.com/products/optimalj/

Czarnecki, K., & Helsen, S. (2003). Classification of model transformation approaches. *Proceedings of the OOPSLA 2003 Workshop on Generative Techniques in the Context of Model-Driven Architecture.* Retrieved from http://swen.uwaterloo.ca/~kczarnec/ECE750T7/czarnecki_helsen.pdf

Denning, P. J., Dennis, J. B., & Qualitz, J. E. (1978). *Machines, languages, and computation.* Prentice-Hall, Inc.

Diao, Y., Hellerstein, J. L., Parekh, S., & Bigus, J. P. (2003). Managing Web server performance with AutoTune agents. *IBM Systems Journal, 42*(1), 136-149.

Forrester, J. W. (1961). *Industry dynamics.* Cambridge, MA: MIT Press.

Gamma, E., Helm, R., Johnson, R., & Vlissides, J. (1995). *Design patterns: Elements of reusable object-oriented software.* Addison-Wesley.

Gough, J. (2002). *Compiling for the .NET common language runtime (CLR).* Prentice Hall PTR.

Hoffner, Y., Field, S., Grefen, P., & Ludwig, H. (2001). Contract-driven creation and operation of virtual enterprise. *Computer Networks, 37*, 111-136.

IBM. (2005a). *Rational Software Architect.* Retrieved from http://www-306.ibm.com/software/awdtools/architect/swarchitect/index.html

IBM. (2005b). *WBI Modeler.* Retrieved from http://www-306.ibm.com/software/integration/wbimodeler/

IBM. (2005c). *The who, what, when, where, why and how of becoming an on demand business.* Retrieved from http://t1d.www-306.cacheibm.com/e-business/ondemand/us/pdf/ExecGuide1214.pdf

IDS-Scheer. (2005). *ARIS design platform.* Retrieved from http://www.ids-scheer.com/international/english/products/aris_design_platform/49623

Janssen, J., & Corporall, H. (1997). Making graphs reducible with controlled node splitting. *ACM Transactions on Programming Languages and Systems, 19*(6), 1031-1052.

Keller, A., Kar, G., Ludwig, H., Dan, A., & Hellerstein, J. L. (2002). Managing dynamic services: A contract based approach to a conceptual architecture. *Proceedings of the 8th IEEE/IFIP Network Operations and Management Symposium (NOMS)* (pp. 513-528).

Kiczales, G., Lamping, J., Mendhekar, A., Maeda, C., Videira Lopes, C., Loingtier, J. M., et al. (1997). Aspect-oriented programming. *Proceedings of the European Conference for Object-Oriented Programming (ECOOP)* (pp. 220-242).

Koehler, J., Hauser, R., Sendall, S., & Wahler, M. (2005). Declarative techniques for model-driven business process integration. *IBM Systems Journal, 44*(1), 47-65.

Kong, R. (2005). *Transform WebSphere business integration modeler process models to BPEL* (Tech. Rep.). IBM Toronto. Retrieved from http://www3.software.ibm.com/ibmdl/pub/software/dw/wes/pdf/0504_kong_transform_bpel.pdf

Kumaran, S., & Nandi, P. (2003). *Adaptive business object: A new component model for business applications* (White paper). IBM T. J. Watson Research Center. Retrieved from http://www.research.ibm.com/people/p/prabir/ABO.pdf

Mantell, K. (2003). *From UML to BPEL: Model driven architecture in a Web services world* (Tech. Rep.). IBM Corp. Retrieved from http://www-128.ibm.com/developerworks/webservices/library/ws-uml2bpel/

Matlab. (2005). Retrieved from http://www.mathworks.com

Microsoft. (2002). *Common language runtime profiling.*

Microsoft. (2004). *Understanding BizTalk Server 2004* (White paper). Retrieved from http://www.msdn.microsoft.com/library/default.asp?url=/library/en-us/BTS2004IS/htm/understanding_abstract_syfs.asp?frame=true

Microsoft. (2005) *.Net.* Retrieved from http://www.microsoft.com/net

Nainani, B. (2004). *Closed loop BPM using standards based tools* (Tech. Rep.). Oracle Corp. Retrieved from http://www.oracle.com/technology/products/ias/bpel/pdf/bpm.closedloop.pdf

Nigam, A., & Caswell, N. S. (2003). Business artifacts: An approach to operational specification. *IBM Systems Journal, 42*(3), 428-445.

Oracle. (2005). *BPEL process manager.* Retrieved from http://www.oracle.com/technology/products/ias/bpel/index.html

SAP NetWeaver. (2006). Retrieved from http://www.sap.com/solutions/netweaver/index.epx

Schach, S. (2005). *Object-oriented and classical software engineering* (6th ed.). McGraw-Hill.

Sterman, J. D. (2000). *Business dynamics: System thinking and modeling for a complex world.* Irwin McGraw-Hill.

Sterman, J. D. (2002). System dynamics modeling: Tools for learning in a complex world. *IEEE Engineering Management Review, 30*(1), 42-52.

Stutz, D., Neward, T., & Shilling, G. (2003). *Shared source CLI: Essentials.* O'Reilly Press.

UML 2.0 superstructure final adopted specification. (2003). Retrieved from http://www.omg.org/cgi-bin/doc?ptc/2003-08-02

Vensim. (2005). Retrieved from http://www.vensim.com/software.html

Zhao, W., Bhattacharya, K., Bryant, B. R., Cao, F., & Hauser, R. (2005). Transforming business process models: Enabling programming at a higher level. *Proceedings of the 2005 IEEE International Conference on Services Computing (SCC)* (pp. 173-180).

Zhao, W., Bryant, B. R., Cao, F., Hauser, R., Bhattacharya, K., & Tao, T. (in press). Transforming business process models in the presence of irreducibility and concurrency. *International Journal of Business Process Integration and Management.*

Zhao, W., Hauser, R., Bhattacharya, K., Bryant, B. R., & Cao, F. (2006). Compiling business processes: Untangling unstructured loops in irreducible flow graphs. *International Journal of Web and Grid Services, 2*(1), 68-91.

Chapter III

Service Portfolio Measurement (SPM): Assessing Financial Performance of Service-Oriented Information Systems

Jan vom Brocke,
European Reseach Center for Information Systems (ERCIS),
University of Münster, Germany

Abstract

This chapter addresses service-oriented information systems from a management perspective. It is evident that running a service-oriented enterprise brings up new challenges for management. Given the technological opportunities, the challenge lies essentially in choosing the right mix of services on the basis of an appropriate architecture. For this purpose, strategic considerations regarding, for example, the company's flexibility have to be justified by financial performance measures. This is particularly evident as long-term economic consequences result from decisions on the service portfolio. Thus, evidence is required about the fact that these decisions are in alignment with the company's financial situation. The total costs of ownership (TCO) caused by a particular service-oriented information system, as well as the return on investment (ROI) gained by it, give examples for appropriate financial performance measures. In this chapter, a measurement system is presented that facilitates the assessment of the various financial consequences within a comprehensive framework. The system is grounded in decision theory and capital budgeting, and it is illustrated by its application within practical examples.

The Challenge of Managing Service-Oriented Information Systems

In today's markets, enterprises are increasingly forced to act flexibly. To do so, there is a distinct trend for enterprises to concentrate on core competences in order to gain strategic competitive advantages. As a precondition, information systems are required that incorporate means to support this flexibility. For this purpose, the concept of enterprise service computing offers promising ways to design a company's information system. In service-oriented architectures (SOAs; Loh & Venkatraman, 1992; Weikum & Vossen, 2002), processes of an information system can be extracted and out-tasked to service providers. According to Keen and McDonald (2000), "Out-tasking...breaks a company into a portfolio of process-centred operations rather than interlocking departments or functions."

The deployment of service-enterprise computing puts companies in a position to concentrate on their core competences by sourcing out parts of a process to service providers. In contrast to conventional outsourcing, out-tasking enables the companies to keep control of the entire process at the same time (vom Brocke & Lindner, 2004). According to Forrester, companies with a service-oriented architecture can reduce costs for the integration of projects and maintenance by at least 30% (Vollmer &

Figure 1. Positioning out-tasking as a sourcing strategy

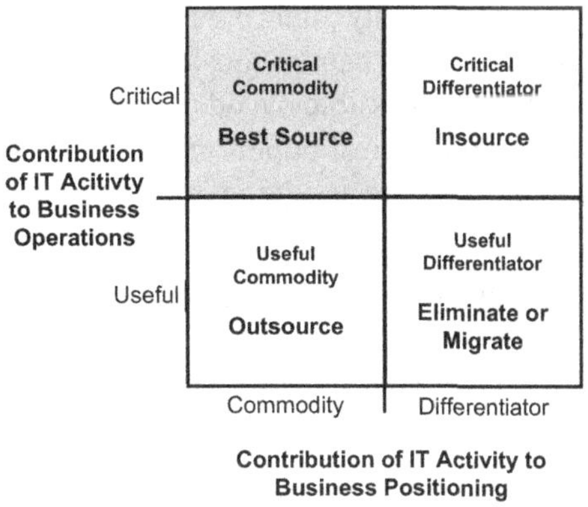

Gilpin, 2004). Major providers of ERP systems incorporate service-oriented architectures in their solutions: Sonic ESB by Sonic Software (Craggs, 2003), mySAP Business Suite by SAP (2004), e-Business on Demand by IBM (2004), and the Application Server by Oracle (2004). As a future trend, Gartner (2002) predicts that by 2007, most company frameworks will have changed to service-oriented architectures (Farber, 2004).

In order to differentiate outsourcing from various sourcing strategies, a framework provided by Lacity, Willcocks, and Feeny (1996) can be applied.

The approach aims at structuring business processes according to characteristics that are relevant for sourcing decisions. In particular, two dimensions are applied: the importance of activities for the operation of business processes and its contribution to strategic business positioning. Against the background of the approach, the following service categories can be differentiated.

- **Critical differentiators:** Tasks of high strategic and operative relevance are suggested to be fully in-sourced. An outsourcing of these tasks would result in a loss of know-how and innovation potential, and would threaten core competences.
- **Useful commodities:** Tasks that are neither strategically nor operatively outstanding are candidates to be fully outsourced. In these situations, the potential of reducing costs by aid of outsourcing can likely be realised.
- **Useful differentiators:** For tasks that are considered to be highly relevant for strategic differentiation but that show little relevance in operational business, neither in- nor outsourcing is recommended. They should be eliminated sooner or later.
- **Critical commodities:** Tasks that are customary in trade but show little strategic relevance should be controlled inside but operated outside the company. That way, the positive effects of specialisation can be realised (Quinn, 1999).

As for the model, out-tasking especially offers potentials for the case of critical commodities. In this respect, out-tasking can be considered as the realisation of a best sourcing strategy that is rendered possible by aid of service-oriented architectures. In order to be able to use technological opportunities for the support of business strategies, new management tasks are arising in information-systems science.

As information systems offer means to outsource services, the question arises of which combination of the various services available should be chosen for the specific needs of a company. For that purpose, appropriate service-portfolio management is required. On the whole, this means that a management process has to be established for the appropriate composition of a corporate service portfolio. In order to support management, methods for performance measurement of a company's service portfolio are required (Kaplan & Norton, 1992). In this chapter, these methods are referred to as service-portfolio measurement (SPM).

In service-portfolio measurement, multiple perspectives of the profitability of service-oriented information systems have to be taken into account. This can, for example, be seen from the work of Kaplan and Norton (1992, 1996). They have illustrated that corporate management should not only reflect financial aspects, but also nonfinancial ones as the performance in these dimensions eventually drives the financial results (Johnson & Kaplan, 1987; Kaplan, 1986; Neely, 2004). Accordingly, preliminary works that may be used in service-portfolio measurement can be differentiated regarding their perspectives on financial and nonfinancial aspects. With respect to both aspects, related works can be found in the field of decision support for outsourcing, which has in part already been transferred to service-oriented computing.

- **Nonfinancial assessments:** Nonfinancial assessments are essentially based on argumentations, partly structured by means of pros and cons lists (Knolmayer, 1997), checklists (Buck-Lew, 1992; Kador, 1990; Kascus & Hale, 1995), analytical hierarchy process models (Putrus, 1992), and flowcharts (Knolmayer). These approaches give a good basis for path-leading decisions on information-systems design. For decision making, however, a stronger methodological foundation is required. Such a foundation can be found in studies assessing the quality of service (QoS) in service-oriented enterprises (Cardoso, Sheth, Miller, Arnold, & Kochut, 2004; Wang, Chen, Wang, Fung, & Uczekaj, 2004). These contributions can well be used for operational service discovery and process control. However, they do not provide decision support for finding the most profitable mix of services within the company's portfolio. For this purpose, financial approaches are additionally required.
- **Financial assessments:** Financial assessments focus on cost analyses, such as special task comparisons (Espinosa & Carmel, 2004), multitask cost comparisons, and holistic cost-risk comparisons (Aubert, Patry, & Rivard, 2002; Bahli & Rivard, 2003; Jurison, 2002). Special work in the field of costs of Web services has been carried out in the field of service detection and composition with a rather operational focus (Cardoso et al., 2004; Day & Deters, 2004). The main disadvantage of these approaches, however, is that they are built on a cost basis and, thereby, apply a measurement system for short-term planning. Decisions on the information-systems architecture, however, drive long-term economic consequences. Apart from system payments, influences on payments related to interests and taxes also have to be taken into account during the information system's life cycle. Thus, methods of capital budgeting have to be applied (Grob, 1993; Seitz & Ellison, 2004; Shapiro, 2004).

A first approach to apply capital budgeting in service-oriented computing has recently been presented (vom Brocke & Lindner, 2004). In this work, the monetary consequences of a service-oriented architecture are evaluated and opposed to those of conventional architectures. This approach can well be used as a methodological basis in order to develop a performance-measurement system for service-oriented information systems. An essential requirement of design is to consider the various different configurations of services in the measurement system that are applicable from different service providers with various conditions in pricing and service levels. This gives ground for decisions on sourcing strategies and thereby facilitates the finding of the most adequate mix of services for an enterprise.

In this chapter, design principles of an appropriate measurement system for the financial performance of service portfolios will be presented. In order to find these principles, a design-science approach is applied (Hevner, March, Park, & Ram, 2004). Hence, the concept of an appropriate measurement system is introduced on the basis of basic principles of decision theory and capital budgeting. The system is then applied in a practical example that serves as a proof of concept. Finally, major results as well as limitations are summed up and further research is pointed out.

A Measurement System for Financial Performance of Service-Oriented Information Systems

Framework

The decision-support system is mainly structured in two dimensions that are integrated: the level and the subject of calculation. The framework of the system is displayed in Figure 2 and will be illustrated in the following.

The system has three levels of evaluation: the process level, the budgeting level, and the corporate level. The process level provides a basis for the entire evaluation. On this level, payments (out-payments) and receivables (in-payments) brought about by the information-system design are analysed. On the budgeting level, additional parameters are taken into account. These are relevant for judging the economic value created by series of payments. Relevant parameters are derived from the specific conditions of funding and taxes that a company has to face. These series of payments are consolidated over time by applying methods from capital budgeting. That way, a survey of financial consequences is created. Finally, on the corporate level, the profitability of the information-system design has to be compared to alternative investments available for the company. Measures like the total cost of ownership (TCO) and the return on investment (ROI) help one to consider relevant parameters for this purpose (Seitz & Ellison, 2004; Shapiro, 2004).

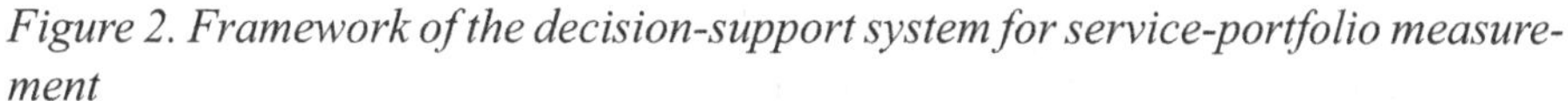

Figure 2. Framework of the decision-support system for service-portfolio measurement

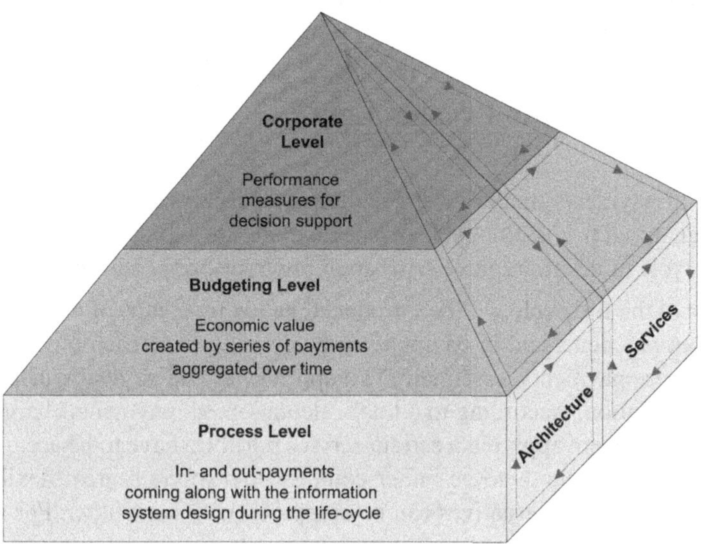

In order to use this methodological framework as a decision-support system for the management of service-oriented information systems, special subjects of measurement have to be taken into account. Essentially, these are the architecture and the services applied in a company's information system. For each subject, different types of payments have to be considered on the process level. With respect to the architecture, payments for development, operation, adaptation, and disintegration of a certain architectural design have to be considered. These payments set a long-term frame for short-term payments that are driven by a certain service portfolio that a company is running on the basis of the architecture. Each service available for a company within a certain process has to be evaluated and specified within the decision-support system. The evaluation is based on payments, taking into account various pricing models of services. This also comprises the operational availability of the service due to various service-level agreements. Thanks to the specific evaluation of each service, varied configurations of the service portfolio can be sampled by management. The alternative configuration can hence be compared, and the most profitable one can be chosen for a company's situation.

Once the payments are planned on the process level, both the architecture and the services are calculated by the same methods on both the budgeting and the corporate level. In the framework of the system, this is indicated by giving the upper levels the shape of a roof. In addition, the same shape shows that the payments are gradually aggregated until the financial performance of the company's information-system design is indicated by common financial measures. Within this integrated design, the decision-support system can also be used partly to support special-interest calculations. The arrows on the right of the framework describe typical measurement processes in which, for example, separate calculations of alternative portfolio configurations are carried out.

Measurement on the Process Level

The series of payments that are chargeable to a decision on the service portfolio should be analysed from a life-cycle perspective. Figure 3 gives a general structure of the life cycle of information systems applying a service portfolio.

The phases of the life cycle will be introduced below by means of analysing characteristic out-payments and in-payments to be considered in each phase. In order to identify these payments, essentially two approaches can be distinguished: total or partial calculation. According to a total calculation, all payments chargeable to an information system applying a certain service portfolio have to be accumulated. The total calculation tends to be rather complex, but offers a great flexibility of calculation as various alternatives can be compared with each other. For a partial calculation, on the contrary, only additional payments that are relevant in comparison to two alternative solutions are calculated. Partial calculation reduces the scale of the computation. However, the assessment is limited to the pair of alternatives selected.

For choosing between a partial or total approach of assessing the payments on the process level, the specific situation of the company has to be considered. In general, different situations have to be analysed regarding the assessment of the architecture and the services. Information about relevant types of payments is given by work in the field of the cost analysis of information systems (Faye Borthick & Roth, 1994; Gartner, 2002; Tam, 1992).

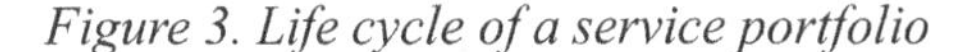
Figure 3. Life cycle of a service portfolio

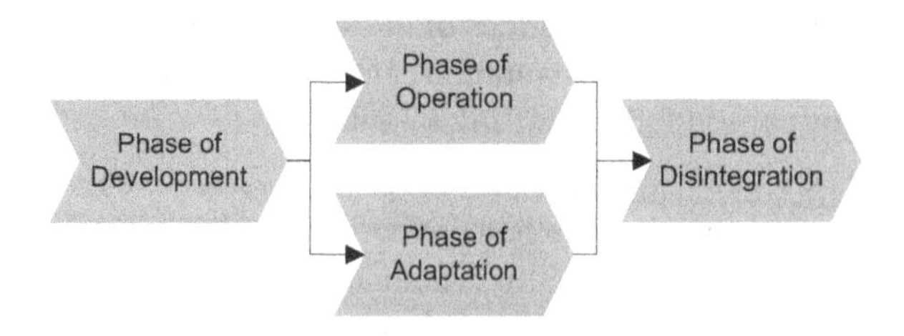

Payments Related to the Architecture

With respect to the information-systems architecture, two typical context situations have to be distinguished: projects for new information systems and projects for the redesigning of information systems. In the case of designing a new system, a total calculation may be undertaken in order to compare a wide range of alternatives. In case of a redesign, on the contrary, payments for setting in place the existing architecture are not relevant for decision making. They have to be considered as so-called sunk costs. Thus, a partial assessment appears to be adequate. As most companies in fact run certain architectures already, they are actually concerned with decisions on migrating toward service-oriented architectures. Hence, a partial calculation will be conducted in the following.

Phase of Development

In analysing the series of payments brought about by a migration toward a service-oriented architecture, payments related to purchasing hardware and software, implementing the architecture, building up know-how, and those for administration and support have to be considered. In-payments will hardly be occurring in this phase. They can result from saving on labour by not implementing services that are outsourced.

Considering the situation of a systems redesign, special aspects have to be considered. As the tasks that are likely to be outsourced have been implemented already, the payments driven by them are no more relevant for the out-tasking decision. They are classified as so-called sunk costs. In addition, further out-payments brought about by the work for redesigning have to be considered. As far as the functionality of the information system is not extended, these payments are totally to be charged to the out-tasking decision.

Phase of Operation

In TCO analyses, costs for the maintenance work on information systems and user support are usually considered during operations (Faye Borthick & Roth, 1994). Against this background, it can be argued that by out-tasking services, total payments for maintenance work are reduced by the equivalent for the work on these services. In contrast, additional payments have to be considered for the maintenance of the interfaces. Examples of this kind of maintenance work are adaptations to new versions of exchange formats on both data and services. Correspondingly, the payments related to support are derived. In addition, the implementation of a service-oriented architecture offers the potential of modernising information systems.

If legacy systems can be replaced by the purchase of an SOA, running costs would consequently be cut. Savings would be rendered possible that would be calculated as in-payments of an operational phase.

Another important effect, however, is neglected in most of the analyses: the efficiency of business processes that are enabled by the information system. In order to calculate these effects properly, Grob and vom Brocke (2005) suggest a method in which process models are used in order to identify relevant in- and out-payments brought about by a certain process design. According to this concept, out-tasking creates the opportunity to reduce the resources that are needed for running the information system. In order to calculate the amount of savings that are chargeable to a service, the values of all relevant resources as well as the rate at which they are used have to be considered.

Further in-payments can be gained during the operation phase by sharing parts of the information-systems architecture with partners. These partners can be found either inside or outside the company. Due to increasing costs and risks of an information system's architecture, these cooperations are increasingly attractive for companies in order to reach economies of scale. Therefore, service-oriented architectures offer promising means for selective service sharing. Further in-payments may be achieved as a wide range of partners can be involved in sharing information-system services.

Moreover, specific payments to the service provider have to be considered during the phase of operation. Both the amount of these payments as well as their distribution throughout the life cycle clearly vary according to the model of pricing that has been agreed upon. Also, payments for general licensing agreements have to be taken into account.

Phase of Adaptation

During the run time of an information system, adaptations will have to be made to the system. These adaptations may be necessary in order to both implement new services as well as to modify existing ones. Examples of drivers of such adaptations are technological innovations and changing demands.

Depending on the information-systems architecture, different financial consequences of these adaptations have to be taken into account. In case parts of the system that are run inside the company are affected, out-payments for the implementation of changes to the system have to be charged. Relevant indicators are both the amount of man months needed as well as the appropriate average cost rate to be calculated. In case out-tasked services are affected, these payments might be saved. However, it should be taken into account that the prices for the services provided might rise.

Moreover, it is likely that changes might have to be made that are not covered by service providers, so services might have to be in-tasked by the company again.

As the need for changes is highly uncertain, methods of probability calculation should be applied here. One approach might be to estimate out-payments for adaptations in each period during the planning horizon, for example, by using the scenario technique. In order to simplify these estimations, different types of adaptations can be defined depending on their duration and the average amount of payment. These types of adaptations can be planned by estimating the probability and the point in time they occur in.

Figure 4. Service-portfolio measurement on the process level assessing SOA

Elements of the Series of Payments with SOA			
Point in Time	0	...	n
Phase of Development			
Out-Payments			
- for building up know-how			
- for implementing SOA as a new architecture			
- due to reorganization work (redesign only)			
In-Payments			
+ due to not implementing services in-house (new design only)			
Phase of Operation			
Out-Payments			
- for additional maintenance work on the interfaces			
- for support (1st / 2nd level)			
- for conducting operation			
- for licenses with service providers			
In-Payments			
+ for shared service			
+ by savings concerning lowering resources needed			
+ by saving for maintenance work on the services			
Phase of Adaptation			
Out-Payments			
- for adaptation work on interfaces			
In-Payments			
- by savings due to adaptation work on services			
Phase of Disintegration			
Out-Payment			
- for replacement like ongoing contracts			
In-Payment			
+ by saving according to idle time costs of the resources			
+ by liquidation of technology			
Series of Payments According to Partial Calculation			

Phase of Disintegration

Finally, how far specific financial consequences of a migration to a service-oriented architecture can be foreseen has to be analysed in the phase of disintegration. A positive effect can result from the reduction of the resources that are needed in order to run the information system. Due to this effect, idle time costs can be saved by disintegrating services. However, potential bounds by the contracts with service providers have to be taken into consideration. These bounds might limit the flexibility of changing partners as well as bring about ongoing payments.

All relevant parameters have to be aggregated in a series of payments. A summary of the consequences discussed above is given in Figure 4.

In addition to the listing of relevant payments, their derivation is characterised by marking the main emphasis of each payment over the planning horizon. These payments reflect the monetary consequences coming along with the decision about a service-oriented architecture on the process level. Further examinations have to be undertaken regarding the selection of services that are combined in a company's service portfolio.

Payments Related to Services

Apart from evaluating monetary consequences of a service-oriented architecture, the decision about the sourcing structure of a service portfolio has to be made on the process level. In doing so, concrete services of a specific quality have to be selected that are offered by providers under certain conditions. As for the quality, various service levels need to be considered.

In contrast to the evaluation of the architecture, the seizing of the services should be carried out according to a total calculation. The greater flexibility that is offered by the total calculation sets the basis for optimising the service portfolio. By evaluating each service entirely regarding its relevant out- and in-payments, various alternative compositions of a service portfolio can be calculated. Payments for the evaluation of services can also be distinguished between the phases of development, operation, adaptation, and disintegration.

Quality of Services

When choosing appropriate services for the company's service portfolio, differences in the quality of the services have to be considered (Kester, 1984; Niessen &

Oldenburg, 1997). As a consequence of the total calculation, it has to be considered that differences in service quality are not to be quantified by savings. Savings can only be calculated by comparing two alternatives, which is done on the corporate level.

Service quality is considered best on a two-fold level in the decision-support system. First, a service needs to be appropriate on a factual level in order to conduct a subprocess according to operational demands. The system only records those services that meet the demands according to the factual specification. Second, various security levels can be defined with which the service will be provided according to the factual specification. In this case, the level of security represents the equivalent value of the breakdown ratio quantified in percentage.

For a more detailed assessment, the monetary consequences of a total breakdown and failures of different extent can be distinguished. In order to record monetary consequences of different security levels, average costs for service breakdown have to be calculated. The amount of the costs can be quantified either on the basis of a process calculation (Grob & vom Brocke, 2005) or by means of estimation (OGC, 2001). The out-payments for the risk of a breakdown of the service can be derived by calculating the product of the failure probability and the amount of personal losses due to breakdowns. In the same way, additional characteristics of the service quality can be incorporated in the decision-support system.

Out-Payments

Apart from mere production costs, transaction costs have to be considered. According to transaction-cost theory, these costs are specifically high when incorporating market-like coordination mechanisms more than hierarchical ones (Coase, 1937; Williamson, 1985). In the decision-support system particularly, coordination costs are relevant. Coordination costs are special transaction costs that refer to internally arising costs for the planning, steering, and control over the work of the external service provider that exceed the actual production of one product (Nam, Rajagopalan, Rao, & Chaudhury, 1995). Parts of these costs can, for example, be explained by the principal-agency theory (Bamberg & Spremann, 1989). Hence, coordination costs essentially comprise costs for control and so-called bonding costs. Bonding costs result from the need of documenting the service relation. Costs for control also refer to those costs caused by the management of the service portfolio. They have to be reduced by appropriate system management. For this purpose, costs of lacking profits due to miscalculation have to be listed also. Allocating payments for coordination estimations has to be carried out, which can partly be grounded on the internal transfer prices of the company's cost-accounting system.

In-Payments

Services that come into consideration for outsourcing can hardly be calculated for specifying in-payments. However, in case an internal cost accounting is carried out in a company, internal transfer price rates may be accounted for. Originally, transfer prices are charged for IT services within a company for coordination purposes. Yet, these rates may as well be allocated as an equivalent for the in-payment of a service.

If appropriate methods for calculating transfer prices are not available, it might also seem reasonable to consider the costs brought about by the service. However, this assumes that appropriate cost information is available concerning single services. Methods for activity-based costs offer promising means to provide this information (Grob & vom Brocke, 2005).

Series of Payments

The most relevant parameters of the monetary consequences discussed above are summarised in Figure 5.

Figure 5. Service-portfolio measurement on the process level assessing services

Elements of the Series of Payments of a Service			
Point in Time	0	...	n
Phase of Development			
Out-Payments			
- for building up relations to services provider			
for implementing the interface for integrating			
- the service			
Phase of Operation			
Out-Payments			
- for production of a service (if in-sourcing only)			
- for co-ordination of service integration			
- for risk of service breakdown			
- for risk of service failure			
In-Payments			
+ from internal calculation of transfer prices			
Phase of Adaptation			
Out-Payments			
- for adaptation work on interfaces			
Phase of Disintegration			
Out-Payment			
- for replacement like ongoing contract			
In-Payment			
+ for replacement like credit			
Series of Payments According to Total Calculation			

On the basis of the series of payments calculated above, additional financial parameters have to be taken into account in order to assess the economic value of the service portfolio based on a service-oriented architecture. These aspects are considered on the budgeting level of the decision-support system, which is described in the following.

Measurement on the Budgeting Level

On the budgeting level of the decision-support system, the financial consequences are measured that arise from providing the payments needed during the life cycle. For their analysis, the method of "visualisation of financial implications" (VOFI) is

Figure 6. Service-portfolio measurement on the budgeting level using VOFI

VOFI for Process Layouts				
Point in Time	0	1	...n...	h
Series of Payments				
Internal Funds				
– Withdrawals				
+ Deposits				
Instalment Loan				
+ Credit Intake				
– Redemption				
– Debitor Interest				
Annuity Loan				
+ Credit Inatake				
– Redemption				
– Debitor Interest				
– Creditor Interest				
Loan in Current Account				
+ Credit Inatake				
– Redemption				
– Creditor Interest				
Financial Investment				
– Reinvestment				
+ Disinvestment				
– Debitor Interest				
Tax Payments				
– Out-Payment				
+ In-Payment				
Accounting Balance	0	0	0	0
Capital Stock				
Balance				
on instalment loan				
on annuity loan				
on current account				
on financial investment				
Net Balance				NBVn

applied (Grob, 1993). Using VOFI, the financial consequences of long-term decisions are structured and calculated by means of spreadsheets that serve as a database for further analyses. Compared to formulae applied by conventional methods (Seitz & Ellison, 2004; Shapiro, 2004) of capital budgeting (e.g., present value or the annuity of an investment project), VOFI mainly has the following advantages for evaluating the financial consequences of out-tasking decisions.

- **Transparency:** The description of the financial consequences of decision making by using spreadsheets helps to show the various underlying impacts on the efficiency measures. By means of VOFI, all payments driven by a decision can be taken into account comprehensively, including various conditions for funding and loaning as well as taxes.
- **Adaptability:** VOFI serves as a reference model for long-term decision making. Due to the explicit description of the financial consequences, it can easily be adapted to special decisions that have to be evaluated. On top of individual rates for loaning and funding, dynamically changing conditions like tax rates can also be calculated. Due to clearly defined interfaces for the in- and outflow of data, also extensions of the framework can be built.

Both the transparency and the adaptability eventually may contribute to the actual use of efficiency calculations in service-portfolio management in practice.

Customising VOFI-specific parameters that are relevant in a certain capital situation (e.g., funding conditions) can be considered. Typical parameters that may serve as a reference are summed up in the VOFI given in Figure 6.

In order to consolidate the various influences on the effectiveness of the decision over time, a periodic update of the capital stock has to be calculated. Starting in Period 0, each period has to be calculated in a way in which there is a balance between in- and out-payments. The following example may illustrate the essential procedure. In the first period, usually an out-payment has to be financed. If the internal funds available are insufficient, a loan has to be taken out. As usual, various conditions for loaning can be agreed upon, and also a combination of various loans can be calculated in the VOFI. Correspondingly, multiple forms of funding can be included.

When calculating the adequate amount of loaning or funding, tax payments have to be considered. In order to calculate tax payments, an auxiliary calculation has to be carried out. Depending on tax laws, relevant parameters are, for example, individual depreciations that are chargeable to the investment, as well as tax rates. In each period, these periodical in- and out-payments have to be balanced. As a checkup, the net funding value, which is defined as the accounting balance of all out- and in-payments, should be zero.

On the basis of these flow figures mentioned above, the capital stock can be updated periodically. For this purpose, the balances of all loans and funds have to be recorded.

The accounting balance of both finally results in the net balance of the total investment. By this algorithm, the value of an investment in a service-oriented architecture can be monitored during the whole life cycle of the information system simply by observing the net balance in each relevant period. For special-interest investigations, additional efficiency measures can be calculated on the basis of VOFI. These measures are calculated on the corporate level of the decision-support system, which is illustrated in the following.

Measurement on the Corporate Level

In order to facilitate the design of a company's service portfolio properly, the various evaluation results have to be aggregated for decision purposes. On the basis of the calculations on the budgeting and process levels described above, a wide range of efficiency measures can be applied from accounting science. For service-portfolio measurement, particularly the TCO and ROI are significant. They are well spread in practice and serve as a means to illustrate the relevant parameters for decision making at the same time.

Total Cost of Ownership

In case the consolidation of the payments was done on the basis of a total calculation, the total costs of ownership brought about by a certain information-system architecture and service portfolio can be calculated. The TCO originally aims at summing up all relevant costs chargeable to an information system throughout its

Figure 7. Service-portfolio measurement on the corporate level using TCO

Total Profit of Ownership	**Value**
Total Surplus of Payments for Service Portfolio + Irregular Revenue – Depreciation	
= Total Monetary Profit I (prior Interest and Further Investments) + Revenues by Further Investments – Interest Expense	
= Total Monetary Profit II (prior Taxes on Profit) + Taxes on Profit	
= Total Monetary Profit III (after Taxes on Profit) – Calculatory Interest + Adjustment of Tax Payments	
= Toal Calculatory Profit	
= Total Cost of Ownership	

life cycle (Ferrin & Plank, 2002). At this stage of calculation, the total payments chargeable to a company's service portfolio during its life cycle can be analysed. Apart from out-payments, in-payments can also be included in the calculation. That way, the total profit of the layout can be computed. The calculation is carried out by summing up payments of each row on the spreadsheet and balancing them like in Figure 7.

The total profit of the investment can be calculated gradually. Starting from the surplus of in- and out-payments, the relevant decrease of assets used has to be computed. Therefore, the sum depreciation rates of all periods are calculated together with irregular in-payments, like in-payments for the liquidation of assets. That way, the so-called total monetary profit (I) is measured. Considering in-payments by further investments made during the life cycle as well as interest expenses, the total monetary profit (II) can be calculated. By charging tax payments on profit, the total monetary profit (III) can be reported. While these measures all show the monetary profit realised by the investment, alternative allocations of the capital also have to be taken into account for consistent decision making. For this purpose, calculatory interest has to be charged quantifying the profit that is assumed to be made by the alternative investment (the so-called opportunity). This total calculatory profit finally represents the total profit (or costs) of ownership calculated on the basis of VOFI.

In a number of cases, no in-payments of the service portfolio are to be charged. This can, for example, be the case when in-payments are hard to identify. In other cases, the in-payments might be invariant so that the out-payments are driving the financial efficiency of alternative layouts. In the computation presented in Figure 7, the in-payments are zero. Hence, the resulting value quantifies the TCO that will mostly be negative as, apart from the liquidation of resources, only out-payments are charged. For reporting the total costs of ownership, the value should consequently be transformed into a positive value.

Using the total cost of ownership, out-tasking is profitable if it serves to reduce the TCO. Moreover, the measure supports designing the appropriate corporate service portfolio. Accordingly, the portfolio should be designed with the aim of minimising the TCO.

Return on Investment

The efficiency of out-tasking decisions can also be evaluated by a measurement of profitability. A widespread measure of this kind in practice is the return on investment (Friedlob & Plewa, 1996). By applying this measure, a ratio is calculated that sets the total profit in relation to the stock of capital provided for the investment. Consequently, the ROI indicates the interest payment made by an investment. While the ROI seems to be suitable for comparing investments of different kinds, it is a

Figure 8. Service-portfolio measurement on the corporate level using ROI

Target: ROI

$$ROI = \sqrt[n]{\frac{NBI_n + EF_0 + CI \cdot (1 - pt)}{IF + EF_0}} - 1 \text{ and } NBI_n + EF_0 + CI \geq 0$$

Benchmark: c

$$c = \sqrt[n]{\frac{NB_n + OI + CI \cdot (1 - pt)}{IF + EF_0}} - 1$$

Decision Support

Investment is financially profitable, if ROI > c

Symbols

ROI	ROI, on a dynamic basis
NBI_n	Net Balance of the Total Investment in t=n
IF	Internal Funds Used
EF_0	External Funds Used in t=0
n	Planning Horizon
pt	Tax Rate on Profit
CI	Total Creditor Interest
OI	Total Interest of the Opportunity
c	Average Capital Cost

static measure and, therefore, is inappropriate to support long-term decisions. This point can be considered by calculating the ROI on the basis of VOFI. In doing so, the ROI is not only a dynamic measure, but it also considers various conditions of loaning and funding as well as taxes. For evaluating the efficiency of an investment, the ROI has to be compared to the average capital cost within planning periods. Figure 8 shows the definition of this ROI as well as the criteria for decision support.

In particular, the ROI can be used for decision support on the design of the information-system architecture, providing the series of payments are partially calculated. The relevant data for the calculation can be taken directly from the VOFI database. In case distributions of funds are planned, these payments have to be considered in the fraction's numerator. That way, the ROI is appropriate for evaluating whether the payments necessary for the migration to a service-oriented architecture are justified by future savings due to the performance of the company's service portfolio.

Up to this stage, the decision-support system has been described from a methodological perspective. In the next section, the system is applied to a concrete situation of a company.

Application of the Measurement System for Financial Performance of Service-Oriented Information Systems

In this passage, the decision-support system for the management of service-oriented information systems will be illustrated by means of an application. The case of the travel agency *TravelSmart Ltd.* may serve as an example. The situation is invented following a practical case.

Introduction

TravelSmart is a travel-management service provider that predominantly offers holiday travels for private and business customers. Round trips only form a small part of its product portfolio. The core competence of the business is the individual configuration of holidays according to the specific needs and wishes of its customers. Therefore, the company's tourism market is limited to the high-end segment. *TravelSmart* uses both digital chains of distribution and classic travel-agency sales. Excellent customer service is the trademark of the company, which differentiates itself clearly from other businesses and competitors.

However, especially in recent years, *TravelSmart* has been exposed to increased pressure by more and more competitors who have been pushing both reduced costs and an amelioration of the service. Until now, IT tasks have only been accomplished by the IT department of the company itself. The company's IT development has achieved excellent tasks, yet the costs for individual IT services have been partly too high compared to the market value.

The business processes of *TravelSmart* are substantially based on information systems for e-commerce as well as travel-agency sales. The goal from management's perspective of the business is now to use, despite the high cohesion, service-oriented architectures. That way, parts of the information-system environment could be given away to external service providers. The focus of the evaluation is whether the supply of a service-oriented architecture would be worthwhile. Furthermore, the service portfolio has to be determined.

The funds needed are covered by internal funds, an installment loan, and a loan in the current account. Figure 9 gives an overview of the financial conditions.

The management of *TravelSmart* has grasped the long-term character of the decision that needs to be taken. It determined a planning horizon of five periods. Thus, the decision-support system described in this chapter was applied.

Figure 9. Financial conditions for the investment in the service portfolio at TravelSmart

Financial Conditions of *TravelSmart*	
Internal Funds	10,000 €
Installment Loan	
Funds provided	10,000 €
Creditor Interest Rate	6%
Disagio	5%
Duration	2 Years
Loan in Current Account	
Creditor Interest Rate	8%
Debitor Interest Rate	6%
Tax Rates	
t = 1	55.0%
t = 2	52.5%
t = 3	50.0%
t = 4	47.5%
t = 5	45.0%

Identification of Services

At the kick-off meeting, the following services were identified for further analysis.

- **Service travel records:** In this business process, extensive and informative records for individual tours are being collected and set for the customer on the basis of data warehouses and internal work flows. This extensive and high-quality processing, and its quick delivery to the customer via e-mail right after the booking of the travel or within a maximum of 2 days in the mail, is a strategic characteristic for differentiation from the competition. This core competence is constantly being extended and ameliorated by means of software solutions in the internal IT department. The question that comes up concerns sourcing as greater adaptations are planned in the near future.
- **Service credit assessment:** The service provides *TravelSmart* with a digital assessment of the solvency of its customers from accredited rating agencies. On the basis of assessment, customers receive generous or restricted payment and financing conditions for the individual travel package. The process is of high operative importance as it gives both the customer and company a higher degree of flexibility.
- **Service order processing:** This business process is conducted by a work-flow-management system that connects numerous application systems in order to support individual steps in the processing for a booking order, and to possibly coordinate the interaction of the responsible employee and the applications.

As a starting point, the sourcing strategy for the services has to be agreed upon. Figure 10 gives the classification of the services in the model from Lacity et al. (1996).

The travel-records service as one result of the analysis should be carried out internally in the future. The strategic impact and the internal further developmental potentials can possibly balance the possible advantages of the outsourcing. The service of credit assessment demonstrates a service that can be supported by the best practices of external service providers, leading to a cost reduction for the company. Also, for the order processing, an appropriate out-tasking alternative should be provided.

Due to experiences gained during the project, the management defines the demands of the services as follows.

- For credit assessment and order processing, a medium availability of 99.5% according to the ITIL standard is required. The error ratio should not exceed 0.01%.
- For the service of providing travel records, quick, high-quality, and highly available functionality is required. The requirement concerning availability is around 99.9%; the maximum error ratio is around 0.001%.

In order to ground the decision on the service portfolios, a calculation of the monetary consequences was carried out according to the decision-support system introduced in this chapter. Hence, an assessment of the payments on the process level was conducted first.

Figure 10. Deriving sourcing strategies at TravelSmart

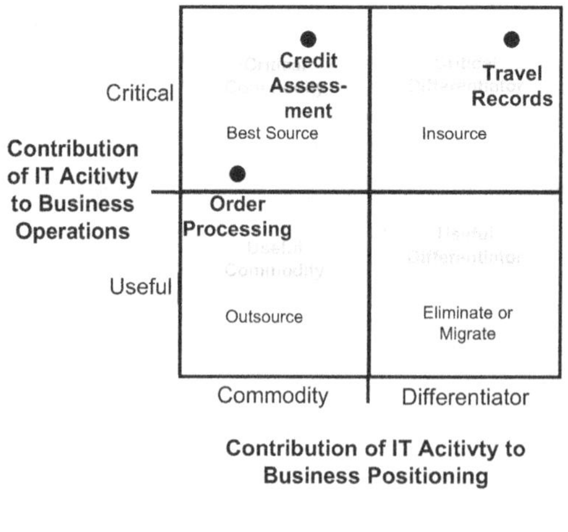

Credit Assess-ment
Travel Records
Critical
Best Source
Insource
Contribution of IT Acitivty to Business Operations
Order Processing
Useful
Outsource
Eliminate or Migrate
Commodity
Differentiator
Contribution of IT Acitivty to Business Positioning

Measuring on the Process Level

Payments Related to the Architecture

The assessment of the payments on the process level led to the following initial situation: The implementation of the SOA goes along with 14,000€ for the provision of hardware and 4,000€ for software (which is an enterprise service bus application). On a time schedule of 5 years, 16,000€ on the whole have to be written off of the capital costs. The write-off has to be considered for the calculation of the foundation of the tax calculation. Furthermore, out-payments for the training of the administrators will cost about 10,000€.

In the company, 1,000€ have to be calculated for maintenance, 4,000€ for the operation, and 2,000€ for first- and second-level support. A periodical increase of 10% can then be expected. Further payments of 3,000€ are caused by license fees for both software and maintenance. At the same time, in-payments of 6,000€ can be expected for the parallel use of the architecture with a partner enterprise. Starting with the fourth period, an intensification is planned that increases in-payments to 9,000€.

Adaptations of the architecture were expected to be necessary. As to the value, 10% of the capital costs of the purchase per period were estimated. In the phase of disintegration, *TravelSmart* can realise in-payments by aid of the further utilisation of hardware (liquidation proceeds of 15,000€) as well as the decrease of idle time of resources due to an early cut of capacities (1,000€). Additional payments for ending contracts are not being calculated. The results are summed up in Figure 11.

The financial efficiency of the investment in SOA is essentially driven by the performance of the service portfolio that can be run on the architecture. Consequently, the payments related to the services available for *TravelSmart* were assessed in the following.

Payments Related to the Services

For the process of credit assessment, two service offers were identified in an announcement for external service providers. The offers were assessed by aid of the payment categories provided in this study. For one of the offers, SOA technology (B1) has been applied; for the other, it was not (B3). The conditions concerning the service offers are compared in Figure 12.

Figure 11. Estimated series of payments with SOA at TravelSmart

Estimated Series of Payments with SOA for *TravelSmart*						
Point in Time	0	1	2	3	4	5
Phase of Development						
Out-Payments						
- for building up know-how	10,000					
- for implementing SOA as a new architecture	18,000					
In-Payments						
+ due to not implementing services in-house						
Phase of Operation						
Out-Payments						
- for additional maintenance work on the interfaces		1,000	1,100	1,210	1,331	1,464
- for support (1st / 2nd level)		2,000	2,200	2,420	2,662	2,928
- for conducting operation		4,000	4,400	4,840	5,324	5,856
- for licenses with service providers		3,000	3,000	3,000	3,000	3,000
In-Payments						
+ for shared service		6,000	6,000	6,000	9,000	9,000
+ by savings concerning lowering resources needed		1,000	1,000	1,000	1,000	1,000
Phase of Adaptation						
Out-Payments						
- for adaptation work on interfaces		1,800	1,800	1,800	1,800	1,800
Phase of Disintegration						
In-Payments						
+ by saving according to idle time costs of the resources						1,000
+ by liquidation of technology						1,500
Series of Payments According to Partial Calculation	-28,000	-4,800	-5,500	-6,270	-4117	-2,549

The values of these figures were grounded in the following considerations.

- By applying standards for Web-service specification, implementations and adaptations of service interfaces can be carried out in an efficient manner. In particular, out-payments for the removal of the service can be reduced because the complexity of the system is shifted toward the supplier. Over and above that, no further architecture components that would have to be removed later on are required.

- The clearly defined and manageable out-tasking interfaces reduce the expenditures connected to the know-how of the service provider.
- Using the best-sourcing strategy, service providers for B1 can be provided that are highly specialised. Thus, scale effects lead to a reduction of production costs and an increase in the quality of the services.
- Coordination costs are lower than before due to standardised interfaces simplifying internal result control.
- Independent from SOA, B1 offers better availability and a lower error rate. That way, out-payments for risks of failure and breakdown are lower.

For the calculation of the series of payments for the investment of the service offer B1, the amount of in-payments caused by internal payments was calculated. With respect to the quantity structure, 7,500 travel bookings (transactions) per period have been forecasted with a dynamic development rate of 10%.

On the basis of this information, the series of payments have been calculated along the process described above. The calculation is presented in Figure 13.

By means of a market analysis, further service offers have been collected. The results are described in the following. For credit assessment, there are now three external service offers available. Two have been implemented on a Web-service basis (B1 and B2), and one has been implemented on the basis of an alternative architecture (B3). According to the process of the travel records, the IT department made a suggestion to migrate toward SOA and to provide the service internally (A1). Apart from that, the service may well be carried out according to the current state of the

Figure 12. Conditions of services for SOA and non-SOA at TravelSmart

Conditions of Services	B1 (SOA)	B3 (Alt.)
Phase of Development		
Out-payments - for building up relations to services provider	1,500 €	3,500 €
- for implementing the interface for integrating the service	1,000 €	3,000 €
- for providing the software components	0 €	4,000 €
Phase of Operation		
Out-payments		
- for production of a service (per transaction)	0.70 €	0.85 €
- for co-ordination of service integration	3,000 €	8,000 €
Average percentage		
- of service availability	99.9%	99.5%
- of service failure	0.001%	0.01%
Phase of Adaptation		
Proportional work on adaptation per period	4.0%	4.0%
Phase of Disintegration		
Out-Payment		
- for replacement	1,000 €	8,000 €

Figure 13. Estimated series of payments related to service B1 on the process level at Travelsmart

Series of Payments related to Service B1						
Point in Time	0	1	2	3	4	5
Phase of Development						
Out-Payments						
- for building up relations to services provider	1,500					
- for implementing the interfaces	1,000					
Phase of Operation						
Amount of Transactions		7,500	8,250	9,075	9,983	10,981
Out-Payments						
- for production of a service		5,250	5,775	6,353	6,988	7,687
- for co-ordination of service integration		3,000	3,000	3,000	3,000	3,000
- for risk of service breakdown		375	413	454	499	549
- for risk of service failure		75	83	91	100	110
In-Payments						
+ from internal calculation of transfer prices		30,000	33,000	36,300	39,930	43,923
Phase of Adaptation						
Out-Payments						
- for adaptation work on interfaces		100	100	100	100	100
Phase of Disintegration						
Out-Payment						
- for replacement						1,000
Series of Payments	**-2,500**	**21,200**	**23,630**	**26,303**	**29,243**	**31,478**

art (A2). For order processing, there are two offers available from the same external service providers (C1, C2). C2 offers 99.7% of availability as opposed to the 99.5% availability offered by C2. The positive evaluation due to lower risk out-payments, however, leads to a more rational offer on an economic basis for C2 although C2's implementation costs are higher. Neither C1 nor C2 apply SOA.

Figure 14. Series of payments related to service offers on the process level at TravelSmart

Series of Payments Related to the Various Service Offers						
Point in Time	0	1	2	3	4	5
Service A1	-3,000	35,746	39,457	43,539	48,028	50,967
Service A2	-17,000	25,926	28,887	32,144	35,726	31,667
Service B1	-2,500	21,200	23,630	26,303	29,243	31,478
Service B2	-7,500	16,700	18,800	21,110	23,651	22,446
Service B3	-10,500	12,580	14,680	16,990	19,531	14,326
Service C1	-11,000	4,962	6,069	7,286	8,624	5,097
Service C2	-13,000	12,263	14,119	16,161	18,407	13,377

For each service, monetary consequences have been assessed following the approach described above. The results are summarised in Figure 14.

These results set the basis for the further analysis of the payments on the budgeting level. This part of the system's application is presented in the next passage.

Measurement on the Budgeting Level

For each potential combination of services, the financial consequences have been calculated on the budgeting level. Comparing the net present value on the planning horizon in t = n, the most efficient combination was identified. This is the combination of A1, B1, and C2. As an example, the computation for this combination is displayed in Figure 15. It comprises the aggregation of the series of payments, the VOFI, and auxiliary calculations for computing tax payments.

During the computation, further VOFI calculations have been carried out for each combination of services regarding their specific requirements on the architecture. As an example, the calculation for a service portfolio without SOA is given in Figure 16.

As part of the various calculations of alternatives, the situation without any changes to the sourcing strategy has to be included according to the measurement system. This very calculation represents the so-called opportunity, which essentially assesses the yield of the internal funds as if they would have been allocated in a financial investment. The calculations for this alternative are given in Figure 17.

An instant comparison of the alternatives can be carried out by checking the net balance value in t = n = 5, which should reach a maximum level. For a more detailed analysis, performance measures can be calculated on the database.

Figure 15. Financial consequences of the investment in the service portfolio with SOA on the budgeting level at TravelSmart

Series of Payments						
Point in Time	0	1	2	3	4	5
Service A1	-3,000	35,746	39,457	43,539	48,028	50,967
Service B1	-2,500	21,200	23,630	26,303	29,243	31,478
Service C2	-13,000	12,263	14,119	16,161	18,407	13,377
Series of Payments for Services	**-18,500**	**69,209**	**77,206**	**86,002**	**95,678**	**95,822**
Series of Payments for Architecture	-28,000	-4,800	-5,500	-6,270	-4,117	-2,549
Total Series of Payments	**-46,500**	**64,409**	**71,706**	**79,732**	**91,561**	**93,274**

Figure 15.continued

VOFI for the Investment with SOA						
Point in Time	0	1	2	3	4	5
Series of Payments	-46,500	64,409	71,706	79,732	91,561	93,274
Internal Funds	10,000					
Installment Loan						
+ Credit Intake	10,000					
– Redemption			10,000			
– Creditor Interest	500	600	600			
Loan in Current Account						
+ Credit Intake	27,000					
– Redemption		27,000				
– Creditor Interest		2,160				
Financial Investment						
– Reinvestment		2,502	25,526	42,307	51,805	56,771
+ Debtor Interest			150	1,682	4,220	7,328
Tax Payments						
– Out-Payments		32,147	35,729	39,107	43,976	43,831
Accounting Balance	0	0	0	0	0	0
Balance on						
Installment Loan	10,000	10,000				
Loan in Current Account	27,000					
Financial Investment		2,502	28,028	70,335	122,141	178,912
Net Balance	**-37,000**	**-7,498**	**28028**	**70,335**	**122,141**	**178,912**

Calculation of Tax Payments					
Point in Time	1	2	3	4	5
Tax Rates	55.0%	52.5%	50.0%	47.5%	45.0%
Surplus of In- and Out-Payments	64,409	71,706	79,732	91,561	93274
– Creditor Interest	2,760	600			
+ Debtor Interest		150	1,682	4,220	7,328
– Depreciation	3,200	3,200	3,200	3,200	3,200
Assessment Base for Taxes	58,449	68,056	78,214	92,582	97,402
In-Payments Out-Payments	32,147	35,729	39,107	43,976	43,831

Calculation of Depreciation					
Point in Time	1	2	3	4	5
Book Value, Beginning of the Year	16,000	12,800	9,600	6,400	3,200
– Depreciation Linear Rate	**3,200**	**3,200**	**3,200**	**3,200**	**3,200**
Book Value, End of the Year	12800	9,600	6,400	3,200	

Measurement on the Corporate Level

On the basis of the detailed assessment on both the budgeting and process levels, performance measures have been calculated in order to support the management

Figure 16. Financial consequences of the investment in the service portfolio without SOA on the budgeting level at TravelSmart

VOFI for the Investment without SOA						
Point in Time	0	1	2	3	4	5
Series of Payments	-40,500	50,769	57,686	65,294	73,664	59,370
Internal Funds	10,000					
Installment Loan						
+ Credit Intake	10,000					
– Redemption			10,000			
– Creditor Interest	500	600	600			
Loan in Current Account						
+ Credit Intake	21,000					
– Redemption		21,000				
– Creditor Interest		1,680				
Financial Investment						
– Reinvestment		820	17,139	33,186	40,284	35,671
+ Debtor Interest			49	1,078	3,069	5,486
Tax Payments						
– Out-Payments		26,669	29,996	33,186	36,448	29,185
Accounting Balance	0	0	0	0	0	0
Balance on						
Installment Loan	10,000	10,000				
Loan in Current Account	21,000					
Financial Investment		820	17,959	51,145	91,429	127,100
Net Balance	**-31,000**	**-9,180**	**17,959**	**51,145**	**91,429**	**127,100**

Calculation of Tax Payments					
Point in Time	1	2	3	4	5
Tax Rates	55.0%	52.5%	50.0%	47.5%	45.0%
Surplus of In- and Out-Payments	50,769	57,686	65,294	73,664	59,370
– Creditor Interest	2,280	600			
+ Debtor Interest		49	1,078	3,060	5,400
– Depreciation	0	0	0	0	0
Assessment Base for Taxes	48,489	57,135	66,372	76,732	64,856
Out-Payments	**26,669**	**29,996**	**33,186**	**36,448**	**29,185**

of *TravelSmart*. Figure 18 gives an excerpt from the entire report that was generated.

In the *TravelSmart* case, the investment in SOA turned out to be profitable. The investment in the migration toward the new technology can be compensated by savings in later phases of the information system's life cycle. In particular, the benefits result from the opportunity of out-tasking services to specialised service providers according to common standards. In addition, out of the various service offers available, the most profitable sourcing strategy for *TravelSmart* could be calculated. By aid of the decision-support system, a combination of the service offers A1, B2, and C2 is suggested. That way, an ROI of 36.43% of the entire investment in SOA is calculated.

Apart from the initial decision on the information system's architecture, an ongoing assessment of the financial efficiency of the information system can be conducted.

Figure 17. Financial consequences of the opportunity on the budgeting level at TravelSmart

VOFI for the Investment with SOA						
Point in Time	0	1	2	3	4	5
Series of Payments	0	0	0	0	0	0
Internal Funds	10,000					
Financial Investment						
– Reinvestment	10,000	270	293	317	343	370
+ Debtor Interest		600	616	634	653	673
Tax Payments						
– Out-Payments		330	324	317	310	303
Accounting Balance	0	0	0	0	0	0
Balance on Financial Investment	10,000	10,270	10,563	10,880	11,222	11,593
Net Balance	**10,000**	**10,270**	**10,563**	**10,880**	**11,222**	**11,593**

Calculation of Tax Payments					
Point in Time	1	2	3	4	5
Tax Rates	55.0%	52.5%	50.0%	47.5%	45.0%
Surplus of In- and Out-Payments	0	0	0	0	0
+ Debtor Interest	600	616	634	653	673
Assessment Base for Taxes	600	616	634	653	673
Out-Payments	**330**	**324**	**317**	**310**	**303**

Figure 18. Measuring the performance of service portfolios on the corporate level at TravelSmart

Service Portfolio Performance on the Corporate Level			
Performance Measures	**With SOA**	Without SOA	Opportunity
Final Value, NBI_n	**178,912**	127,100	11,593
Pay-Off-Period	**2**	2	-
Return on Investment, ROI	**36.43%**	31.87%	3.00%

For that purpose, it might prove positive that the decision-support system provides common performance measures for corporate management. That way, the investment in the company's information-systems design can easily be compared to other potential investments within the scope of management.

Conclusion and Outlook

Given the technological achievements in designing service-oriented information systems, the problem of assembling the right service portfolio according to a company's needs becomes relevant. Consequently, software engineering will increasingly evolve from programming to decision making in a service-oriented enterprise. In order to meet this challenge, decision-support systems for service-portfolio measurement were the subject of this chapter. In order to consider long-term economic consequences that come along with decisions on the system's design, a measurement system for the financial performance of service-oriented information systems has been designed. The system is based on principles of decision theory and capital budgeting which are applied for service-portfolio measurement within a comprehensive framework. That way, common business measures are calculated that indicate the financial performance of the decisions in comparison to alternative investments.

The decision-support system described focuses on the profitability of an investment in the company's service portfolio from a financial perspective. In addition, nonmonetary consequences of the service-portfolio design also have to be taken into account. Hence, future work will concentrate on the enlargement of the system from various perspectives that may be relevant. Perspectives of great importance should address, for example, customer relations or security aspects. These perspectives can, for the most part, be built on the basis of preliminary works in the field of nonfinancial assessments of services. The system presented in this chapter may serve as one major component of such an entire decision-support system that is needed for managing the service-oriented enterprise. Hence, the future challenge will mainly lie in extending the system by embedding components for nonfinancial assessments. The work on multidimensional performance measurement, introduced above, might set a suitable frame for this integration.

Acknowledgments

This publication is based on work done within the research cluster of Internet economy at the University of Muenster. The author wishes to thank the German Federal Ministry of Education and Research (BMBF) for financial support (Grant No. 01AK704). Additionally, many thanks go to Mario Thaten, McKinsey & Company, Inc., for the insightful discussion on the case study.

References

Aubert, B., Patry, S., & Rivard, S. (2002). Managing IT outsourcing risk: Lessons learned. In *Information systems outsourcing in the new economy.* Heidelberg, Germany.

Bahli, B., & Rivard, S. (2003). The information technology outsourcing risk: A transaction cost and agency theory-based perspective. *Journal of Information Management*, 211-221.

Bamberg, G., & Spremann, K. (1989). *Agency theory, information and incentives* (2nd ed.). Berlin, Germany.

Buck-Lew, M. (1992). To outsource or not? *International Journal of Information Management, 12*(1), 3-20.

Cardoso, J., Sheth, A. P., Miller, J. A., Arnold, J., & Kochut, K. (2004). Quality of service for workflows and Web service processes. *Journal of Web Semantics, 3*(1), 281-308.

Coase, R. H. (1937). The nature of the firm. *Economica, 4*(11), 386-405.

Craggs, S. (2003). *Best-of-breed ESBS: Identifying best-of-breed characteristics.* Retrieved from http://www.sonicsoftware.com/products/whitepapers/docs/best_of_breed_esbs.pdf

Day, J., & Deters, R. (2004). *Selecting the best Web service.* Paper presented at the 2004 Conference of the Centre for Advanced Studies on Collaborative Research, Markham, Ontario, Canada.

Espinosa, J. A., & Carmel, E. (2004). *The effect of time separation on coordination costs in global software teams: A dyad model.* Paper presented at the 37th Hawaii International Conference on System Sciences.

Farber, D. (2004). *All roads lead to SOA.*

Faye Borthick, A., & Roth, H. P. (1994). Understanding client/server computing. *Management Accounting*, 36-41.

Ferrin, B. G., & Plank, R. E. (2002). Total cost of ownership models: An exploratory study. *The Journal of Supply Chain Management, 38*(3), 18-29.

Friedlob, G. T., & Plewa, F. J. (1996). *Understanding return on investment.* John Wiley & Sons.

Gartner. (2002). *TCO manager.* Retrieved October 9, 2005, from http://gartner11.gartnerweb.com/bp/static/tcomanhome.html

Grob, H. L. (1993). *Capital budgeting with financial plans: An introduction.* Wiesbaden, Germany.

Grob, H. L., & vom Brocke, J. (2005). Measuring financial efficiency of process layouts. In J. Becker, M. Rosemann, & M. Kugeler (Eds.), *Process management: A guide for the design of business processes* (2nd ed.). Berlin, Germany: Springer.

Hevner, A. R., March, S. T., Park, J., & Ram, S. (2004). Design science in information systems research. *MIS Quarterly, 28*(1), 75-105.

IBM. (2004). *E-business on demand.* Retrieved from http://www-5.ibm.com/services/de/ondemand/solutions_bizproc.html

Johnson, T. H., & Kaplan, R. S. (1987). *Relevance lost: The rise and fall of management accounting.* Boston.

Jurison, J. (2002). Applying traditional risk-return analysis to strategic IT outsourcing decisions. In R. Hirschheim, A. Heinzl, & J. Dibbern (Eds.), *Information systems outsourcing: Enduring themes, emergent patterns and future direction* (pp. 177-186). Berlin, Germany.

Kador, J. (1990). The dollars and sense of outsourcing. *Candle Computer Report, 12*(8), 1-5.

Kaplan, R. S. (1986). Must CIM be justified by faith alone? *Harvard Business Review, 64*(2), 87-95.

Kaplan, R. S., & Norton, D. P. (1992). The balanced scorecard: Measures that drive performance. *Harvard Business Review, 70*(1), 71-79.

Kaplan, R. S., & Norton, D. P. (1996). Using the balanced scorecard as a strategic management system. *Harvard Business Review, 74*(1), 75-85.

Kascus, M. A., & Hale, D. (1995). *Outsourcing cataloging, authority work, and physical processing: A checklist of considerations.* Chicago.

Keen, P., & McDonald, M. (2000). *The eprocess edge: Creating customer value and business wealth in the Internet era.* Berkeley.

Kester, W. C. (1984). Today's options for tomorrow's growth *Harvard Business Review*, 153-161.

Knolmayer, G. A. (1997). Hierarchical planning procedure supporting the selection of service providers in outtasking decisions. In H. Krallmann (Ed.), *Wirtschaftsinformatik '97* (pp. 99-119). Heidelberg, Germany.

Lacity, M. C., Willcocks, L. P., & Feeny, D. F. (1996). The value of selective IT sourcing. *Sloan Management Review, 37*(3), 13-25.

Loh, L., & Venkatraman, N. (1992). Determinants of information technology outsourcing: A cross-sectional analysis. *Journal of Management Information Systems, 9*(1), 7-24.

Nam, K., Rajagopalan, S., Rao, H. R., & Chaudhury, A. (1995). Dimensions of outsourcing: A transactions costs framework. In M. Khosrow-Pour (Ed.), *Managing information technology investments with outsourcing* (pp. 104-128). Harrisburg, PA: Idea Group Publishing.

Neely, A. (2004). The challenges of performance measurement. *Management Decision, 42*(8), 1017-1023.

Niessen, J., & Oldenburg, P. (1997). Service level management: Customer focused. In *IT infrastructure library.* Norwich: The Stationery Office.

OGC. (2001). Service delivery. In *IT infrastructure library.* Norwich: The Stationery Office.

Oracle. (2004). *Oracle application server.* Retrieved from http://otn.oracle.com/products/integration/pdf/integration_tech_wp.pdf

Putrus, R. S. (1992). Outsourcing analysis and justification using AHP. *Information Strategy, 9*(1), 31-36.

Quinn, J. B. (1999). Strategic outsourcing: Leveraging knowledge capabilities. *Sloan Management Review, 40*(4), 9-21.

SAP. (2004). *Sap-solutions.* Retrieved from http://www.sap.com/solutions/business-suite/index.aspx

Seitz, N., & Ellison, M. (2004). *Capital budgeting and long-term financing decisions* (3rd ed.). Farmington Hills, MI.

Shapiro, A. C. (2004). *Capital budgeting and investment analysis.* Prentice Hall.

Tam, K. Y. (1992). Capital budgeting in information systems development. *Information & Management, 23*(6), 345-357.

Vollmer, K., & Gilpin, M. (2004). *Integration in a service-oriented world.*

vom Brocke, J., & Lindner, M. A. (2004). *Service portfolio measurement: A framework for evaluating the financial consequences of out-tasking decisions.* Paper presented at the ICSOC04 Second International Conference on Service Oriented Computing, New York.

Wang, G., Chen, A., Wang, C., Fung, C., & Uczekaj, S. (2004). *Integrated quality of service (QoS) management in service-oriented enterprise architectures.* Paper presented at the 8th IEEE International Enterprise Distributed Object Computing Conference (EDOC'04).

Weikum, G., & Vossen, G. (2002). *Transactional information systems: Theory, algorithms and the practice of concurrency control and recovery.* San Francisco.

Williamson, O. E. (1985). *The economic institutions of capitalism.* New York: Tree Press.

Section II

Enterprise Service Computing: Requirements

Chapter IV

Requirements Engineering for Integrating the Enterprise

Raghvinder S. Sangwan, Penn State Great Valley, USA

Abstract

In an era of global economy, an enterprise must demonstrate agility in order to stay competitive. Agility requires continuous monitoring of the ever-changing business landscape and quick adaptation to that change. Often times, this means businesses must merge to form strategic partnerships allowing them to provide new products and services. Such partnerships create the need for critical information to flow seamlessly across the newly formed enterprise and be available on demand for effective collaboration and decision making. However, the legacy business information systems that each partner brings into the newly formed enterprise typically have a very narrow focus serving the needs of a single business unit within an enterprise. As such, it becomes necessary to integrate multiple different systems before the right information can be delivered to the right person at the right time. Integrating disparate systems from a technical perspective is not hard to achieve since the Web-services standard is fairly mature and provides an open infrastructure for software systems to interoperate. One must, however, first understand the need and level of cooperation and collaboration among the different segments of an enterprise, its suppliers, and its customers in order for this integration to be effective. This chapter motivates the need for model-driven requirements engineering for enterprise integration, reviews the research to date on model-driven requirements engineering, and examines a case study on integrating health-care providers to form integrated health networks to gain insight into challenges and issues.

Introduction

The global economy and electronic commerce are creating new opportunities for conducting business. It is becoming commonplace for organizations to form strategic partnerships for collaboration with other organizations and creating value propositions that will give them a competitive edge. Information technology can play a key role in such endeavors, enabling disparate segments of the newly created enterprise to work together effectively by providing the right information to the right person at the right time. Such a seamless flow of information has become possible through the use of Internet and Web services. The technologies have in fact created innovative ways of conducting business. Interactions are now possible between the following entities.

- **Customers and businesses:** This allows customers access to business services and goods over the Internet in the comfortable setting of their home at virtually any time of the day, and allows businesses to disseminate promotional information to their customers. Customer-to-business (C2B) and business-to-customer (B2C) applications have become almost essential for any business to stay competitive.
- **Businesses and suppliers:** This allows the creation of electronic supply chains. These supply chains automate the flow of goods, raw materials, and parts, creating a sort of assembly line with each participating business entity along the way adding some value to create the final product. Business-to-business (B2B) applications make it possible to optimize this flow so that warehouse space and business capital can be used in the most efficient manner and the product demand can be fulfilled in real time.

Technology, however, is only an enabler; it creates the opportunity but cannot provide guidance for how the business is to be conducted to take advantage of the opportunity. It, therefore, becomes important to understand the new business model for achieving the business goals set forth in the initial intent of the organizations that merged to form a strategic partnership.

A business model shows the structure and dynamics of an organization. The structure of an organization is captured through its workers, their roles, and their responsibilities. The dynamics are captured through its processes. Together they show how an enterprise operates, providing a mechanism to clearly think through these operational aspects and determine if and how they add any business value (Leffingwell & Widrig, 2000).

A business model is an organization's most valuable core asset. It is a highly reusable strategic artifact that can be used for multiple purposes.

1. **Building an application:** It can serve as a basis for understanding the functional requirements of an application that will automate the business processes within a particular segment of an enterprise.
2. **Building a family of applications:** It can serve as the basis for a product-line architecture, providing a single unified model of the business information (called a domain model) and services common across a family of products within a product line. It also serves as an input for functional requirements for the individual products.
3. **Business-process engineering:** It can be used for exploring new business opportunities and the functional requirements of the business information system to support such opportunities.
4. **Business-process reengineering:** It can be used for streamlining existing processes, realigning them to take advantage of the industry-wide best practices.

The focus of this chapter is on using the business model as a means for understanding the extent of collaboration and cooperation within and between different segments of an enterprise, and generating the requirements for an integrated enterprise information system that helps an enterprise meet its strategic objectives. This model-driven requirements-engineering approach (Berenbach, 2003, 2004a, 2004b) has numerous advantages over the more traditional document-driven requirements-engineering processes that use loosely structured natural language as a mechanism for specifying requirements. It makes it easier to analyze and version control the individual requirements, and maintain traceability among these requirements, their design specifications, and the test cases. It also becomes easier to assess the size and progress of an integration project as well as the completeness, consistency, and quality of its requirements.

Model-Driven Requirements Engineering: A Motivating Example

UML (unified modeling language) is a visual language that was initially designed to model aspects of software-intensive systems. However, it has been extended over the years to include modeling elements that can be used among other things to create models of a business. A business model consists of the following key artifacts (Kruchten, 2004).

- **Business-vision document:** This is a textual document that describes the objectives and goals of the business model under consideration. The goals become the rationale or basis for justifying the business processes described in the business use case model.
- **Business use-case model:** This is a UML-based model that shows the processes of the business in terms of business use cases. It also shows business actors that initiate these use cases. Actors represent entities such as customers that are external to a business. Use cases represent sequences of actions within a business process that fulfill the request of an actor initiating a particular use case.
- **Business-analysis model:** This is a UML-based model that shows the details of the business processes by elaborating each business use case. The elaboration shows how business workers interact with business entities to fulfill the request of an actor initiating a given business use case. The business entities represent key concepts within the business domain that encapsulate business information essential to the fulfillment of responsibilities embodied within a business use case.

Figure 1. The business context showing the business actors

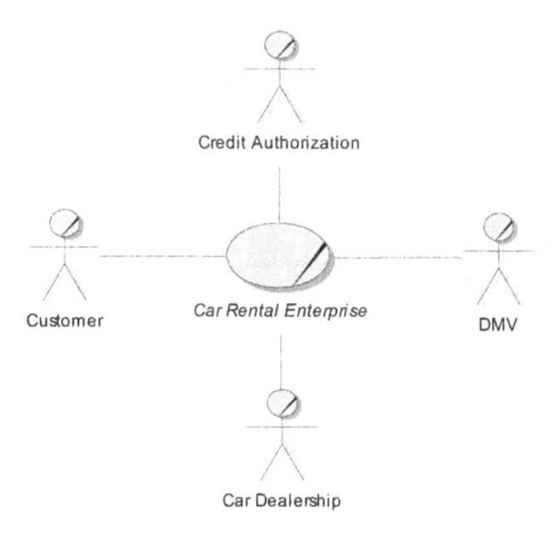

Business workers and business entities can be organized into business systems representing organizational units within an enterprise.

Consider as an example a car-rental enterprise. We may start at the most abstract level with the business context showing the enterprise surrounded by the business actors: the external entities that interact with the business. This is shown in Figure 1.

It is useful to create a context diagram since it allows one to think in terms of the external entities a business must interact with and enumerate the business use cases from their perspective. It is these entities that generate business events to which a business must respond in an efficient and effective manner (Robertson & Robertson, 1999). In this example, the external entities include a customer who is interested in renting a car, a credit-authorization system used for verifying customer credit, a DMV system to check a customer's driving record, and a car dealership that maintains and supplies cars for the car-rental enterprise.

Figure 2. Business use cases for the business actors

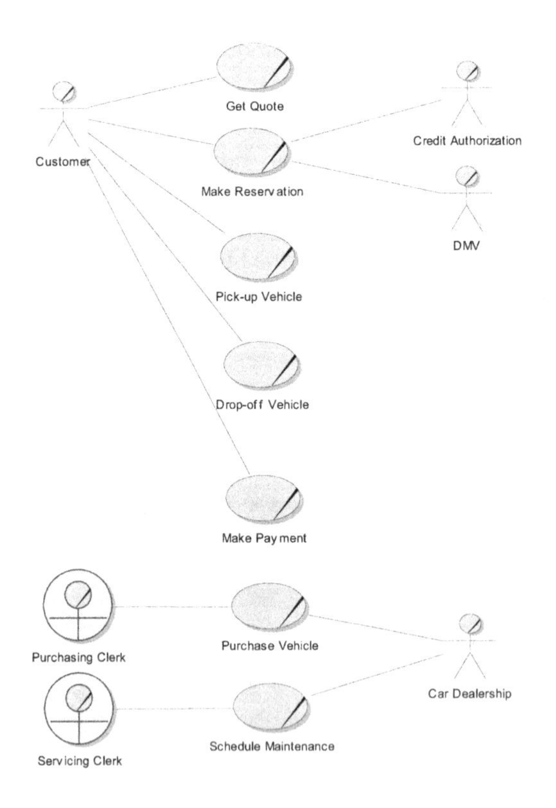

Business Use-Case Model

The business context can be elaborated to create a business use-case model to show the business use cases for the business actors. Figure 2 shows this model for the car-rental enterprise.

Each use case represents a business process that must be implemented within the enterprise to service the business event related to its external entities.

Business-Analysis Model

We show the details of a business process through interactions between business workers and business entities. Figure 3 shows part of such a model for the car-rental enterprise, elaborating only the "Make Reservation" business use case. The complete model would show details for all the business use cases.

The business entities are important concepts in a business domain that represent physical and conceptual resources. They typically encapsulate important business

Figure 3. Business-object model showing interactions between business workers, entities, and actors

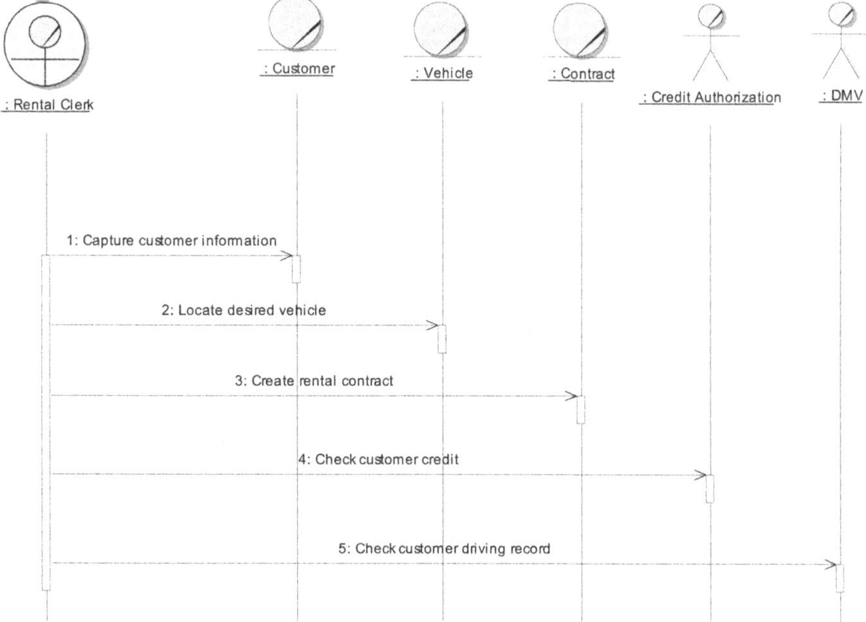

information along with the associated business logic and rules. In this example, the customer, vehicle, and contract are business entities.

Once a business model has been created, it can be used for creating a model of a software system desired by the enterprise. A software system may automate some business processes and in doing so may require integration with other software systems that are either internal or external to an enterprise. In the example of the car-rental enterprise, we may choose to automate all the business processes shown in the business use-case model. The "Make Reservation" business use case requires integration with the "Credit Authorization" and "DMV" systems; there may also be an opportunity for the "Purchase Vehicle" and "Maintain Vehicle" business use cases to integrate with the business information system at the "Car Dealership." The dealership business information system could receive orders for new vehicles and maintenance schedules for existing fleets of cars from the car-rental enterprise's system and in turn may integrate with information systems of its suppliers, forming an e-supply-chain where demand is generated and fulfilled in real time.

Figure 4. System use-case model

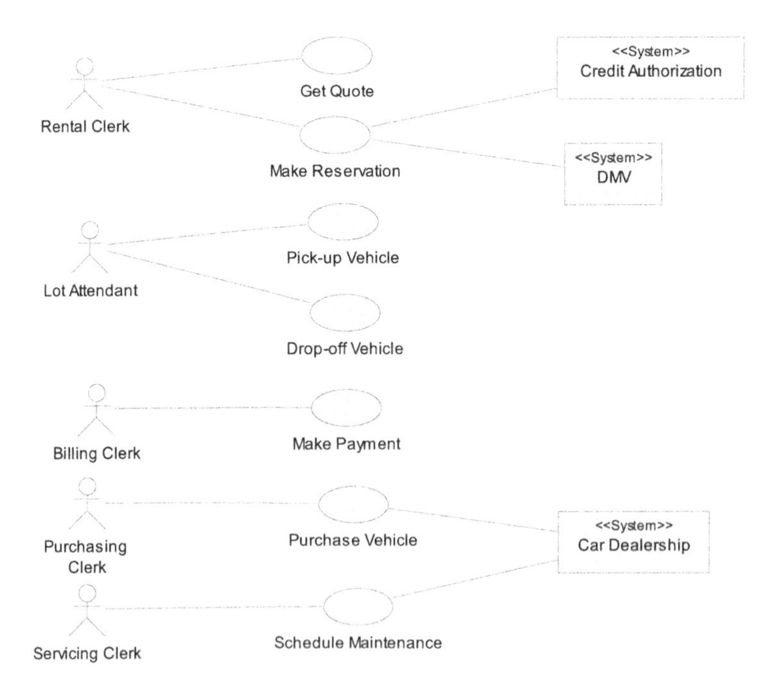

Generating Requirements for the Integrated Enterprise

A business model has a close correspondence with a software-system model. The business use cases become the system use cases, and the business workers become the system actors that initiate these use cases. The business entities in the business-analysis model and how they are interrelated to each other can be captured in a domain model for the software system and also become the entities in the system-analysis model. Figure 4 shows the system use-case model.

Notice how the customer from the business use-case model has been replaced by the rental clerk, the lot attendant, and the billing clerk. These roles represented business workers in the business use-case model but will now be interacting on the customer's behalf with the software system to be created.

The detailed operations within a system use case are first shown using a system sequence diagram treating the software system as a black box. Figure 5, for instance, shows a system sequence diagram for the "Make Reservation" use case.

Each system operation within the system sequence diagram becomes a software requirement that must now be designed and implemented in order to satisfy the system use case, which can be traced back to the business process "Make Reservation" that is being automated. This essentially completes the path for generating software requirements from a business model. Tools are available on the market such that not only can traceability be maintained between all the related artifacts, but models can be analyzed to generate software requirements from business models automatically.

Figure 5. System sequence diagram for "Make Reservation" use case

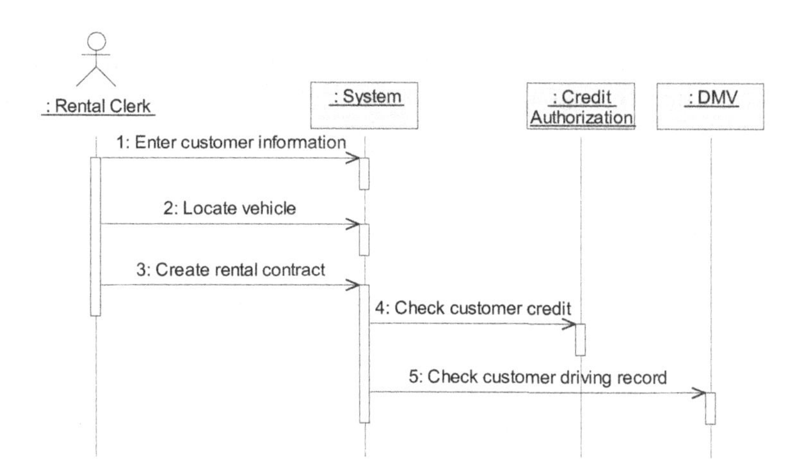

Figure 6. Partial domain model

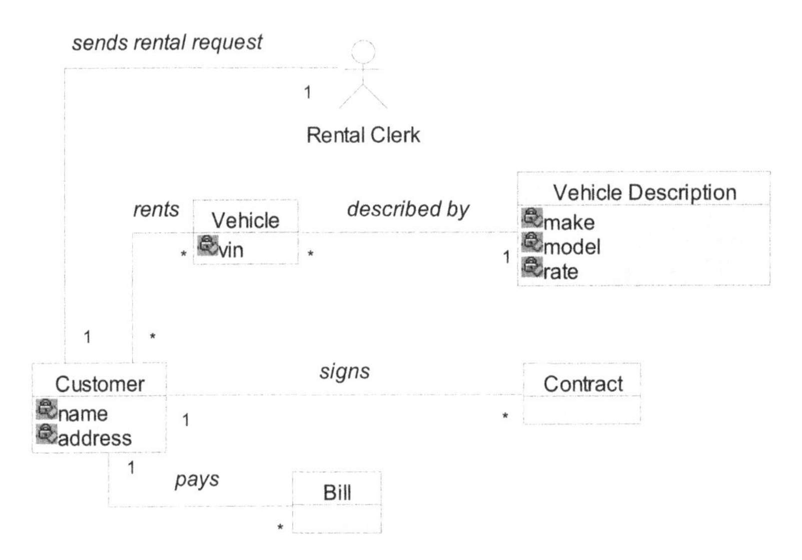

One final word on the business entities—Figure 6 shows the business entities from the business-analysis model in the domain model for the software system to be created. It is only a partial domain model obtained from the details of the "Make Reservation" business use case. The figure only shows a customer's demographic information such as name and address, the vehicle identification number (VIN) for the vehicle, and the vehicle's description such as its make, model, and rental rate. However, other attributes as they are discovered should also be added to the entities in the domain model as they represent critical business information. Interrelationships among these entities are also shown.

These entities will be used to describe the details of the system operations shown in the system sequence diagram. For example, in the description of the operation "Locate vehicle," the postcondition may be stated as follows:

The desired *vehicle* is located and its *vehicle description* is retrieved.

Complexity in Large Models

As a more complex case, consider the health-care industry. It has experienced a gradual change in its business model over the years. Independent hospitals, clinics, laboratories, and physician group practices have merged to form large integrated health networks (IHNs). Patients can now walk into any institution within such a network for treatment and be easily referred to and treated at other specialized facilities within the same network. Once their electronic patient record and account is set up with one institution, there is never a need to establish a separate such record with other institutions within the network. This account serves as a single repository for all patient visits to any part of the network, making it possible for the patient to receive one statement rather than a multitude of statements from each different provider within the network. Creating a health information system (HIS) for such an integrated health network can be a quite challenging and expensive undertaking. It therefore behooves one to do due diligence to the requirements for this system. Model-driven requirements engineering can provide the necessary formalism and rigor.

To provide a sense of the complexity of an IHN, Figure 7 shows the business context.

Figure 7. The business context for an IHN

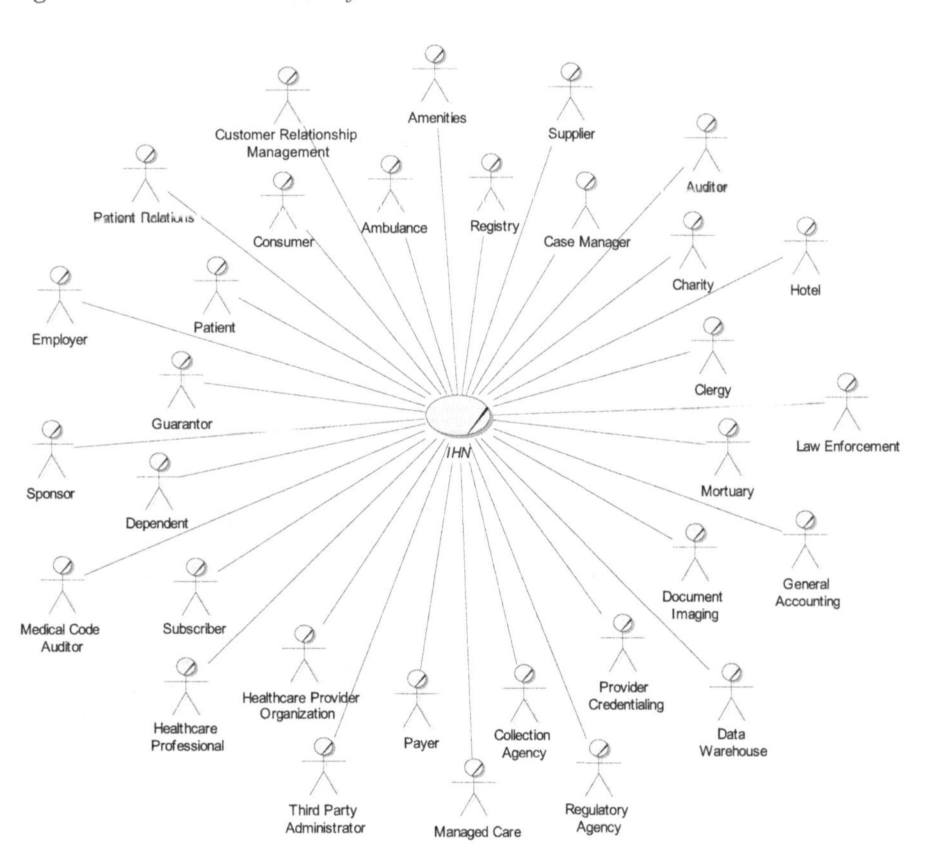

The number of external entities and, therefore, the business events they generate is several orders of magnitude higher compared to the example introduced earlier. The business-context diagram captures this complexity succinctly and provides a structured way to proceed with the creation of a business use-case model, its analysis model, the system use-case model, the system sequence diagrams, and, finally, the generation of requirements for an HIS. For brevity, we do not show the entire process as it is similar to the example introduced earlier. A leading provider of health-care information systems for which this effort was undertaken resulted in a massive model with more than 1,100 business use cases and their associated elaboration artifacts.

We, however, use a typical scenario for an emergency room (ER) patient brought into a health-care facility by an emergency medical team (EMT) upon receiving a 911 call to highlight a few important requirements-modeling issues. The following steps occur during this scenario (Sangwan & Qiu, 2005).

- The EMT identifies the patient and performs a preliminary diagnosis.
- The appropriate health-care facility is notified to prepare for the arrival of the patient.
- The patient is transported to the health-care facility.
- The patient is checked into the health-care facility.
- The medical staff does a triage and prioritizes the treatment plan for the patient.
- The patient is stabilized before the treatment can begin.
- The patient is diagnosed.
- The patient is treated.
- Arrangements are made for aftercare and follow-up.
- The patient is discharged.

If the patient requires further treatment, the appropriate health-care facility within the IHN is notified; otherwise, the patient is transported back home.

Two interesting issues arise when creating a requirements model in this situation.

- **Different flavors of a business service:** The emergency-room check-in business service is very different from a check-in at a doctor's office. The patient may not be in a condition to provide any information at all, whereas in a doctor's office it is expected that a patient provide the necessary demographic and insurance information along with the co-pay amount.

- **Different representations of a business entity:** While a person may be admitted to a facility as a patient, in the financial world, he or she may act as a guarantor responsible for making payments for the services provided during the emergency-room visit. Patient and guarantor are different roles played by the same business entity.

There is, therefore, a need for modeling this variability. Marshall (2000) provides an approach for handling similar situations.

Conclusion

This chapter made an argument for the importance of model-driven requirements engineering in enterprise integration. The business model used in this approach not only helps one understand the structure and dynamics of a business, but also provides a mechanism for investigating opportunities for business-process engineering and reengineering. This includes investigating scenarios for e-commerce and e-supply-chains. Models for software systems needed to take advantage of these opportunities can then be created from the business models to fulfill software requirements generated from these models. The chapter demonstrated this using a car-rental enterprise as a motivating example and a case study on creating a health-care information system for integrated health networks.

References

Berenbach, B. (2003). The automated extraction of requirements from UML models. In *Proceedings of the 11th Annual IEEE International Requirements Engineering Conference (RE'03)* (pp. 287-288).

Berenbach, B. (2004a). The evaluation of large, complex UML analysis and design models. In *Proceedings of the 26th International Conference on Software Engineering (ICSE 2004)* (pp. 232-241).

Berenbach, B. (2004b). Towards a unified model for requirements engineering. In *Proceedings of the Fourth International Workshop on Adoption-Centric Software Engineering (ACSE 2004)* (pp. 26-29).

Booch, G., Rumbaugh, J., & Jacobson, I. (2005). *The unified modeling language user guide* (2nd ed.). Boston: Addison-Wesley.

Fowler, M. (2004). *UML distilled* (3rd ed.). Boston: Addison-Wesley.

Kruchten, P. (2004). *The rational unified process: An introduction* (3rd ed.). Boston: Addison-Wesley.

Leffingwell, D., & Widrig, D. (2000). *Managing software requirements: A unified approach.* Boston: Addison-Wesley.

Marshall, C. (2000). *Enterprise modeling with UML.* Boston: Addison-Wesley.

Robertson, S., & Robertson, J. (1999). *Mastering the requirements process.* Boston: Addison-Wesley.

Sangwan, R., & Qiu, R. (2005). Using RFID tags for tracking patients, charts and medical equipment within an integrated health delivery network. In *Proceedings of the International Conference on Networking, Sensing and Control* (pp. 1070-1074).

Chapter V

Mobile Workforce Management in a Service-Oriented Enterprise: Capturing Concepts and Requirements in a Multi-Agent Infrastructure

Dickson K.W. Chiu, Dickson Computer Systems, Hong Kong

S.C. Cheung, Hong Kong University of Science and Technology, Hong Kong

Ho-fung Leung, The Chinese University of Hong Kong, Hong Kong

Abstract

In a service-oriented enterprise, the professional workforce such as salespersons and support staff tends to be mobile with the recent advances in mobile technologies. There are increasing demands for the support of mobile workforce management (MWM) across multiple platforms in order to integrate the disparate business functions of the mobile professional workforce and management with a unified infrastructure, together with the provision of personalized assistance and automation. Typically, MWM involves tight collaboration, negotiation, and sophisticated business-domain knowledge, and thus can be facilitated with the use of intelligent software agents. As mobile devices become more powerful, intelligent software agents can now be deployed on these devices and hence are also subject to mobility. Therefore, a multiagent information-system (MAIS) infrastructure provides a suitable paradigm to capture the concepts and requirements of an MWM as well as a phased development and deployment. In this book chapter, we illustrate our approach with a case study at a large telecommunication enterprise. We show how to formulate a scalable, flexible, and intelligent MAIS with agent clusters. Each agent cluster comprises several types of agents to achieve the goal of each phase of the workforce-management process, namely, task formulation, matchmaking, brokering, commuting, and service.

Introduction

The advancement of mobile technologies has resulted in an increasing demand for the support of mobile-workforce management (MWM) across multiple platforms anytime and anywhere. Examples include supply-chain logistics, group calendars, dynamic human-resources planning, and postal services. Existing solutions and proposals often treat the workforce as passive-moving resources and cannot cope with the current requirements for the knowledge-based economy and services, such as technical-support teams (e.g., computer- or network-support engineers and technicians).

Recent advances in hardware and software technologies have created a plethora of mobile devices with a wide range of communication, computing, and storage capabilities. New mobile applications running on these devices provide users with easy access to remote services at anytime and anywhere. Moreover, as mobile devices become more powerful, the adoption of mobile computing is imminent. The Internet is quickly evolving toward a wireless one, but the wireless Internet will not be a simple add-on to the wired Internet. New challenging problems arise from the handling of mobility, handsets with reduced screens, and varying bandwidth. Moreover, the business processes involving the workforce tends to get complicated with requirements from both within the organization's management and external Web services (e.g., tracking and logistics integration). New mobile applications running on these devices provide users easy access to remote services regardless of where they are, and will soon take advantage of the ubiquity of wireless networking to create new virtual worlds. Therefore, the main challenge of MWM is to provide an effective integration of the ever-increasing disparate business functions in a unified platform not only to management, but also to the mobile professional workforce.

An additional challenge to MWM in service-oriented enterprises (such as telecom and computer vendors) is the provision of personalized assistance and automation to the mobile professional workforce, whose members each have different capabilities, expertise, and support requirements. Often, consultations and collaborations are required for a task. Because of their professional capabilities and responsibilities, members of the workforce have their own job preferences and scheduling that cannot be flexibly managed in a centralized manner. As mobile devices become more powerful, peer-to-peer mobile computing becomes an important computation paradigm. In particular, intelligent software agents can now run on these mobile devices and can adequately provide personalized assistance to the mobile workforce. Under the individual's instructions and preferences, these agents can be delegated to help in the negotiating and planning of personalized tasks and schedules, thereby augmenting the user's interactive decisions. In addition, agent-based solutions are scalable and flexible, supporting variable granularities for the grouping of workforce management.

We have been working on some related pilot studies related to MWM, such as constraint-based negotiation (Chiu, Cheung, et al., 2004), m-service (mobile-service) adaptation (Chiu, Cheung, Kafeza, & Leung, 2003), and alert management for medical professionals (Chiu, Kwok, et al., 2004). Based on these results, we proceed to a larger scale case study, and the contributions of this chapter are as follows. First we formulate a scalable, flexible, and intelligent multiagent information-system (MAIS) infrastructure for MWM with agent clusters in a service-oriented enterprise. Then we propose the use of agent clusters, each comprising several types of agents to achieve the goal of each phase of the workforce-management process, namely, task formulation, matchmaking, brokering, commuting, and service. Next we formulate a methodology for the analysis and design of MWM in the context of enterprise service integration with MAIS. Finally, we illustrate our approach with an MWM case study in a large service-oriented telecom enterprise, highlighting typical requirements and detailing architectural design considerations. This book chapter is an extension of our previous work (Chiu, Cheung, & Leung, 2005). It refines our previous MAIS infrastructure and relates that to the believe-desire-intention (BDI) agent architecture (Rao & Georgeff, 1995). The application of the refined MAIS infrastructure is illustrated by a case study based on a large service-oriented telecom enterprise.

The rest of the chapter is organized as follows. First we introduce background and related work. Next we explain an overview of an MAIS and a development methodology for MWM. After this, we highlight the MWM process requirements. The next section details our MAIS architecture and implementation framework. Then we evaluate our approach from different stakeholders' perspectives. We conclude this chapter with our plans for further research.

Background

Users under mobile or wireless computing environments are no longer constrained by working at a fixed and known location where wired connection is available. Users of a workforce-management system can collaborate at anywhere and anytime. This facilitates timely and location-aware decision making. Although a mobile system shares many characteristics with a distributed system, it imposes new challenges (Barbara, 1999) to computing applications, including workforce management. First, communication between parties in a mobile system is no longer symmetric. The downstream data rates are much wider than upstream data rates. Some two to three orders of magnitude differences are generally expected. As such, mobile applications need to be designed with care to minimize the upstream data transfer. Second, mobile communication channels are more liable to disconnection and data-rate frustration. Message exchanges should be designed to be as idempotent as possible. As a result, mobile process flows must support exception handling and be able to adapt to environmental changes. Third, the screen sizes of mobile devices are usually small and vary across different models. This affects how information can be effectively disseminated and displayed to users. Fourth, mobile or wireless networks are ad hoc in nature. A wireless connection infrastructure typically consists of thousands of mobile nodes whose communication channels can be dynamically reconfigured. To reduce overheads, channel reconfiguration generally requires limited network management and administration. The availability of mobile ad hoc networking technology imposes challenges to effective multihop routing, mobile data management, congestion control, and dynamic quality-of-services support. The autonomy of mobile nodes is desired (Shi, Yang, Xiang, & Wu, 1998). Fourth, mobile nodes have stringent constraints on computational resources and power. Expensive computations as required by asymmetric encryption or video encoding should not be performed frequently.

Advanced work-flow-management systems (WFMSs) are mostly Web enabled. Recently, researchers in work-flow technologies have been exploring cross-organizational work flows to model these activities, such as Grefen, Aberer, Hoffner, and Ludwig (2000), Kim, Kang, Kim, Bae, and Ju (2000), and the Workflow Management Coalition (1995, 1999). In addition, advanced WFMSs can provide various services such as coordination, interfacing, maintaining a process repository, process (work flow) adaptation and evolution, matchmaking, exception handling, data and rule bases, and so on, with many opportunities for reuse. With the advance in mobile and wireless technologies, mobile workforce management has become more and more decentralized, with involved components becoming increasingly autonomous, and location and situation awareness being incorporated into system design (Karageorgos, Thompson, & Mehandjiev, 2002; Lee, Buckland, & Shepherdson, 2003; Thompson & Odgers, 2000).

A business process is carried out through a set of one or more interdependent activities, which collectively realize a business objective or policy goal. Work flow is the computerized facilitation or automation of a business process. WFMSs can assist in the specification, decomposition, coordination, scheduling, execution, and monitoring of work flows. In addition to streamlining and improving routine business processes, WFMSs help in documenting and reflecting upon business processes. Often, traditional WFMSs can only coordinate work flows within a single organization. However, contemporary WFMSs can now interact with various types of distributed agents over the Internet.

Intelligent agents are considered autonomous entities with abilities to execute tasks independently. He, Jennings, and Leung (2003) present a comprehensive survey on agent-mediated e-commerce. An agent should be proactive and subject to personalization, with a high degree of autonomy. In particular, due to the different limitations on different platforms, users may need different options in agent delegation. Prior research studies usually focus on the technical issues in a domain-specific application. For example, Lo and Kersten (1999) present an integrated negotiation environment by using software-agent technologies for supporting negotiators. However, all of these works did not support their models on different platforms.

This problem is further complicated by the dynamicity of the mobile e-commerce environment brought about by wireless communication channels and portable computing devices. Mobile-agent technology is a promising solution to the problem (Kowalczyk et al., 2003). Various studies have been made to integrate mobile and wireless technologies into agents (Bailey & Bakos, 1997; Kotz & Gray, 1999; Kowalczyk & Bui, 2000; Lomuscio, Wooldridge, & Jennings, 2000; Papaioannou, 2000).

However, the problem of MWM and the deployment of agents for this purpose are rarely studied. Research in mobile computing mainly focuses on the enabling technologies at communication layers instead of the deployment of applications such as MWM on the application layer. Guido, Roberto, Tria, and Bisio (1998) point out some MWM issues and evaluation criteria, but the details are no longer up to date because of the fast-evolving technologies. Jing, Huff, Hurwitz, Sinha, Robinson, and Feblowitz (2000) present a system called WHAM (workflow enhancements for mobility) to support the mobile workforce and applications in work-flow environments, with emphasis on a two-level (central and local) resource-management approach. Both groups did not consider distributed agent-based, flexible, multiplatform business-process interactions or any collaboration support. Although there have been studies on related technologies for MWM, there have not been in-depth studies on how to integrate these technologies for a scalable MWM MAIS.

The emergence of MAIS dates back to Sycara and Zeng (1996), who discuss the issues in the coordination of multiple intelligent software agents. In general, an MAIS provides a platform to bring together the multiple types of expertise for any decision making (Luo, Liu, & Davis, 2002). For example, F. R. Lin, Tan, and Shaw (1998) present an MAIS with four main components: agents, tasks, organizations, and information infrastructure for modeling the order-fulfillment process in a supply-chain network. Furthermore, F. R. Lin and Pai (2000) discuss the implementation of

MAIS based on a multiagent simulation platform called Swarm. Next, Shakshuki, Ghenniwa, and Kamel (2000) present an MAIS architecture in which each agent is autonomous, cooperative, coordinated, intelligent, rational, and able to communicate with other agents to fulfill the users' needs. Choy, Srinivasan, and Cheu (2003) propose the use of mobile agents to aid in meeting the critical requirement of universal access in an efficient manner. Chiu et al. (2003) also propose the use of a three-tier view-based methodology for adapting human-agent collaborative systems for multiple mobile platforms. In order to ensure interoperability of an MAIS, standardization on different levels is highly required (Gerst, 2003). Thus, based on all these prior works, our proposed MAIS framework adapts and coordinates agents with standardized mobile technologies for MWM.

E-collaboration (Bafoutsou & Mentzas, 2001), being a foundation of WFM, supports communication, coordination, and cooperation for a set of geographically dispersed users. Thus, e-collaboration requires a framework based on strategy, organization, processes, and information technology. Furthermore, Rutkowski, Vogel, Genuchten, Bemelmans, and Favier (2002) address the importance of structuring activities for balancing electronic communication during e-collaboration to prevent and solve conflicts. For logic-based collaboration, Bui (1987) describes various protocols for multicriteria group-decision support in an organization. Bui, Bodart, and Ma (1998) further propose a formal language based on first-order logic to support and document argumentation, claims, decisions, negotiation, and coordination in network-based organizations. In this context, a constrain-based collaboration can be modeled as a specific case of the Action-Resource Based Argumentation Support (ARBAS) language.

Wegner, Paul, Thamm, and Thelemann (1996) present a multiagent collaboration algorithm using the concepts of belief, desire, and intention. In addition, Fraile, Paredis, Wang, and Khosla (1999) present a negotiation, collaboration, and cooperation model for supporting a team of distributed agents to achieve the goals of assembly tasks. However, this paper mainly focuses on the overall integration of MWM support with MAIS.

Another foundation of MFM is meeting scheduling. There are some commercial products, but they are just calendars or simple diaries with special features, such as availability checkers and meeting reminders (Garrido, Brena, & Sycara, 1996). Shitani, Ito, and Sycara (2000) highlight a negotiation approach among agents for a distributed meeting scheduler based on the multiattribute-utility theory. Lamsweerde, Darimont, and Massonet (1995) discuss a goal-directed elaboration of requirements for a meeting scheduler, but do not discuss any implementation frameworks. Sandip (1997) summarizes an agent-based system for an automated distribution meeting scheduler, but it is not based on BDI agent architecture. However, all these systems cannot support manual interactions in the decision process or any mobile support issues.

In summary, none of the existing works consider an MAIS infrastructure for MWM as a solution for integration and personalized workforce support. Scattered efforts have looked into subproblems but are inadequate for an integrated solution. There is neither any work describing a concrete implementation framework and methodology by means of a portfolio of contemporary enabling technologies.

MAIS Infrastructure

An MAIS provides an infrastructure for the exchange of information among multiple agents as well as users under a predefined collaboration protocol. Agents in the MAIS are distributed and autonomous, each carrying out actions based on their own strategies. In this section, we explain our MAIS infrastructure and metamodel in which the computational model of an agent can be described using a BDI framework. Then, we summarize our methodology for the design and analysis of an MAIS for MWM.

MAIS Layered Infrastructure for MWM

Figure 1 summarizes our layered infrastructure for MWM. Conventionally, services and collaboration are driven solely by human representatives. This could be a tedious, repetitive, and error-prone process, especially when the professional workforces have to commute frequently. Furthermore, agents facilitate the protection of pri-

Figure 1. A layered infrastructure for MWM

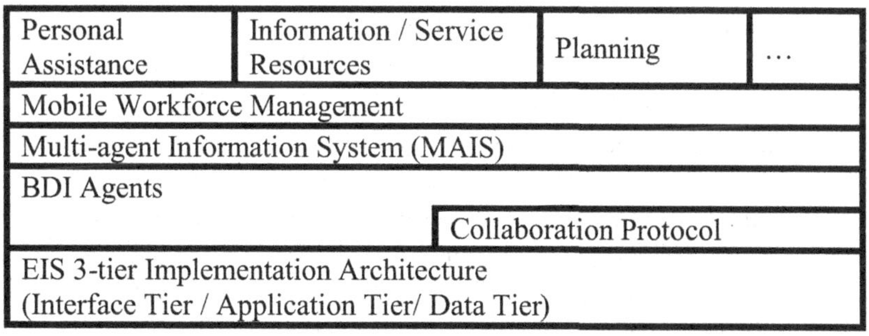

Personal Assistance | Information / Service Resources | Planning | ...
Mobile Workforce Management
Multi-agent Information System (MAIS)
BDI Agents
Collaboration Protocol
EIS 3-tier Implementation Architecture
(Interface Tier / Application Tier/ Data Tier)

Figure 2. BDI conceptual model

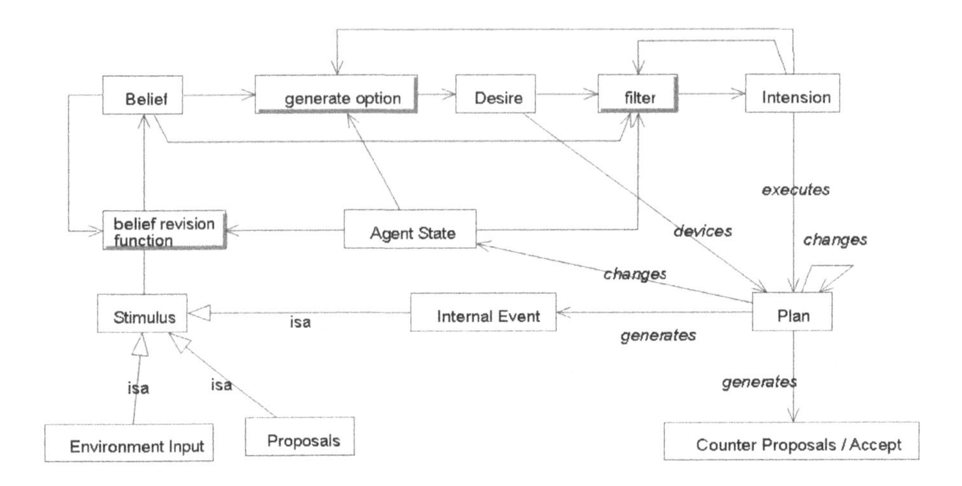

vacy and security. The provision of computerized personal assistance to individual users across organizations by means of agents is a sensible choice. These agents, acting on behalf of their delegators, collaborate through both wired and wireless Internet, forming a dynamic MAIS over an enterprise information system (EIS). Such repeatable processes can be adequately supported, and the cost of developing the infrastructure is well justified.

The BDI framework is a well-established computational model for deliberative intelligent agents, as summarized in Figure 2. A BDI agent constantly monitors the changes in the environment and updates its information accordingly. Possible goals are then generated, from which intentions to be pursued are identified. A sequence of actions will be performed to achieve the intentions. BDI agents are proactive by taking initiatives to achieve their goals, yet adaptive by reacting to the changes in the environment in a timely manner. They are also able to accumulate experience from previous interactions with the environment and other agents.

Internet applications are generally developed with a three-tier architecture comprising the front, application, and data tiers. Though the use of a three-tier architecture in the agent community is relatively new, it is a well-accepted pattern to provide flexibility in each tier (Chiu et al., 2003) and is absolutely required in the expansion of e-collaboration support. Such flexibility is particularly important to the front tier, which often involves the support of different solutions on multiple platforms. In our architecture, users may either interact manually with other collaborators or delegate an agent to make decisions on their behalf. Thus, users without agent support can still participate through flexible user interfaces for multiple platforms.

Figure 3. Metamodel of an MAIS in a UML (unified modeling language) class diagram

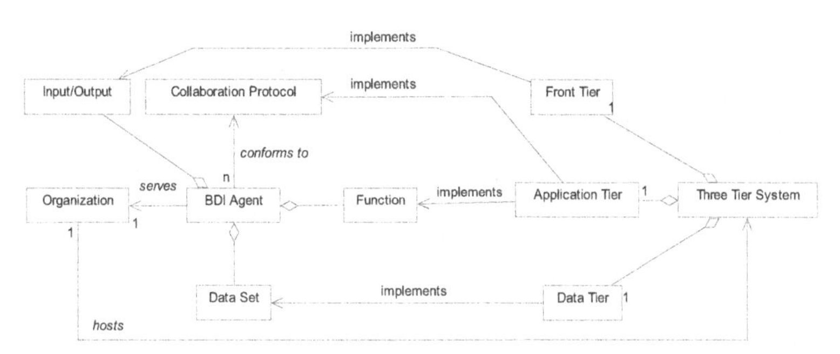

MAIS Metamodel

Figure 3 describes the metamodel of an MAIS system in a class diagram of UML (Object Management Group [OMG], 2001), which is widely used for visualizing, specifying, constructing, and documenting the artifacts of a software-intensive system. It summarizes our mapping between the components of a BDI agent to individual tiers of a three-tier system hosted by an organization. A BDI agent is made up of three major components: input and output, functions, and data sets. It acts on behalf of a user in an organization and interacts with other agents according to a predefined collaboration protocol. The agent receives inputs and generates outputs through the front tier. The agent's functions and the protocol logic can be implemented at the application tier. The data tier can be used to implement the various data sets of an agent.

A BDI computational model is composed of three main data sets: belief, desire, and intention. Information or data are passed from one data set to another through the application of some functions. Once a stimulus is sensed as input, the belief-revision function (BRF) converts it to a belief. The desire set is updated by generating some options based on the data in the belief set. Options in the desire set are then filtered to become the new intentions of the agent, and a corresponding plan of action can then be generated. As such, the BDI agent mimics an assistant for decisions on behalf of a human user, which is particularly useful for collaborations.

Though an agent can receive signals from the environment (such as user location), the stimulus inputs are mainly incoming requests and responses from other agents and users. These inputs are usually associated with a set of constraints and/or options (solutions) to a proposal. As a result, the belief set contains several sets of constraints representing the requirements of a proposal. All solutions or even future options should satisfy these sets of constraints. As such, acceptable workforce service and collaboration arrangements are solved by mapping the constraints generated to the well-known constraint-satisfaction problem (CSP; Tsang, 1993), where efficient solvers are available.

MAIS Analysis and Design Methodology for MWM

Based on our previous experience in constraint-based negotiation, m-service adaptation, and alert management for mobile medical professionals, we proceed in this study to generalize and scale up our framework to a MAIS for MWM. We advocate the system analysis and design methodology to be carried out in two parts. Part 1 deals with the overall architectural design. That is, we have to analyze high-level requirements and formulate an enterprise MAIS infrastructure and system integration aspects that are specific for a particular purpose (MWM here) and to a particular domain (service-oriented enterprises here). The application of MWM for service-oriented enterprises has not been studied before and is therefore the focus of this chapter. The steps for Part 1 are as follows.

1. Identify different categories of services and objectives for the workforce in the enterprise. The identification can make use of available service ontologies, such as those defined in Semantic Web services.
2. Identify the life cycle (i.e., different phases) for the management of a typical service task, from task request to completion.
3. For each phase, identify the major agent to represent it and then the interactions required based on the process requirements.
4. Further identify minor agents that assist the major agents in carrying out these functionalities. As a result, clusters of different types of agents (instead of a single monolithic pool of agents) constitute the MAIS.
5. Identify the interactions required for each minor agent type.
6. Design the basic logics for all these agents.
7. Identify the (mobile) platforms to be supported and where to host different types of agents. See if any adaptation is required.

Only after successful high-level requirement studies and the design of the overall architecture can we proceed to the next part. Part 2 deals with the detail design of agents, and the methodology has been preliminarily studied in our previous work (Chiu, Cheung, et al., 2004). It should be noted that the actual detailed design for each type of agent in the MWM domain has high potentials for further research because of its emerging adoptions. Here, we summarize the steps as follows for conveying a more complete picture of the required effort.

1. Design and adapt the user interface required for users to input their preferences. Customize displays to individual users and platforms.
2. Determine how user preferences are mapped into constraints and exchange them in a standardized format.
3. Consider automated decision support with agents. Identify the stimulus, collaboration parameters, and output actions to be performed by a BDI agent.
4. Partition the collaboration parameters into three data sets: belief, desire, and intention. Formulate a data subschema for each of these data sets. Implement the schema at the data tier.
5. Derive transformations amongst the three data sets. Implement these transformations at the application tier.
6. Enhance the performance and intelligence of the agents with various heuristics gathering during the testing and pilot phase of the project.

MWM Requirements Overview

This study is based on the requirements of a large service-oriented telecom enterprise, in which sales, technical, and professional workforces are mobile. We first highlight the requirements of the users and management before introducing the service-task categories. Then we present the workforce services and process overview.

User and Management Requirements

The main target users of the MWM are the mobile sales, technical, and professional workforces. Their main job functions are to carry out quality consultations and customer services, with commitments in improving customer relationships (thereby increasing sales). Users employ MWM systems to assist their work. The provision of anytime and anywhere connections is essential because the workforce tends to become mobile, especially for professionals such as physicians, service engineers, and sales executives as well as other staff who need to travel. In particular, the flexibility of supporting multiple front-end devices increases users' choice of hardware and therefore their means of connectivity. Agent automation helps reduce tedious collaboration tasks that are often repeated, including meeting scheduling as well as negotiations with standardized parameter (Chiu, Cheung, Hung, et al., 2005).

For management, it is expected that the MWM can integrate disparate heterogeneous organizational applications. In addition, MWM can locate mobile workforce members and therefore improve staff communications. Though this may not be in the interest of the workforce, the MWM infrastructure helps management to control and manage them, such as for location-dependent job allocation. Also, agents help improve the quality and consistency of decision results through preprogrammed intelligence through the BDI-agent architecture. In addition, an integration approach reduces development costs through software reuse and the time required for development.

Service-Task Categories

To effectively support mobile workforces in fulfilling their tasks and in particular services, we have to understand different types of service requirements. We analyze the characteristics of tasks, and each task may have one or more of the following characteristics.

A collaboration task requires more than several workforce members; that is, the availability of more than one person at the same time. As such, there is a subproblem similar to a well-known and nontrivial collaboration problem: meeting scheduling. In practice, scheduling is a time-consuming and tedious task. It involves intensive communications among multiple persons, taking into account many factors or constraints. In our daily life, meeting scheduling is often performed by ourselves or by our secretaries via telephone or e-mail. Most of the time, each attendee has some uncertain and incomplete knowledge about the preferences and the diaries of the other attendees. Historically, meeting scheduling emerged as a classic problem in artificial intelligence and MAIS.

An on-site task requires the workforce member(s) to travel to a specific location. This is typical for sales representatives, construction-site supervisors, field engineers, medical professionals, and so on. They often need to visit numerous locations in a day. Thus, a route advisory system (possibly supported by a third party or public services) can help them find the viable routes to their destinations. This could also help the organizations save time and costs by providing the fastest and most economical routes, respectively. However, if an organization has its own transportation vehicles for their workforce, further integration of the vehicles with the workforce-management system is required.

A personal task requires one or more specific members of the workforce to fulfill the tasks (say, because of job continuity). Otherwise, a flexible task allows the capability requirements of the task to be specified instead so that the system can select the best possible candidate(s).

A remote task requires communications support. The user, workforce, or agent involved has to be connected to the EIS or portals from remote sites for effective work. Information transcoding or even process adaptation may be required (Chiu et al., 2003).

Workforce Services and Processes Overview

Tracing the overall process from the placement of a customer service call or visit plan to its completion, we identify the following phases of a typical MWM service task.

1. The task-formulation phase concerns the creation of a task request and its specification from various sources inside and outside the enterprise.
2. The matchmaking phase concerns the tactical identification of the possible workforce capable of the task and ranks a subset of them for consideration in the brokering phase by using protocols such as the contract net (Smith, 1980).
3. The brokering phase concerns negotiation with a short list of the workforce to pick the best available one for a suitable appointment time according to schedule, location, and preferences.
4. The commuting phase concerns the travel of the workforce (if necessary), their vehicles (if any), and their locations.
5. The service phase concerns the actual execution of the task and the necessary support for the remote workforce.

Figure 4. MAIS architecture overview for MWM

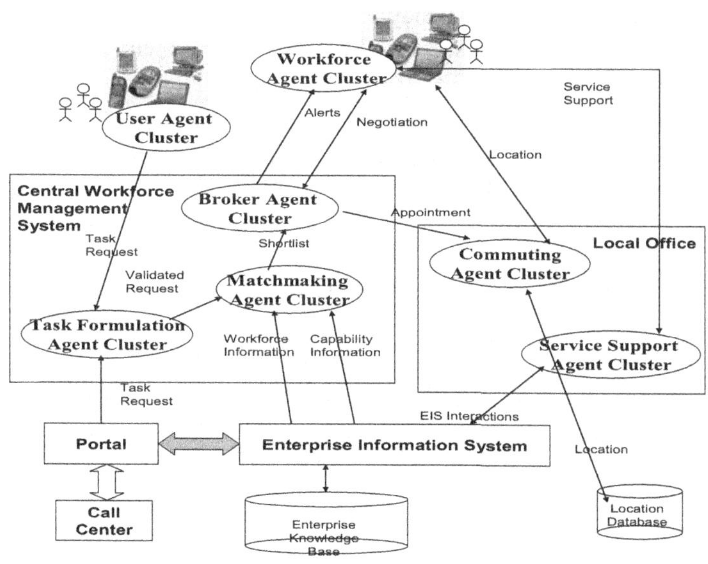

System Architecture and Implementation

We employ an MAIS design in our infrastructure for MWM (as shown in Figure 4) because of the requirements of mobile users as well as application flexibility and scalability. In particular, local offices can maintain their own commuting agent clusters and service-support agent clusters. This approach not only off-loads the central MWM, but also facilitates the maintenance of local knowledge. We discuss in subsections below each major agent cluster, which corresponds to various phases of a task-management life cycle.

Task-Formulation Agent Cluster

The task-formulation agents are assisted by a cluster of agents (as summarized in Table 1) to carry out the functions for the task-formation phase of the MWM process. There are many possible sources for a service-task request, such as (a) call centers, (b) customer Web portals, (c) management orders, (d) regular service schedulers, (e) service follow-ups, (f) customer relationship management (CRM) systems, (g) EIS triggers, and so on. Because of the diversity of request formats from existing systems, request-translation agents are built as the front end for each of these sources to map these requests into a common compatible format. Important request attributes include the task category expressed in the enterprise's ontology, urgency, importance, budget, resource requirement, location, requestor, related customer, and so on. However, requests from call centers and Web portals are often (problem) case reports, and are currently diagnosed by customer-services specialists and engineers. To reduce cost and increase efficiency, automation with report-diagnosis agents should be developed in the next stage of the system deployment.

Task-validation agents then attempt to fill in unspecified or implied attributes for requests with various heuristics. For example, the urgency and importance of tasks are often not specified clearly. These requests are then validated against a rule base

Table 1. Key agents in the task-formulation agent cluster

Agents	Functions
Task Formulation	Main agent formulates a task requirement from requests
Request Translation	Translate requests into a common format
Task Validation	Fill in unspecified or implied task attributes and check task validity
Requirement Negotiation	Negotiate requests with requestors
Report Diagnosis	Diagnose requests and calls

of constraints, which specifies the rules and policies of the enterprise as well as various units. In particular, request authorization and budget are validated. Rejected tasks are passed back to the requestors clearly stating the violations and problems so that they can revise their request effectively. In the next stage of the system deployment, requirement-negotiation agents should be developed to handle failed requests in a more effective manner.

Validated task requests are recorded in the enterprise case base and then forwarded to the matchmaking agents for the next phase as well as the monitor agents for monitoring.

Matchmaking Agent Cluster

The matchmaking agents are assisted by a cluster of agents (as summarized in Table 2) for the matchmaking phase of the MWM process. They identify the possible workforce members who are capable of carrying out the task. The overall approach is based on our earlier work on capability and role modeling for work flows (Chiu et al., 1999). We separate the matchmaking and brokering phases because they deal with the operational and tactical allocation of workforces, respectively. In particular, a centralized management personal schedule of all workforce members together with intelligent task allocations based on such massive information is infeasible.

After receiving a validated task request, a capability-analysis agent analyzes the request to identify the detailed breakdown of the capability requirements according to the enterprise's ontology.

If the task request has personal specification, the capability of the workforce member is validated against the elicited capability requirements. Otherwise, the matchmaking agent has to select a preliminary short list of candidates from the workforce database according to the capability requirements. Though we do not consider complete schedules of the workforce here, we can still filter the workforce according to duty rosters and avoid those marked busy. Some preliminary algorithms are presented

Table 2. Key agents in the matchmaking agent cluster

Agents	Functions
Matchmaking	Main agent identifies the possible workforce members capable of carrying out the task
Capability Analysis	Break down the capability requirements of a task
Location	Determine the location of workforce members
Cost Evaluation	Estimate cost of service

by Chiu et al. (1999). However, for a comprehensive MWM, we have to consider the workforce locations as well as the possibilities of a composition workforce for a task. For example, we have to decide between two engineers who each have one of the two required capabilities or a senior engineer with the two capabilities.

Workforce locations are determined by locations agents, which retrieve location contexts that are streamed continuously from spatial sensitive signals, such as those delivered by the global positioning system (GPS). More sophisticated location agents may also compute position and orientation based on signal strengths received from nearby wireless transmission stations for mobile communications.

The preliminary short list of the workforce is then passed to the cost-evaluation agent for estimating the cost of service. If there are many valid combinations, the final short list of combinations has to be pruned according to various search heuristics. In addition, the salary cost, the travel cost, and the cost incurred due to the travel time must be considered among other heuristic components of cost functions. The final short list, together with the estimated costs, is then forwarded to the broker agents for the next phase.

Broker Agent Cluster

The broker agents are assisted by a cluster of agents (as summarized in Table 3) for the brokering phase of the MWM process. They have to negotiate with the short list of workforce to pick the best available one for a suitable appointment time, according to schedule, location, and preferences.

An appointment agent first obtains the update locations of the short-listed workforce and then contacts them for a possible appointment. As detailed in our recent work (Chiu et al., 2003), we advocate the use of constraints for time and place negotiation because the collection of complete personal schedules can be avoided for efficiency, privacy, and communication costs. An additional requirement over our previous protocol is that not all the contacted workforces are appointed; preferences should be given to those at the top of the short list. It should be noted that in the case of a personal task, we still have to carry out the negotiation for selecting the best appoint-

Table 3. Key agents in the broker agent cluster

Agents	Functions
Broker	Main agent negotiates and picks the best available short-listed workforce
Appointment	Negotiate with workforce members for appointment
Alert	Keep track of alert messages to the workforce members

ment time and location. In addition, if customers are involved, customer preferences should be considered a priority over those of the workforce.

Furthermore, alert agents tackle messaging on behalf of appointment agents for task requests with an alert mechanism (Chiu, Kwok, et al., 2004). This is particularly important if a target user (such as an external customer) without agent support is involved in the appointment. Manual responses have to be tracked. In the case of no reply, the alert agent has to resend the message and/or inform the appointment agent to raise the urgency or consider other alternatives.

When an appointment is confirmed, the workforce members go into the commuting phase if traveling is required; otherwise, they go directly into the service phase.

Commuting Agent Cluster

The commuting agent cluster (as summarized in Table 4) plays a main role in the commuting phase. These agents take care of the traveling needs of the workforce if they have to travel to work on site. Location agents track the location of each workforce member.

Transport-advisory agents search for a suitable route from public transport for a commuting workforce member. In developing countries or crowded large cities, even professional workforce members may not have their own vehicles unless they are senior employees or are usually traveling with a lot of equipment. For those with vehicles owned by the enterprise, vehicle agents help plan the route for the vehicles to their service destinations and track vehicle locations. Transport-advisory agents also consult nearby vehicle agents for possibilities of picking up colleagues to take them to their destinations.

If the workforce and vehicles are mobile in a large metropolis, the main challenge is performance and efficiency because of the large number of public transport routes (for example, more than 1,000 in Hong Kong) and locations. In addition, both travel time and cost may have to be considered. We are working on an agent-based mobile

Table 4. Key agents in the commuting agent cluster

Agents	Functions
Commuting	Main agent manages the traveling needs of workforce members
Location	Determine the location of workforce members
Transport Advisory	Search for a suitable route from public transport for a commuting workforce member
Vehicle	Manage vehicles used by workforce members

Table 5. Key agents in the service-support agent cluster

Agents	Functions
Collaboration Session	Maintain widgets for collaborating workforce members
Remote EIS	Connect to EIS for required information
Monitor	Keep track of the progress of tasks

route-advisory system for public-transport networks to address this problem (Chiu, Lee, et al., 2005).

Service-Support Agent Cluster

The service-support agent cluster (as summarized in Table 5) supports the workforce in the service phase. These agents take care of the communication needs of the workforce when connection to remote collaborators or systems is required.

Collaboration-session agents maintain widgets such as shared desktops and blackboards for collaborating workforce members. For example, the workforce on the same project may edit the same document concurrently even if they are located in different sites. These collaboration-session agents must therefore coordinate and consistently reflect changes in shared widget properties. In the past, there were two approaches to coordinating communications among individual groupware widgets. The first approach is to have the coordination handled by specialized applications. This inevitably complicates the application logistics and limits the reuse of groupware widgets. The second approach is to provide a set of generic groupware widgets with built-in logistics for communications among individual widgets. However, prebuilt groupware widgets may not be easily customized to suit various user needs. Generic groupware widgets tend to be bulky and accompanied by many unused features.

With collaboration-session agents, many groupware widgets can just mirror the functionality of their single-user counterparts, except for the additional logistics to synchronize shared properties. Collaboration-session agents communicate with one another through registered connections to update the widgets of all the users in the collaboration section with the changes in the shared properties, say, with a callback mechanism. Thus, widget designers only need to determine the set of properties by which a group of widgets should be synchronized and to what extent they are synchronized. These properties collectively define the coupling among a group of widgets as a coupling portfolio among the collaboration-session agents. The dynamic modification of coupling portfolios is thus supported by on-the-fly reconfiguration of multicast groups through negotiation among these agents.

Remote EIS agents enable the workforce to connect to the EIS for information relevant to their task. Security is the main concern and therefore EIS agents act as guards and filters to allow only the authorized users to connect to the authorized EIS resources. Additional filtering is necessary to screen sensitive information for security as well as for conserving bandwidth.

Monitor agents keep track of the progress of all tasks. In particular, they are interested in when a workforce member or group commits to a task and when a task is completed. If deadlines are missed or exceptions are reported by the workforce or their agents, the monitor agents will report the cases to relevant supervisors or management.

Workforce and User Agent Cluster

Each workforce member has a workforce agent cluster (as summarized in Table 6) to assist with daily work. As the workforce (especially senior ones) can schedule meetings and arrange work for their subordinates, workforce members are also users. For external users or customers, we only need to limit them to a subset of agents and functions from a security perspective. Thus, we discuss these two types of agent clusters together.

Calendar agents maintain their personal schedules and act on their user's behalf for appointment negotiation. Reminder agents help the calendar agents to remind users of their upcoming schedules (especially the important ones) and urgent alerts received from alert agents. Preference agents provide interfaces for users to input their requests and preferences.

Interface agents transform the extended markup language (XML) output from other agents to the current user platform with XML stylesheet language (XSL) technologies. For example, different hypertext markup language (HTML) outputs are generated for Web browsers on desktop PCs (personal computers) and PDAs (personal digital assistants), while wireless access protocol (WAP) markup language (WML) outputs are generated for mobile phones (Y. B. Lin & Chlamtac, 2000).

Table 6. Key agents in the workforce and user agent cluster

Agents	Functions
Calendar	Maintain individual's schedule and negotiates appointment
Reminder	Remind of upcoming events and interact with alert agents
Preference	Maintain individual's preferences
Interface	Transform input and output to conform with individual's device

Discussions

In this section, we evaluate the applicability of our implementation framework and methodology with respect to the major stakeholders, including users, management, and system developers. The issues considered are based on the research framework on nomadic computing proposed by Lyytinen and Yoo (2002).

User's Perspective

Users employ MWM systems to assist in their work. In particular, agents help improve the reliability and robustness of workforce collaboration by retrying upon unsuccessful attempts, searching for alternatives, and so on. Agent-based adaptation of collaboration-protocol design for different operating environments improves the ease of use. This helps overcome the impact of expanding system functionalities and operation environments. Our proposed infrastructure also increases the chances to connect to the EIS and to interoperate with systems of other organizations. Thus, the main problem of integration and personalized assistance can be archived.

Our proposed way of applying constraint technologies helps achieve a balance of the performance and privacy of the workforce because they need not send all or part of their private information to a designated agent. This also avoids too much unnecessary data being sent, which wastes bandwidth and is not suitable for mobile users or agents.

Management's Perspective

A major concern of management is the costs against the benefits of the MWM system. In particular, if any of the improvements to the workforce as discussed in the previous subsection can significantly help improve their productivities, the costs can be justified. MWM provides tangible benefits for organizations by allowing information sharing among the mobile workforce. In addition, MWM usually implies the ability of locating mobile workforce members, therefore improving staff communications. Though this may not be in the interest of the workforce, the MWM infrastructure helps management to control and manage them, such as for location-dependent job allocation.

The incorporation of MWM helps improve customer relationships due to improved communications and service. Indirectly, business opportunities may increase, too. The disparity of heterogeneous organizational applications has created inflexible boundaries for communicating and sharing information and services among the mobile workforce, management, and customers. Therefore, MWM provides a standardized way to share the information through information agents and services among various heterogeneous applications.

Agents help improve the quality and consistency of decision results through pre-programmed intelligence. The BDI-agent architecture mimics the human practical deliberation process by clearly differentiating among the mental modalities of beliefs, desires, and intentions. Flexibility and adaptation are achieved by the agent's means- and ends-revising capabilities. As such, costs to program into the agent the operation and even the management knowledge elicited are minimized. Also, expertise to handle practical problems can be incorporated into the options function to generate desires and the filter function to determine intentions.

As for cost factors, our approach is suitable for the adaptation of existing systems by wrapping them with communication and information agents. Through software reuse, a reduction in not only the total development cost but also training and support cost can be achieved. For security, as explained in the previous subsection, constraints help reduce the need of revealing unnecessary information in collaborations and therefore improve security.

System Developer's Perspective

System developers are often concerned about the system-development costs and subsequent maintenance efforts. These concerns can be addressed by systematic, fine-grained requirements elicitation of the functions of various agent types. Thus, with loosely coupled and tightly coherent intelligent software modules encapsulated in agents, system complexity can be managed. Agents are highly reusable and can be maintained with relative ease. Furthermore, it should be noted that the use of XSL technologies and database views as the main mechanism for user-interface adaptation by presentation agents facilitates software maintenance at the application tier. This can significantly shorten the system-development time, meeting management expectations in a competitive environment.

Recent advances in technologies have resulted in fast-evolving mobile-device models and standards. MWM systems require much greater extents of adaptations to keep up. Agents are readily adaptable to cope with new technologies and can further help reduce uncertainties through adequate testing and experimentations of new technologies.

Some system functions have been implemented using entry-level PDAs, HP iPAQs®, each equipped with a 200MHz StrongArm® processor and 32MB SDRAM. The implementation aimed at exploring the feasibility of supporting agents on PDA platforms. The BDI-agent model and the associated constraint solver were written in Microsoft®-embedded Visual C++® and executed under Windows CE®. We found that an agent could comfortably solve 100 constraints with 200 to 300 variables in a second. This is a comfortable problem size for daily applications. As such, there is no need to rely on powerful computational servers to solve these constraints. In fact, a distributed solution of agents favors not only privacy, but also scalability. It further eases the programming of captured knowledge as explained in the previous subsections.

As constraints can be used to express general planning problems, including those involving higher order logic (Tsang, 1993), we anticipate that this approach can be applied in different domains for solving different problems related to MWM.

Conclusion

This chapter has presented a pragmatic approach to developing an MWM system with an MAIS infrastructure. We have also explained a metamodel of MAIS and a layer-infrastructure framework that supports multiple platforms (in particular, wireless mobile ones) and their integration with the EIS. We have summarized our experience in the analysis and design of an MAIS for MWM. We have also explained an overview of MVM requirements and process life cycle. We have further detailed the design of each agent cluster corresponding to each phase of the MWM process life cycle. Finally, we have explained the merits and applicability of our approach from the perspectives of major system stakeholders. As such, we are addressing the main challenge of MWM for a service-oriented enterprise, which is the integration of disparate business functions for the mobile professional workforce and management with a unified infrastructure, together with the provision of personalized assistance and automation.

With this solid foundation, we can proceed to study or reexamine the technical and management perspectives of each phase and functions of the MWM process in detail. We also anticipate this framework can serve as a reference model for this new MWM application area. We believe that only after task management for mobile workforces has been adequately studied can the problem of managing a complete mobile work flow be tackled.

Acknowledgments

The work described in this chapter was supported by a grant from the Research Grants Council of the Hong Kong Special Administrative Region, China (Project No. CUHK4190/03E).

References

Bafoutsou, G.., & Mentzas, G. (2001. A comparative analysis of Web-based collaborative systems. In *Proceedings of the 12th International Workshop on Database and Expert Systems Applications* (pp. 496-500).

Bailey, J., & Bakos, Y. (1997). An exploratory study of the emerging role of electronic intermediaries. *International Journal of Electronic Commerce, 1*(3), 7-20.

Barbara, D. (1999). Mobile computing and databases: A survey. *IEEE Transactions on Knowledge and Data Engineering, 11*(1), 108-117.

Bui, T. X. (1987). *Co-oP: A group decision support system for cooperative multiple criteria group decision making* (LNCS 290). Berlin, Germany: Springer-Verlag.

Bui, T. X., Bodart, F., & Ma, P.-C. (1998). ARBAS: A formal language to support argumentation in network-based organization. *Journal of Management Information Systems, 14*(3), 223-240.

Chiu, D. K. W., Cheung, S. C., Hung, P. C. K. Chiu, S. Y. Y., & Chung, K. K. (2005). Developing e-negotiation process support with a meta modeling approach in a Web services environment. *Decision Support Systems, 40*(1), 51-69.

Chiu, D. K. W., Cheung, S. C., Hung, P. C. K., & Leung, H.-F. (2004). Constraint-based negotiation in a multi-agent information system with multiple platform support. In *Proceedings of the 37th Hawaii International Conference on System Sciences (HICSS37),* Waikoloa, Big Island, III [CD-ROM]. IEEE Computer Society Press.

Chiu, D. K. W., Cheung, S. C., Kafeza, E., & Leung, H.-F. (2003). A three-tier view methodology for adapting M-services. *IEEE Transactions on System, Man and Cybernetics, Part A, 33*(6), 725-741.

Chiu, D. K. W., Cheung, S. C., & Leung, H.-F. (2005). A multi-agent infrastructure for mobile workforce management in a service oriented enterprise. *Proceedings of the 38th Hawaii International Conference on System Sciences (HICSS38),* Waikoloa, Big Island, HI [CD-ROM]. IEEE Press.

Chiu, D. K. W., Kwok, B., Wong, R., Kafeza, E., & Cheung, S. C. (2004). Alert driven e-services management. In *Proceedings of the 37th Hawaii International Conference on System Sciences (HICSS37),* Waikoloa, Big Island, HI [CD-ROM]. IEEE Computer Society Press.

Chiu, D. K. W., Lee, O., & Leung, H.-F. (2005). A multi-modal agent based mobile route advisory system for public transport network. In *Proceedings of the 38th Hawaii International Conference on System Sciences (HICSS38),* Waikoloa, Big Island, HI [CD-ROM]. IEEE Computer Society Press.

Chiu, D. K. W., Li, Q., & Karlapalem, K. (1999). A meta modeling approach for workflow management system supporting exception handling. *Information Systems, 24*(2), 159-184.

Choy, M. C., Srinivasan, D., & Cheu, R. L. (2003). Cooperative, hybrid agent architecture for real-time traffic signal control. *IEEE Transactions on System, Man and Cybernetics, Part A, 33*(5), 597-607.

Fraile, J.-C., Paredis, C. J. J., Wang, C.-H., & Khosla, P. K. (1999). Agent-based planning and control of a multi-manipulator assembly system. In *Proceedings of the IEEE International Conference on Robotics and Automation, 2*, 1219-1225.

Garrido, L., Brena, R., & Sycara, K. (1996). Cognitive modeling and group adaptation in intelligent multi-agent meeting scheduling. In *Proceedings of the 1st Iberoamerican Workshop on Distributed Artificial Intelligence and Multi-Agent Systems* (pp. 55-72).

Gerst, M. H. (2003). The role of standardisation in the context of e-collaboration: ASNAP shot. In *Proceedings of the 3rd Conference on Standardization and Innovation in Information Technology* (pp. 113-119).

Grefen, P., Aberer, K., Hoffner, Y., & Ludwig, H. (2000). CrossFlow: Cross-organizational workflow management in dynamic virtual enterprises. *International Journal of Computer Systems Science & Engineering, 15*(5), 277-290.

Guido, B., Roberto, G., Tria, P. di, & Bisio, R. (1998). Workforce management (WFM) issues. In *Proceedings of the IEEE Network Operations and Management Symposium (NOMS 98), 2*, 473-482.

He, M., Jennings, N. R., & Leung, H.-F. (2003). On agent-mediated electronic commerce. *IEEE TKDE, 15*(4), 985-1003.

Jing, J., Huff, K., Hurwitz, B., Sinha, H., Robinson, B., & Feblowitz, M. (2000). WHAM: Supporting mobile workforce and applications in workflow environments. In *Proceedings of the 10th International Workshop on Research Issues in Data Engineering (RIDE 2000)* (pp. 31-38).

Karageorgos, M., Thompson, S., & Mehandjiev, N. (2002). Agent-based system design for B2B electronic commerce. *International Journal of Electronic Commerce, 7*(1), 59-90.

Kim, Y., Kang, S., Kim, D., Bae, J., & Ju, K. (2000). WW-flow: Web-based workflow management with runtime encapsulation. *IEEE Internet Computing, 4*(3), 56-64.

Kotz, D., & Gray, R. (1999). Mobile agents and the future of the Internet. *ACM Operating Systems Review, 33*(3), 7-13.

Kowalczyk, R., & Bui, V. (2000). On constraint-based reasoning in e-negotiation agents. In F. Dignum & U. Cortés (Eds.), *Agent mediated electronic commerce III* (LNAI 2003, pp. 31-46). London: Springer-Verlag.

Kowalczyk, R., Ulieru, M., & Unland, R. (2003). Integrating mobile and intelligent agents in advanced e-commerce: A survey. In *Proceedings of Agent Technology Workshops 2002* (LNAI 2592, pp. 295-313). Berlin, Germany: Springer-Verlag.

Lamsweerde, A. van, Darimont, R., & Massonet, P. (1995). Goal-directed elaboration of requirements for a meeting scheduler: Problems and lessons learnt. In *Proceedings of the 2nd IEEE International Symposium on Requirements Engineering (RE '95)* (pp. 194-203).

Lee, H., Buckland, M. A., & Shepherdson, J. W. (2003). A multi-agent system to support location-based group decision making in mobile teams. *BT Technology Journal, 21*(1), 105-113.

Lin, F.-R., & Pai, Y.-H. (2000). Using multi-agent simulation and learning to design new business processes. *IEEE Transactions on System, Man and Cybernetics, Part A, 30*(3), 380-384.

Lin, F.-R., Tan, G. W., & Shaw, M. J. (1998). Modeling supply-chain networks by a multi-agent system. In *Proceedings of HICSS31* (Vol. 5, pp. 105-114).

Lin, Y.-B., & Chlamtac, I. (2000). *Wireless and mobile network architectures*. New York: John Wiley & Sons.

Lo, G., & Kersten, G. K. (1999). Negotiation in electronic commerce: Integrating negotiation support and software agent technologies. In *Proceedings of the 29th Atlantic Schools of Business Conference* [CD-ROM].

Lomuscio, A., Wooldridge, M., & Jennings, N. (2000). A classification scheme for negotiation in electronic commerce. In F. Dignum & C. Sierra (Eds.), *Agent-mediated electronic commerce: A European perspective* (LNCS 1991, pp. 19-33). London: Springer-Verlag.

Luo, Y., Liu, K., & Davis, D. N. (2002). A multi-agent decision support system for stock trading. *IEEE Network, 16*(1), 20-27.

Lyytinen, K., & Yoo, Y. (2002). Research commentary: The next wave of nomadic computing. *Information Systems Research, 13*(4), 377-388.

Object Management Group (OMG). (2001). *Foreword UML specification 1.4*. Retrieved from http://www.omg.org/cgi-bin/doc?formal/01-09-67

Papaioannou, T. (2000). Mobile information agents for cyberspace: State of the art and visions. In *Proceedings of Cooperating Information Agents (CIA-2000)* [CD-ROM].

Rao, A. S., & Georgeff, M. P. (1995). BDI agents: From theory to practice. *Proceedings of the 1st International Conference on Multiagent Systems* (pp. 312-319).

Rutkowski, A. F., Vogel, D. R., Genuchten, M. van, Bemelmans, T. M. A., & Favier, M. (2002). E-collaboration: The reality of virtuality. *IEEE TPC, 45*(4), 219-230.

Sandip, S. (1997, July-August). Developing an automated distributed meeting scheduler. *IEEE Expert, 12*(4), 41-45.

Shakshuki, E., Ghenniwa, H., & Kamel, M. (2000). A multi-agent system architecture for information gathering. In *Proceedings of the 11th International Workshop on Database and Expert Systems Applications* (pp. 732-736).

Shi, M., Yang, G., Xiang, Y., & Wu, S. (1998). Workflow management systems: A survey. In *Proceedings of the IEEE International Conference on Communication Technology*, S33.05.01-S33.05.06.

Shitani, S., Ito, T., & Sycara, K. (2000). Multiple negotiations among agents for a distributed meeting scheduler. In *Proceedings of the 4th International Conference on MultiAgent Systems* (pp. 435-436).

Smith, R. G. (1980). The contract net protocol: High-level communication and control in a distributed problem solver. *IEEE Transactions on Computers, C-29*(12), 1104-1113.

Stroulia, E., & Hatch, M. P. (2003). An intelligent-agent architecture for flexible service integration on the Web. *IEEE Transactions on System, Man and Cybernetics, Part C, 33*(4), 468-479.

Sycara, K., & Zeng, D. (1996). Coordination of multiple intelligent software agents. *International Journal of Cooperative Information Systems, 5*(2-3), 181-212.

Thompson, S. G., & Odgers, B. R. (2000). Collaborative personal agents for team working. In *Proceedings of the 2000 Artificial Intelligence and Simulation of Behavior (AISB) Symposium* (pp. 49-61).

Tsang, E. (1993). *Foundations of constraint satisfaction.* London; San Diego, CA: Academic Press.

Wegner, L., Paul, M., Thamm, J., & Thelemann, S. (1996). Applications: A visual interface for synchronous collaboration and negotiated transactions. *Proceedings of the Workshop on Advanced Visual Interfaces* (pp. 156-165).

Workflow Management Coalition. (1995). *The workflow reference model* (WFMC-TC-1003, 19-Jan-95, 1.1). Retrieved from http://www.wfmc.org/standards/model.htm

Workflow Management Coalition. (1999). *Terminology and glossary* (WFMC-TC-1011, 3.0). Retrieved from http://www.wfmc.org/standards/model.htm

Section III

Enterprise Service Computing: Modeling

Chapter VI

Designing Enterprise Applications Using Model-Driven Service-Oriented Architectures

Marten van Sinderen, University of Twente, The Netherlands

João Paulo Andrade Almeida, Telematica Instituut, The Netherlands

Luís Ferreira Pires, University of Twente, The Netherlands

Dick Quartel, University of Twente, The Netherlands

Abstract

This chapter aims at characterizing the concepts that underlie a model-driven service-oriented approach to the design of enterprise applications. Enterprise applications are subject to continuous change and adaptation since they are meant to support the dynamic arrangement of the business processes of an enterprise. Service-oriented computing (SOC) promises to deliver the methods and technologies to facilitate the development and maintenance of enterprise applications. The model-driven architecture (MDA), fostered by the Object Management Group (OMG), is increasingly gaining support as an approach to manage system and software complexity in distributed-application design. Service-oriented computing and the MDA have some common goals; namely, they both strive to facilitate the development and maintenance of distributed enterprise applications, although they achieve these goals in different ways. This chapter discusses a combination of these approaches and discusses the benefits of this combination.

Introduction

Enterprise applications are subject to continuous change and adaptation since they are meant to support the dynamic arrangement of the business processes of an enterprise. For example, an enterprise may decide to outsource a business process for efficiency reasons, or different processes may be integrated to provide a new product. Therefore, it is necessary to devise methods and technologies that can cope with this dynamic characteristic of enterprise applications in a cost-effective manner.

Service-oriented computing (SOC) promises to deliver the methods and technologies to facilitate the development and maintenance of enterprise applications (Papazoglou & Georgakopoulos, 2003). This should promote the introduction of richer and more advanced applications, thereby offering new business opportunities. Other foreseen benefits are the shortening of application-development time by reusing available applications, and the creation of a service market, where enterprises make it their business to offer generic and reusable services that can be used as application building blocks. The service-oriented paradigm is in essence characterized by the explicit identification and description of the externally observable behavior, or service, of an application. Applications can then be linked based on the description of their externally observable behavior. According to this paradigm, developers in principle do not need to have any knowledge about the internal functioning and the technology-dependent implementation of the applications being linked. Often, the term service-oriented architecture (SOA) is used to refer to the architectural principles that underlie the communication of applications through their services (Erl, 2005).

The model-driven architecture (MDA), fostered by the Object Management Group (OMG), is increasingly gaining support as an approach to manage system and software complexity in distributed-application design. MDA defines a set of basic concepts such as model, metamodel, and transformation, and proposes a classification of models that offer different abstractions (OMG, 2003). Model-driven engineering (MDE) builds on this foundation to introduce the notion of software-development process by organizing models in the modeling space (Kent, 2002). MDE focuses first on the functionality and behavior of a distributed application, which results in platform-independent models (PIMs) of the application that abstract from the technologies and platforms used to implement the application. Subsequent steps lead to a mapping from PIMs to an implementation, possibly via one or more platform-specific models (PSMs). The main advantages of software development based on MDA—software stability, software quality, and return on investment—stem from the possibility to derive different implementations (via different PSMs) from the same PIM, and to automate to some extent the model-transformation process.

Service-oriented computing and model-driven engineering have some common goals; they both strive to facilitate the development and maintenance of distributed enterprise applications, although they achieve these goals in different ways. This chapter aims at characterizing the concepts that underlie a model-driven service-oriented approach to the design of enterprise applications and discusses the benefits of such an approach.

This chapter is further structured as follows. First we discuss the service concept, its relevance to systems design, and its application in service-oriented architectures. Then we introduce the concept of an interaction system in order to explain and distinguish between two important paradigms that play a role in service-oriented design. Next the chapter discusses the concepts of platform, platform-independence, and abstract platform, and how these fit into the model-driven engineering approach. All this serves as a background to the next section; which outlines our model-driven service-oriented approach in terms of a set of related milestones. After this we illustrate the model-driven service-oriented approach with a simple case study, namely, an auction system. Finally the chapter summarizes our findings and presents some concluding remarks.

Service Concept

The concept of service has long been recognized as a powerful means to achieve a separation of concerns and adequate abstraction in distributed-systems design, especially for data communication (Vissers & Logrippo, 1985). In recent years, the advent of service-oriented architectures has renewed the attention for this concept, which is particularly relevant to enterprise service computing (Papazoglou & Georgakopoulos, 2003).

This section discusses the service concept and its application in systems design, with a focus on service-oriented architectures.

Systems and Services

The Webster's Dictionary online (http://www.m-w.com) provides a definition of system particularly applicable to distributed systems: "a system is a regularly interacting or interdependent group of items forming a unified whole." This definition

Figure 1. External and internal system perspectives: (a) system parts and (b) services

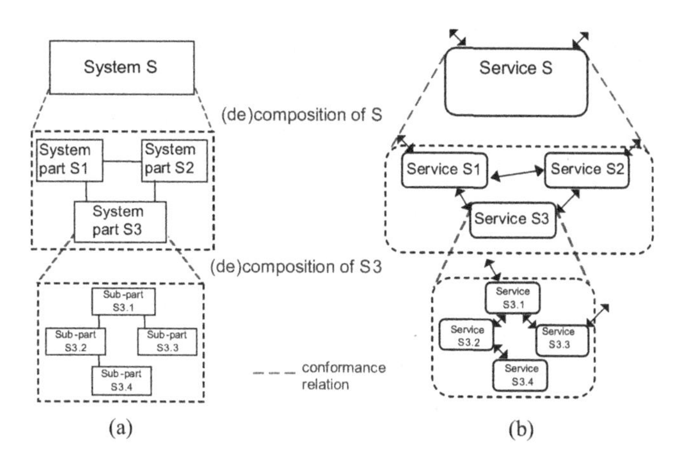

indicates two different perspectives of a system: an integrated and a distributed perspective. The integrated perspective considers a system as a whole or black box. This perspective only defines what function a system performs for its environment. The distributed perspective defines how this function is performed by an internal structure in terms of system parts (or subsystems) and their relationships.

The integrated perspective of a system coincides with the concept of service. A service is a design that defines the observable behavior of a system in terms of the interactions that may occur at the interfaces between the system and its environment and the relationships between these interactions. A service does not disclose details of an internal organization that may be given to implementations of the system (Vissers, Scollo, Sinderen, & Brinksma, 1991).

By considering each part of a system as a system itself, the external and internal perspectives can be applied again to the system parts. This results in a process of repeated or recursive decomposition, yielding several levels of decomposition, also called levels of abstraction. Figure 1 illustrates schematically this process. In the recursive decomposition process, the service concept can represent the behavior of various kinds of entities, such as value chains, business enterprises, software applications, application components, middleware platforms, or communication networks.

Service Concept in Service-Oriented Architectures

A service in service-oriented architectures is a "self-describing, open component that supports rapid, low-cost composition of distributed applications" (Papazoglou &

Georgakopoulos, 2003, p. 26). Services in service-oriented architectures are offered by service providers, who are responsible for the service implementations and supply the service descriptions. These descriptions are published in service registries either by the service providers or on their behalf, and are used by potential service users to search and select the services that they want to use. This implies that a service description should contain the service capabilities, but also the technical information necessary to actually access and use the service. Although many authors associate service-oriented architectures only with Web services, there is nowadays a consensus that Web services is just one possible alternative to implement a service-oriented architecture (Erl, 2005).

The concept of service as introduced in service-oriented architectures fits in our more general definition given above. We have shown in Quartel, Dijkman, and Sinderen (2004) that service-oriented architectures are currently being approached in a technology-driven way. Most developments focus on the technology that enables enterprises to describe, publish, and compose application services, and to enable communication between applications of different enterprises according to their service descriptions. This chapter shows how modeling languages, design methods, and techniques can be put together to support model-driven service-oriented design. The service concept can be used as a starting point for design according to different paradigms. The service concept also supports platform independence.

Design Paradigms

We introduce the concept of interaction system here in order to distinguish design paradigms in terms of how each paradigm addresses the design of interactions between system parts. An interaction system refers to the specification of the way in which system parts interact with each other, that is, affect each other's behavior in a defined way through the exchange of information. An interaction system involves a portion of the function of each system part and the connecting structure that is available for the system parts to exchange information, as depicted in Figure 2a.

Figure 2. Interaction system from (a) distributed (system) and (b) external (service) perspectives

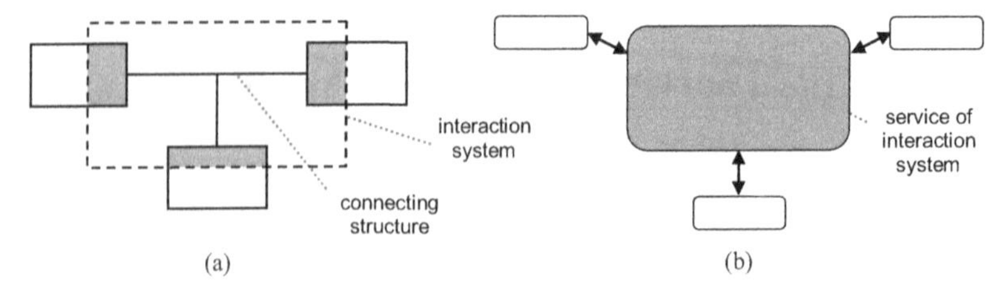

An interaction system is a system in itself, and therefore the external behavior of an interaction system can be defined as a service, as depicted in Figure 2b. The service of an interaction system can be taken as a starting point for the design of interactions between system parts. As such, it accomplishes an important separation of concerns, namely those related to the design of applications of the service and those related to the implementation of the service.

Protocol-Centered Paradigm

In the protocol-centered paradigm, service users interact through a set of interaction patterns offered by a service provider. The set of interaction patterns, or required service, is chosen to suit the interaction needs of the service users, and therefore corresponds to an arbitrary level of functionality that may or may not be easy to implement. A service provider can be decomposed into (or composed of) a layer of protocol entities and an underlying (lower level) service provider, where the protocol entities interact through the lower level service provider in order to provide the required service. Figure 3 illustrates the model of a system according to the protocol-centered paradigm.

The level of functionality offered by the lower level service provider may require further (de)composition, yielding a set of hierarchical protocol layers and a final lower level service provider that corresponds to an available technical construct (e.g., a network product).

Protocol entities communicate with each other by exchanging messages, often called protocol data units (PDUs), through a lower level service. PDUs define the syntax and semantics for unambiguous understanding of the information exchanged between protocol entities. The behavior of a protocol entity defines the service primitives between this entity and the service users, the service primitives between the protocol entity and the lower level service, and the relationships between these primitives. The protocol entities cooperate in order to provide the requested service (Sharp, 1994).

Figure 3. System according to the protocol-centered paradigm

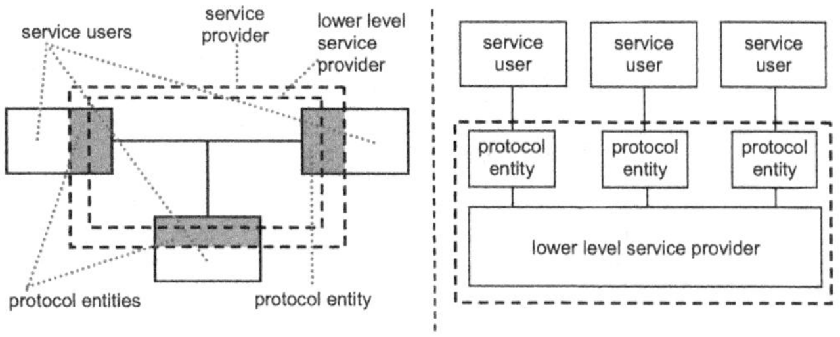

Protocols can be defined at various layers, from the physical layer to the application layer. Lower level services typically provide physical interconnection and (reliable or unreliable) data transfer between protocol entities. Interaction patterns between the protocol entities vary from connectionless data transfer (e.g., "send and pray") to complex control facilities (e.g., handshaking with three-party negotiation). This design paradigm has inspired some (communication) reference architectures, such as the open systems interconnection reference model (OSI RM) and the transmission-control protocol/Internet protocol (TCP/IP) suite.

Middleware-Centered Paradigm

In the middleware-centered paradigm, objects or components interact through a limited set of interaction patterns offered by a middleware platform. The middleware platform is an available technology construct that facilitates application development by hiding the distribution aspects of communication. Any component can offer its service to other (previously unknown) remote components if its service definition is made available (published) in terms of interactions that have a predefined mapping onto the interaction patterns of the middleware. This is the case since service interfaces and protocol elements can be derived from the service definition, thereby geographically extending the service to other components and allowing remote access to the service. In order to distinguish between the possible transient roles of components using and offering a service, we use the terms consumer component and provider component, respectively. Figure 4 illustrates the model of a system according to the middleware-centered paradigm.

A provider component may entail complex functionality, and therefore may require further decomposition and/or distribution over multiple computing nodes. Decomposition is achieved by conceiving the provider component as a composition of a consumer component and one or more extended service providers, where each extended service provider offers an extended service as discussed above. This process continues until the final components can each be conveniently run on a single computing node and offer a useful unit of functionality.

Figure 4. System according to the middleware-centered paradigm

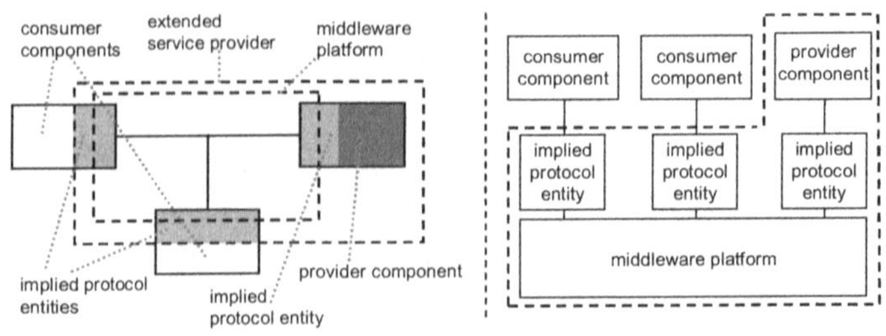

There are several different types of middleware platforms, each one offering different types of interaction patterns between objects or components. The middleware-centered paradigm can be further characterized according to the types of interaction patterns supported by the platform. Examples of these patterns are request-response, message passing, and message queues. Examples of available middleware platforms are common object request broker architecture/CORBA component model (CORBA/CCM) (OMG, 2002a, 2004), .NET (Microsoft Corporation, 2001), and Web services (World Wide Web Consortium [W3C], 2001, 2003).

The middleware-centered paradigm promotes the reuse of a middleware infrastructure. In addition, it enables the development of provider components in isolation of other consumer components, where the latter can use the services of the former through the supported interaction patterns of the middleware. This is different from the protocol-centered paradigm, where (protocol) entities are developed in combination to form a single protocol. An interesting observation with respect to the middleware-centered paradigm is that it somehow depends on the protocol-centered paradigm: Interactions between components are supported by the middleware, which transforms the interactions into (implicit) protocols, provides generic services that are used to make the interactions distribution transparent, and internally uses a network infrastructure to accomplish data transfer (Sinderen & Ferreira Pires, 1997).

Methods for application development that are inspired by the middleware-centered paradigm often consist of partitioning the application into application parts and defining the interconnection aspects by defining interfaces between parts (e.g., by using object-oriented techniques and abstracting from distribution aspects). The available constructs to build interfaces are constrained by the interaction patterns supported by the targeted middleware platform. Examples of these constructs are operation invocation, event sources and sinks, and message queues.

The protocol-centered and middleware-centered paradigms tackle the support of interactions between system parts in different ways, and consequently are suited to different types of applications. For enterprise service computing, the middleware-centered paradigm is more appropriate, although there are situations where protocol design plays an important role. The middleware-centered paradigm is assumed (to some extent) by both SOA/SOC and MDA/MDE.

Model-Driven Architecture Approach

The concept of platform independence has been introduced in the MDA approach in order to shield the design of applications from the choice of platform and to guarantee the reuse of designs across different platforms.

The term platform is used to refer to a system that entails all kinds of technological and engineering details to provide some functionality, which applications can use through defined interfaces and specified usage patterns (OMG, 2003). For the purpose of this chapter, we assume that a platform corresponds to some specific middleware technology.

Platform independence is a quality of a model that relates to the extent to which the model is independent of the characteristics of a particular platform (OMG, 2003), including its specific interfaces and usage patterns. In this chapter, we assume that models are used to specify both the behavior and structure of a system, and that several platform-independent models may be used in conjunction to specify a design. A consequence of the use of platform-independent models is the ability to refine the design or implement it on a number of target middleware platforms.

Ideally, one could strive for platform-independent models that are absolutely neutral with respect to all different classes of middleware technologies. However, we foresee that at a certain point in the development trajectory, different sets of platform-independent modeling concepts may be used, each of which is needed only with respect to specific classes of target middleware platforms. Figure 5 illustrates an MDE trajectory in which such a highly abstract and neutral PIM is depicted as the starting point of the trajectory. In Figure 5, the platform-independent models facilitate the transformation to two particular classes of middleware platforms, namely, the request-response (object based) and asynchronous-messaging (message oriented) platforms.

Figure 5. An MDE trajectory

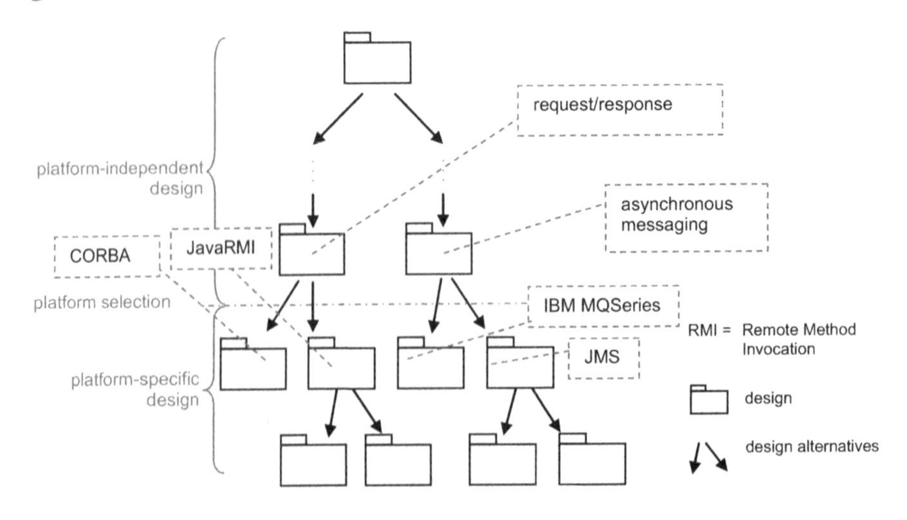

In an MDE trajectory, a designer should clearly define the abstraction levels at which PIMs and PSMs have to be defined. The choices of platforms should also be made explicit in each step in the design trajectory. Furthermore, the choice of design concepts for PIMs should be carefully considered, taking into account the common characteristics of the target platforms and the complexity of the transformations-that are necessary in order to generate PSMs from PIMs. The concept of abstract platform, as proposed initially in Almeida, Sinderen, Ferreira Pires, and Quartel (2003), supports the explicit identification of the level of platform independence that is needed for a model. An abstract platform defines the platform characteristics that are relevant for applications at a given platform-independent level. These characteristics are a compromise between what an application developer ideally wants from a platform and what target platforms actually can provide. The PIM of an application depends on an abstract platform model in the same way as the PSM of an application depends on a concrete platform model (Almeida, Dijkman, Sinderen, & Ferreira Pires, 2004).

Model-Driven Service-Oriented Design

In order to exploit the service concept in an MDA-based design process, we define the following milestones.

1. **Service definition.** The service definition sets the boundaries of the enterprise (sub)system to be designed. Services are specified at a level of abstraction at which the supporting infrastructure is not considered. In our case, the infrastructure is the middleware platform, and therefore service specifications are middleware-platform independent by definition. The service concept defines a platform-independent level that is also paradigm independent in the sense that both the protocol-centered and middleware-centered paradigms may be used in subsequent design steps. In addition, the service concept allows very different types of middleware platforms to be used for the implementation of the application service, not just similar ones. Service definitions are positioned at the top of the design trajectory depicted in Figure 5. Service users associated with (using) the service, and necessarily relying on the service definition, may consequently be defined at the same level of platform independence.
2. **Platform-independent service design.** The platform-independent service design consists of the platform-independent service logic, which is structured in terms of application-part functions, and an abstract platform definition. The choice of abstract platform must consider the portability requirements since it defines the characteristics of the platform upon which application-part services

may rely. The level of abstraction at which the platform-independent service logic is specified depends on the abstract platform definition. Figure 6 illustrates the design trajectory with the service-definition and platform-independent service-design milestones.

3. **Platform-specific service design.** The platform-independent service design is transformed into a platform-specific service design, which is structured in terms of platform-specific application-part functions and a concrete platform definition. This transformation may be straightforward when the selected platform corresponds (directly) to the abstract platform definition. We show in the next section that when the abstract platform is defined as a service, it is also possible to apply service design recursively. The recursive application of service design leads to the introduction of some target platform-specific abstract platform logic to be composed with the target platform. The abstract platform service is then refined into a composition of a target platform and the abstract platform logic.

The service-definition milestone can be considered the starting point for PIM design, or alternatively, the end result of business design in which business processes are articulated and their computer support (automation) is identified without revealing any details on how the automation is accomplished. In the latter case, the service definition corresponds to the computer support. MDA introduces the notion of the computation-independent model (CIM; OMG, 2003), which can be considered as

Figure 6. Service decomposition into underlying interaction system provided by abstract platform

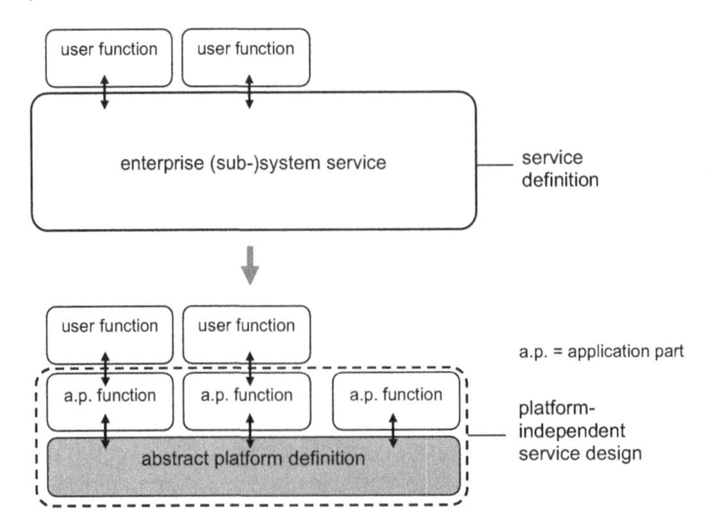

a kind of business model. However, since the notion of CIM and its relationship to PIM are not yet well understood, we refrain from discussing milestones that could be related to CIMs in this chapter.

The platform-independent service-design milestone corresponds to a PIM, whereas the platform-specific service design corresponds to a PSM. The platform-independent service-design milestone does not mention which paradigm—protocol centered or middleware centered—is used to achieve this result. This is because, depending on the type of application, either paradigm can be used. If application parts have a complex interaction and are tightly coupled, they should be designed in combination and the protocol-centered paradigm should be used (with the abstract platform as the lower level service provider). In this case, application parts are normally not described as separate services, but their combined behavior in terms of protocol messages can be abstracted as a service. However, if application parts offer general useful functions that can be invoked through simple interactions supported by existing middleware, they can be defined in isolation and the middleware-centered paradigm should be preferred. In this case, each application-part function in a provider role can be defined as a service. In both paradigms, application parts can be considered as service components that cooperate to provide the required service (according to the service-definition milestone).

The description of the milestones suggests a top-down design trajectory, starting from service definition to service design. However, this does not exclude the use of bottom-up knowledge. Bottom-up experience is what allows designers to reuse middleware infrastructures, by defining an abstract platform that can be realized in terms of these concrete middleware platforms, and to find appropriate service designs that implement the required service.

Case Study: Auction System

In order to illustrate the use of a service definition and its refinement in a design trajectory, we introduce our case study, namely, an auction system. In this case study, service users (participants) participate in auctions in which products are sold to the highest bidder. Any participant may auction off and bid for products. In order to simplify the case study, we consider the number of participants to be fixed and predefined.

Auction Service

The auction service must be specified in such a way that interaction requirements between participants are satisfied without unnecessarily constraining implementation freedom. This freedom includes the structure of the auction service provider (the system that eventually supports the auction service) and other technology aspects such as middleware platforms, operating systems, and programming languages. Therefore, services are described in abstract terms by interactions and their relationships. An interaction in a service definition is a common unit of activity performed by a service user and a service provider; it occurs at the local service interface that directly connects these two system parts (Vissers & Logrippo, 1985; Vissers et al., 1991).

The auction service relates the following interactions:

- *offer* and *offer_ind*, both with attributes *product_id* and *opening_value*. The *product_id* uniquely identifies a product being auctioned.
- *bid* and *bid_ind*, both with attributes *product_id* and *bid_value*.
- *outcome_ind*, with attributes *product_id*, *final_value*, and *participant_id*. The *participant_id* uniquely identifies a participant in the auction. In the case of this interaction, it identifies the winning bidder.

These interactions occur at the interfaces between the auction service provider and the participants. The occurrence of an interaction results in the establishment of values for its attributes. In addition to the attributes listed above, for each interaction, the *participant_id* is implied by the location where the interaction occurs.

A convenient technique for specifying a service is to define the service as a conjunction of different constraints on interactions (Vissers et al., 1991). In particular, a useful structuring principle is that of identifying local and end-to-end (or remote) constraints.

The local constraint (related to each participant) for this service is:

- *bid*, which can only occur after *offer_ind* and before *outcome_ind* (for a given *product_id*).

The remote constraints (between participants) for this service are the following:

- The occurrence of *offer* is followed by an occurrence of *offer_ind* for each participant in the auction.
- The occurrence of *bid* is followed by an occurrence of *bid_ind* for each participant in the auction.

Figure 7. Auction service

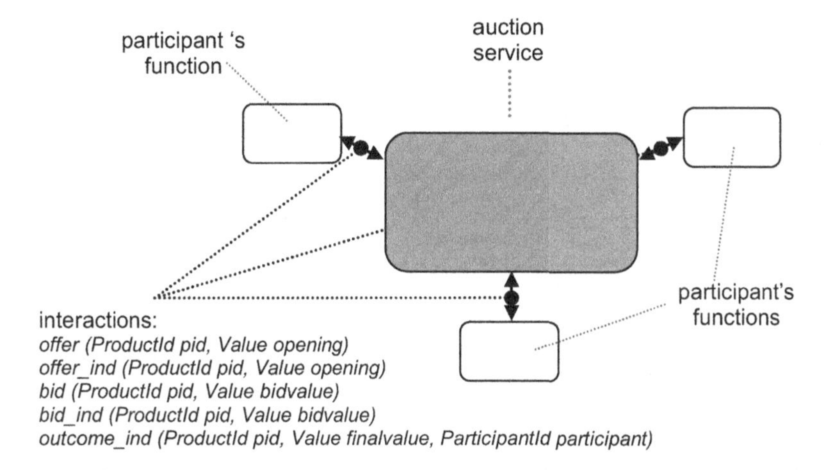

- *outcome_ind* occurs Δt seconds after the last *bid_ind* occurs (for a given *product_id*).

Figure 7 illustrates the auction service.

Design Trajectories

Using the auction service as defined earlier as a starting point, we follow the design trajectory for two different abstract platforms: an abstract platform that supports message exchange and an abstract platform that supports the request-response pattern. The platform-independent design based on the former abstract platform can be characterized as protocol centered, and the one based on the latter abstract platform can be characterized as middleware centered. We consider different design solutions for the auction service, illustrating that the service specification is to a large extent implementation independent. For each platform-independent design obtained, we consider realizations in two concrete platforms: CORBA (OMG, 2004) and the Java Message Service (JMS) point-to-point domain (Sun Microsystems, 2002). Figure 8 illustrates the design trajectories followed in our examples.

The recursive application of service decomposition is shown in the design trajectory on the rightmost side of Figure 8. The service of the request-response abstract platform is further decomposed for realization in the JMS platform. In the following, we discuss each of the alternative trajectories depicted in Figure 8.

Figure 8. Examples of alternative trajectories

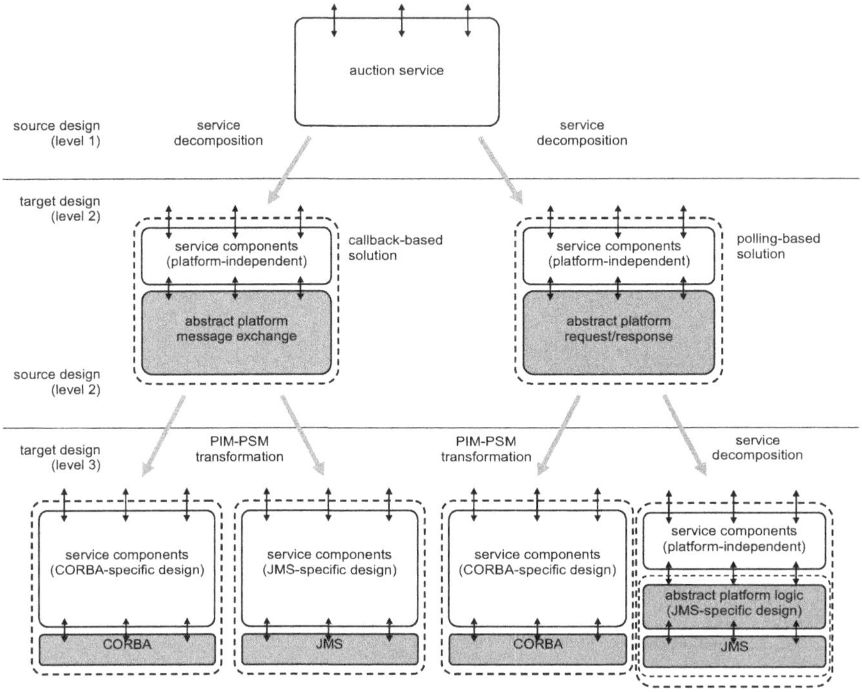

Callback-Based Solution with Message-Exchange Abstract Platform

Abstract Platform: Message Exchange

Initially, let us consider an abstract platform that supports message exchange. We identify two interactions that are related by the abstract platform:

- *send*, with attributes *destination* and *payload*, and
- *receive*, with attribute *payload*.

An occurrence of *receive* follows an occurrence of *send*. The interaction *receive* is executed at the location specified by the attribute destination of *send*. The attribute *payload* represents the information to be sent. The value of the attribute *payload* for an occurrence of *receive* is the value of the attribute *payload* for the related occurrence of *send*.

Platform-Independent Design

The abstract platform is used in our callback-based solution to exchange messages between participant service components and a controller service component. The structure of the platform-independent auction service design is depicted in Figure 9.

The controller service component centralizes the coordination of the auction. When a participant offers a product for auction by executing the interaction *offer*, the participant service component sends a *register_offer* message to the controller. This is done using the abstract platform through the *send* interaction, which is followed by the occurrence of the *receive* interaction on the interface of the controller. A *register_offer* message carries the identification of a product, the opening value for the product, and the identification of the participant (in this case the seller). The controller sends *offer_callback* messages to all other participant service components, informing the identification of the product being auctioned and its opening value. This results in the occurrence of the *offer_ind* interaction for every participant (excluding the seller).

When a participant bids for a product by executing the interaction *bid*, the participant service component sends a *register_offer* message to the controller. This message carries the identification of a product, the bid, and the identification of the participant (in this case the bidder). The controller sends *bid_callback* messages to all other participant service components. This results in the occurrence of the *bid_ind* interaction for every participant (excluding the bidder). Eventually, when no bids are registered with the controller for a period of time, the controller sends

Figure 9. Structure of the callback-based auction service provider

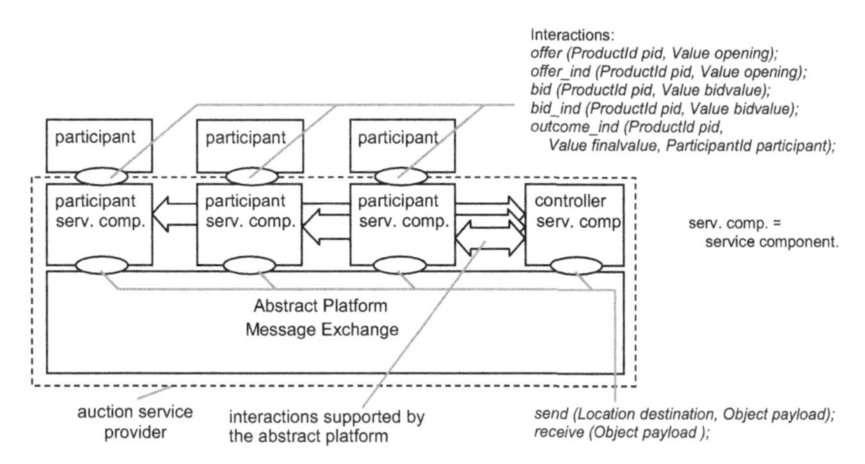

Figure 10. The auction of a product (platform-independent service design)

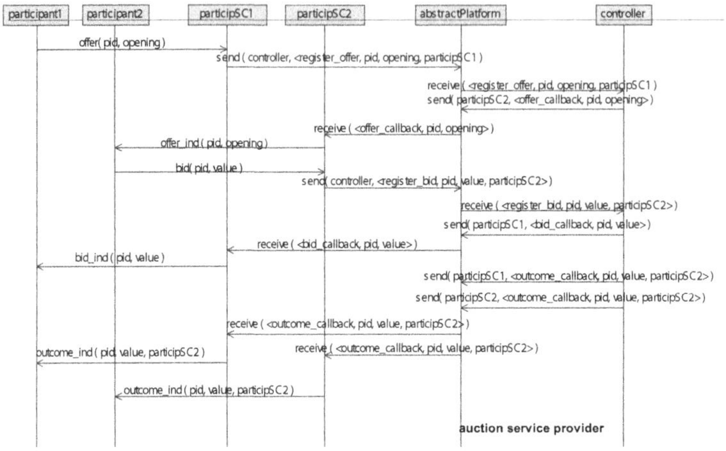

Figure 11. Offer of a product and outcome of the auction (JMS-specific realization)

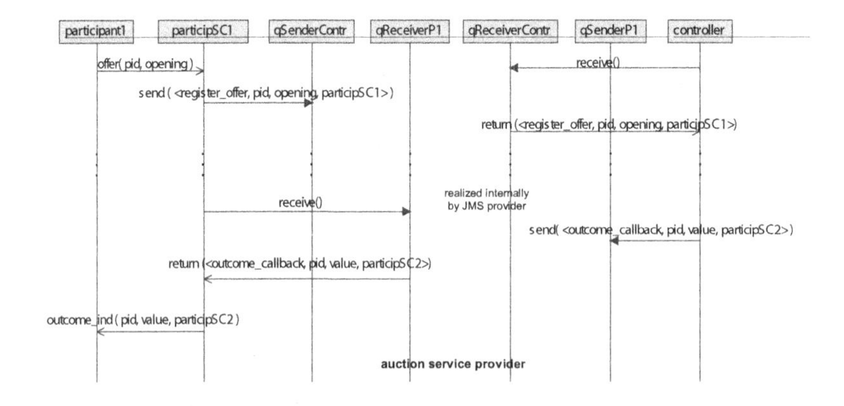

an *outcome_callback* message to all participants, resulting in the occurrence of the *outcome_ind* for every participant.

Figure 10 shows a sequence diagram representing an instance of the execution of the auction service in which *participant1* offers a product that is bought by *participant2*.

Figure 12. The auction of a product (CORBA-specific realization)

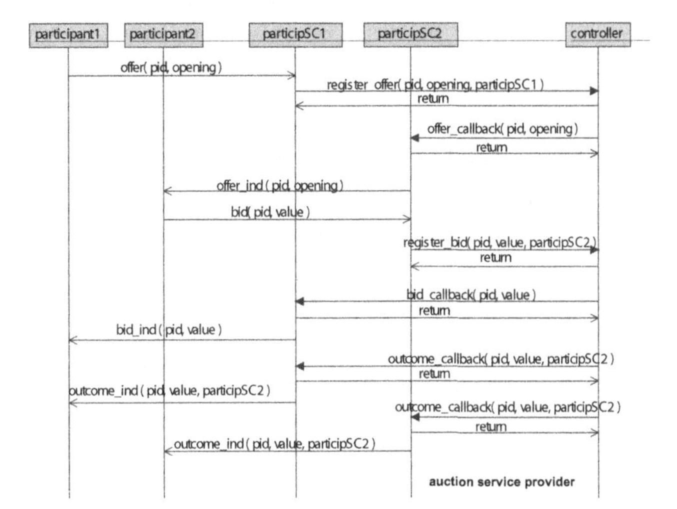

Realization

A realization of the platform-independent design in the JMS platform is straightforward. The service provided by Java Message Service (JMS) corresponds directly to the service provided by the defined abstract platform. Figure 11 shows a sequence diagram of the realization in JMS, which corresponds to part of the sequence diagram depicted in Figure 10. In the JMS platform, the destination of a message is addressed by a queue identifier. In this realization, there is a queue for messages destined to the controller and a queue for messages destined to each participant service component. The addressing of the destination for a message is done through the selection of a queue and the instantiation of a message producer for the queue (*qSenderContr* for the queue directed to *controller*, and *qSenderP1* for the queue directed to *participant1*). The boundaries of the auction service provider are preserved, as well as the behavior of the auction service provider.

The realization in the CORBA platform can be obtained through a simple transformation: Message exchange is realized through an operation invocation with no return parameters. An example auction in our realization in the CORBA platform is shown in Figure 12. This sequence diagram corresponds to the sequence diagram depicted in Figure 10; again, the boundaries of the auction service provider are preserved.

For the CORBA realization, we could have also considered the use of the CORBA Notification Service (OMG, 2002b) in a similar way as we have used JMS to accomplish message exchange. This illustrates our observation that there are many possible ways to realize a platform-independent design even for a particular concrete platform.

Polling-Based Solution with Request-Response Abstract Platform

Abstract Platform: Request-Response

Let us consider an abstract platform that supports the request-response pattern. We identify four interactions that are related to each other through the abstract platform.

- *request*, with attributes *target*, *operation*, and *argument_list*. The attributes represent, respectively, the identifier of the target object, the identifier of the requested operation, and the argument list for the request.
- *request_ind*, with attributes *operation* and *argument_list*
- *response*, with attribute *return_parameters*, which represents the list of return parameters
- *response_ind*, with attribute *return_parameters*

Figure 13. Structure of the callback-based auction service provider

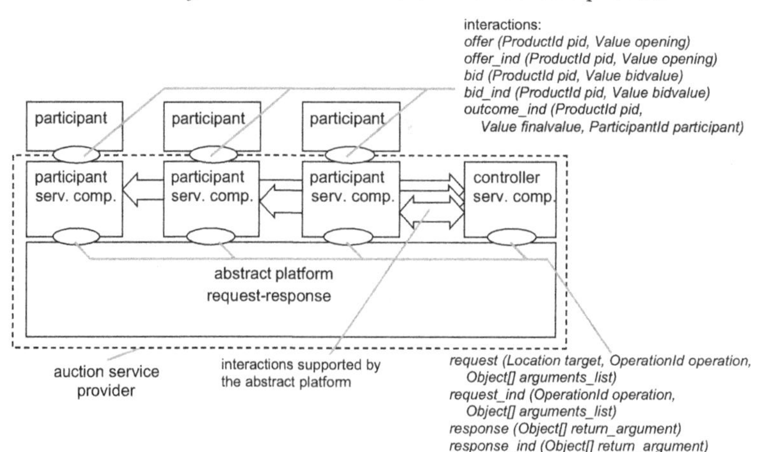

The occurrence of *request_ind* follows the occurrence of *request*, the occurrence of *response* follows the occurrence of *request_ind*, and the occurrence of *response_ind* follows the occurrence of *response*.

This is a generalization of the service provided by existing request-response platforms. These platforms provide some infrastructure to generate customized stubs that in conjunction with the middleware core provide specializations of the service as presented in this section.

Platform-Independent Design

The abstract platform is used in our polling-based solution to enable the participant service components to issue invocations to the controller. The structure of the platform-independent design is depicted in Figure 13, which is identical to Figure 9 except for the abstract platform and its primitive interactions.

In this solution, the controller service component also centralizes the coordination of the auction. Similarly to the callback-based solution, participants may register product offers and bids with the controller. This is done by invoking the operations *register_offer* and *register_bid*. The participants poll the controller for offers and bids by invoking its operations *get_current_offers* and *get_current_bids*, which return sets of current offers and bids, respectively. The participants also poll the

Figure 14. Indication of the outcome of an auction (platform-independent service design)

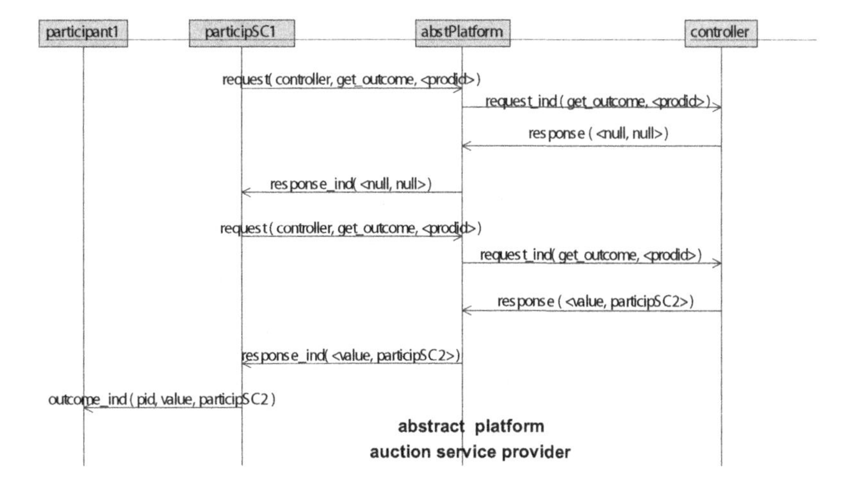

controller for the outcome of the auction for a particular product with the operation *get_outcome*. Figure 14 depicts the activities involved in the indication of the outcome of an auction.

Realization

A realization of the platform-independent design in terms of the CORBA platform is straightforward. The realization in terms of the JMS platform deserves more attention since this platform does not support the request-response pattern directly.

In the JMS-specific realization step, the abstract platform service definition is used as a starting point for a recursive application of service decomposition. Figure 15 illustrates the activities involved in the indication of the outcome of an auction, in a realization with the abstract platform in terms of the JMS platform. The boundaries of the abstract platform are preserved in this realization. The occurrence of a *request* interaction results in the sending of a *request* message to the controller, containing the identification of the request, the name of the operation to be invoked, and the parameters for the operation. Likewise, the occurrence of a *response* interaction at the controller side results in the sending of a *response* message to one of the participant service components. The identification of the request is used by the abstract platform service components to correlate *request* and *response* messages.

Figure 15. Indication of the outcome of an auction (JMS-specific realization)

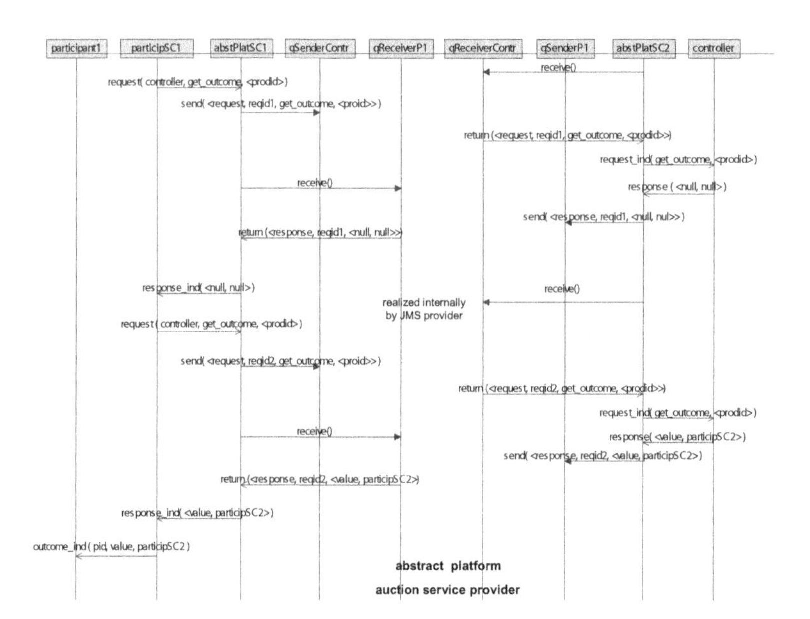

Conclusion

Service-oriented computing has been recognized as an approach that can significantly facilitate the development and maintenance of distributed enterprise applications. The same holds for model-driven engineering. While SOC aims at the creation of a service market consisting of offerings of generic and reusable services, MDE focuses on the creation and use of platform-independent and reusable artifacts in the design process.

The central role of the service concept is explicit in service-oriented architectures. However, it is currently mainly used in a technology-driven fashion, having a strong bias toward Web-services technology. By discussing two design paradigms—protocol centered and middleware centered—we have shown how the service concept can be used in two different ways as a technology-independent starting point for top-down development.

The service concept is not explicit in MDA. We argue that it should be given a more prominent role there since it allows one to precisely define the notion of platform independence adopted for a design. Often, some platform characteristics are assumed implicitly in platform-independent designs. We therefore propose the architectural concept of the abstract platform to completely and explicitly define platform characteristics at the abstraction level that comply with the platform-independent design at hand.

We propose a model-driven service-oriented design approach that combines these findings. The approach identifies the use of the service concept in different milestones of the model-driven engineering trajectory. It supports a top-down development, but does not exclude using bottom-up knowledge. As such, this approach allows designers to derive abstractions that remain stable and that have feasible implementations based on knowledge from the solution space.

The model-driven service-oriented approach is illustrated in this chapter with a simple example: the design of an auction system. The example shows that an auction service can be defined as a stable abstraction that shields the service users from the service's implementation. Options to be considered for the implementation can be different with respect to both design solutions (e.g., callback and polling based) and assumed platform characteristics (e.g., message exchange and request-response). The example shows that an abstract platform can be defined that corresponds to each set of platform characteristics, therefore shielding the platform-independent service design from the concrete platforms that can be used (e.g., CORBA and JMS).

The scope of this chapter did not allow us to give attention to information modeling and language support. Our approach has focused on the behavioral aspects of design, and therefore we have not explored the issues related to information representation and structuring. Nevertheless, we acknowledge that these issues are also important for model-driven service-oriented design. Design artifacts have to be expressed in some suitable modeling language, and UML (unified modeling language) is widely accepted as a de facto modeling-language standard in this area. In particular, Cariou, Beugnard, and Jézéquel (2002) have explored the notion of medium, which corresponds to our notion of application interaction system, with UML as the prime candidate language to represent such media. Other languages may be preferred if they better suit specific design requirements, for example, to represent behavioral aspects at related high levels of abstraction (Quartel et al., 2004).

Finally, we acknowledge that there are many other approaches that propose a separation of concerns similar to or different from the one supported by our approach. Notably, aspect-oriented development can be considered a prominent approach, which is apparently completely different from the one proposed here. However, Atkinson and Kühne (2003) show how the benefits of aspect-oriented development can be exploited in model-driven and middleware-based approaches.

References

Almeida, J. P. A., Dijkman, R., Sinderen, M. van, & Ferreira Pires, L. (2004). On the notion of abstract platform in MDA development. In *Proceedings of the Eighth IEEE International Enterprise Distributed Object Conference* (pp. 253-263).

Almeida, J. P. A., Sinderen, M. van, Ferreira Pires, L., & Quartel, D. (2003). A systematic approach to platform-independent design based on the service concept. In *Proceedings of the 7th IEEE International Conference on Enterprise Distributed Object Computing* (pp. 112-123).

Atkinson, C., & Kühne, T. (2003). Aspect-oriented development with stratified frameworks. *IEEE Software, 20*(1), 81-89.

Cariou, E., Beugnard, A., & Jézéquel, J. M. (2002). An architecture and a process for implementing distributed collaborations. In *Proceedings of the Sixth IEEE International Conference on Enterprise Distributed Object Computing* (pp. 212-223).

Erl, T. (2005). *Service-oriented architecture: Concepts, technology, and design.* Crafordsville, IN: Pearson Education, Inc.

Kent, S. (2002). Model-driven engineering. In M. J. Butler, L. Petre, & K. Sere (Eds.), *Integrated Formal Methods: Proceedings of the Third International Conference (IFM 2002)* (LNCS 2335, pp. 286-298). Berlin; Heidelberg, Germany: Springer.

Microsoft Corporation. (2001). *Microsoft .NET remoting: A technical overview.* Retrieved August 15, 2005, from http://msdn.microsoft.com/library/default.asp?url=/library/en-us/dndotnet/html/hawkremoting.asp

Object Management Group (OMG). (2002a, June). *CORBA components* (formal/02-06-65). Needham, MA: Object Management Group, Inc.

Object Management Group (OMG). (2002b, August). *Notification service specification, v1.0.1* (formal/02-08-04). Needham, MA: Object Management Group, Inc.

Object Management Group (OMG). (2003). *MDA guide* (omg/2003-06-01). Needham, MA: Object Management Group, Inc.

Object Management Group (OMG). (2004). *Common object request broker architecture: Core specification* (formal/04-03-12). Needham, MA: Object Management Group, Inc.

Papazoglou, M. P., & Georgakopoulos, D. (2003). Service-oriented computing. *Communications of the ACM, 46*(10), 24-28.

Quartel, D. A. C., Dijkman, R. M., & Sinderen, M. J. van. (2004). Methodological support for service-oriented design with ISDL. *Service-Oriented Computing: ICSOC 2004, Second International Conference* (pp. 1-10).

Sharp, R. (1994). *Principles of protocol design.* Upper Saddle River, NJ: Prentice-Hall.

Sinderen, M. van, & Ferreira Pires, L. (1997). Protocols versus objects: Can models for telecommunications and distributed processing coexist? *Proceedings of the Sixth IEEE Computer Society Workshop on Future Trends of Distributed Computing Systems* (pp. 8-13).

Sun Microsystems. (2002). *Java(TM) message service specification final release 1.1.* Retrieved August 15, 2005, from http://java.sun.com/products/jms/docs.html

Vissers, C. A., & Logrippo, L. (1985). The importance of the service concept in the design of data communications protocols. In *Proceedings of the IFIP WG6.1 Fifth International Conference on Protocol Specification, Testing and Verification V* (pp. 3-17).

Vissers, C. A., Scollo, G., Sinderen, M. J. van, & Brinksma, E. (1991). Specification styles in distributed systems design and verification. *Theoretical Computer Science, 89*, 179-206.

World Wide Web Consortium (W3C). (2001). *Web services description language (WSDL) 1.1.* Retrieved August 15, 2005, from http://www.w3.org/TR/wsdl

World Wide Web Consortium (W3C). (2003). *SOAP version 1.2 part 1: Messaging framework.* Retrieved August 15, 2005, from http://www.w3.org/TR/soap12-part1/

Chapter VII

A Composite Application Model for Building Enterprise Information Systems in a Connected World

Jean-Jacques Dubray, SAP Labs, USA

Abstract

The Web, as a ubiquitous distributed computing platform, has changed dramatically the way we build information systems, evolving from monolithic applications to an open model that enables real-time and federated information access, unifying the user experience across business processes. The industry has coined a new term for this latest evolution: connected systems. Unlike distributed systems, they are not just about distributing workload or ensuring fail-over, but rather about leveraging connectivity to enable specialized software agents to perform units of work cooperatively and opportunistically by exposing and consuming each other's services to fulfill a common goal. To reach their fullest benefits, connected systems require a new application model that relies exclusively on the consumption and composition of autonomous services. This new blueprint is poised to reshape the information systems' architecture, infrastructure, delivery technologies, programming languages, deployment, and management models. The goal of this chapter is to help you understand why and how IT should evolve the enterprise architecture toward a service-oriented composite application model.

Introduction

Connectivity has been at the foundation of human innovation and progress for the last 5,000 years. Transportation and communication infrastructures have enabled a specialization and composition of human activities empowering each economic agent to use and contribute the best of its abilities. In the last hundred years, this movement has accelerated, and today, vertical industrial conglomerates have all but disappeared under the economic pressure of an agile, layered, and dynamic fabric of enterprises of all sizes offering composable services to each other. Indeed, this fabric is itself creating tremendous competitive strains on its constituents by globally propagating innovations and optimizations, creating a constant need to reengineer design, sourcing, production, delivery, marketing, and support processes. Furthermore, technological advances have shown their ability to wipe out century-old industries within a few years. In this now global fabric, an enterprise must secure a decisive capacity to innovate, adapt, and optimize or else one of its competitors will quickly gain the ability to sell an equivalent product in its markets to its customers.

Paradoxically, the advent of the richest and fastest communication network combined with the use of the most powerful computers and high levels of automation have revealed a crying lack of adaptability and composability of IT organizations, hindering new business models and relationships while slowing productivity gains. The software industry has efficiently produced a software-construction model based on assemblies of technologies and frameworks; however, the systems produced with these technologies are not composable as requirement changes or new systems need to be built. As a result, IT, one of the major vectors of change for the past 30 years, can no longer be perceived as much of a competitive differentiator as the costs of innovating, adapting, or optimizing cannot often match the business cycles of an organization.

In the last 3 years, the software industry has started a major overhaul of the concepts and technologies used to build information systems to both adapt to the connected world and restore IT's leadership in driving business value. The foundation of this evolution is service-oriented architectures (SOAs), and its flagship is composability, that is, enabling the enterprise to build assets that can be reused in contexts unknown at the time they were designed. This ability of reusing assets in new business contexts is expected to improve the response time and the cost of developing or adapting solutions. Composability can be achieved both at the hardware level, with the concept of grids (Garritano, 2003), and the software level in several dimensions as user-interface, business-process, and information composition.

This chapter focuses on this new paradigm whereby IT assets can be constructed to be autonomous and composable based on a new computing model. The first section looks at the existing application models and their limitations in a connected world. The second section reviews the new principles of composition that are introduced by service orientation. The last section introduces how these principles enable composite application models.

Current Issues with the Application-Programming Model

The Economics of IT

Software engineering has made tremendous advances in the last 30 years, with each one conquered through new levels of modeling and abstractions. The most spectacular of all are probably relational databases, object-oriented programming, and graphical user interfaces (GUIs). These three abstractions and their related technologies have

Figure 1. A typical data model in modern IT

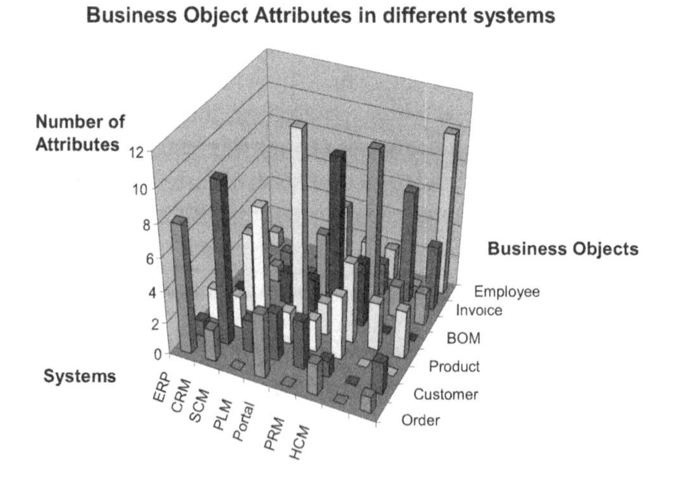

Figure 2. Cost and value trends of adding one new system in an IT organization vs. the number of information systems in this organization

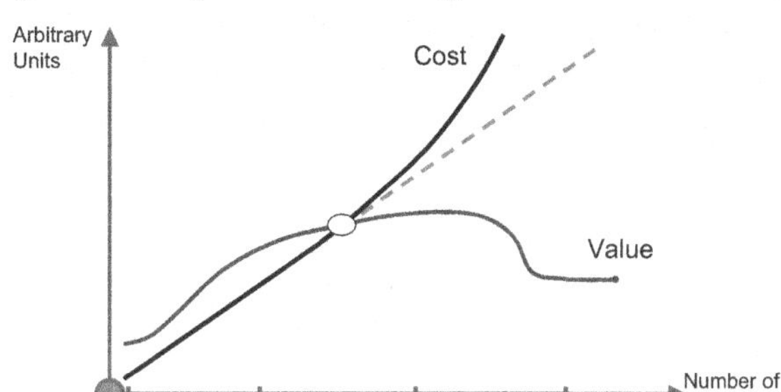

increased dramatically the productivity of developers, who have delivered millions of mission-critical applications and solutions automating all but business functions and roles. For each system individually, a clear return on investment (ROI) justified its construction. However, the levels of automation we have reached have also created a landscape where the data model of any given organization spreads across many systems. As an illustration, Figure 1 represents the attributes of business entities such as the customer, order, bill of material (BOM), and so forth in different functional systems (ERP, CRM, SCM, etc.).

In many organizations, Figure 1 is substantially larger due to geographically dispersed organizations, mergers, and acquisitions. Furthermore, these systems are built over periods of time, during which infrastructure technologies evolve, creating a de facto broad and complex technology landscape. As new systems get added, the cost of integrating and evolving systems added to the cost of managing secure and available information systems will grow probably to the point where it consumes entirely any given IT budget, leaving little room for innovation.

Let us understand why by looking at the value of adding a single new information system to an organization vs. the number of systems that are already supporting this organization (Figure 2). At first, the value increases rapidly because organizations automate the most productive business processes. Over time, lower value processes are automated, though the complexity of adding these new systems remains constant or even increases. It may also happen that adding yet another system could potentially decrease the value of existing systems: It is not uncommon to find information workers that utilize many applications to perform their day-to-day activities. This often lowers productivity because they need to switch contexts, increasing the risk of inconsistencies between similar data inputs, increasing training costs, and so forth.

Figure 2 summarizes the plausible trends of the cost and value of adding new systems vs. the number of systems in a given IT organization (this figure is not based on real data). The reason why IT yields less and less competitive advantage today is because most organizations have reached the crossover point and have entered a situation where their financial margins can no longer be improved by IT projects. Even though many organizations still find innovative ways to improve their businesses, the costs, risks, and complexity of the existing IT landscape prevents most of the projects to go forward.

Shortcomings of Traditional Application Models

The architecture of information systems is the product of an evolution influenced by several factors such as providing access to larger and larger communities of users; enabling scalable, fail-safe hardware configurations and manageable application deployments; simplifying integration with other systems; improving developer productivity; and so forth.

Amongst these factors, code composition has always been a constant focus. Many frameworks or class libraries have enjoyed large commercial success by accelerating the delivery and improving the quality and performance of many solutions. However, code-composition mechanisms such as inheritance or component technologies have proven to be extremely brittle as new releases create inevitable API (application programming interface) changes while being unable to produce assets that themselves could be reused. For instance, the Struts framework (McClanahan & Husted, 2005) is today one of the most reused elements of the implementation of Java-based Web applications. Yet, the assets produced with Struts are not specifically reusable.

Figure 3. State-of-the-art Web-application architecture and technologies

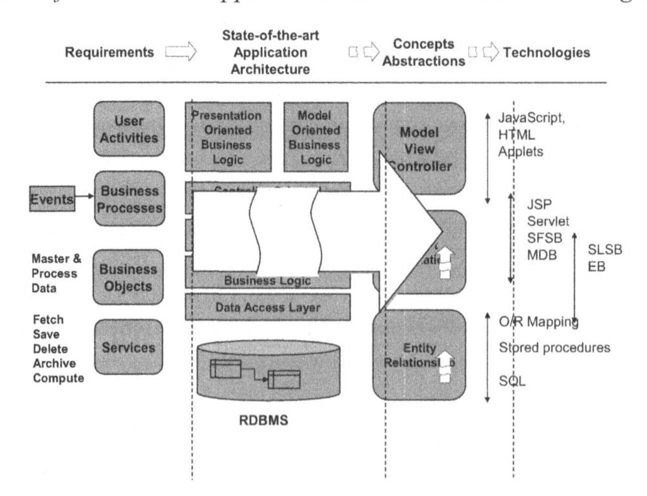

Figure 3 features a simplified representation of the conceptual (requirements), logical (architecture and abstractions), and physical (technologies) views of modern software construction. These three views are misaligned. While the requirements are often expressed in terms of user activities, business processes, events, services, and business objects, none of these concepts can readily be associated to the layering of the architecture or the technologies. This misalignment drives the implementation toward spreading code associated to business entities, processes, and user activities across all layers, often written in different and incompatible technologies, creating debugging and maintenance nightmares. This type of software-construction model was optimized for the separation of developer roles such that a given project can be divided among HTML (hypertext markup language) and JavaScript specialists, SQL specialists, and so on rather than providing a natural translation of requirements into executable artifacts.

For instance, even the abstract foundation of all modern application models, the model-view-controller (MVC) pattern, is poorly aligned with the conceptual level. This pattern was invented in 1978 at Xerox PARC by Reenskaug (2003). MVC does not provide explicit user-activity boundaries: The code is "unaware" of when a user activity starts, ends, or enters any other intermediary state. When a user activity spans more than one view, which is now frequent, the developer needs to implement ad hoc state machines to manage the navigation between the different views, creating a coupling between controllers. This particular problem has far-reaching consequences, for instance, considering an activity-based security model. Similarly, the MVC conceptual model does not provide any facility to specify business processes explicitly, leading again to dedicated code spanning multiple controllers that act together to perform a given process based on the persisted state of the business-process instances captured in ad hoc process-model elements. When business processes are hard-coded in controllers, they become difficult to manage, monitor, change, or compose within and across applications. At the model level, the pattern does not provide a conceptual framework to create meaningful and reusable domain abstractions. Rather than representing a specific business entity, domain objects are often created to support specific views directly bound to the physical data model. As a result, MVC-model objects often couple the user interface and the physical data structure.

The MVC pattern is an excellent technical pattern that can be applied successfully to implement GUI-based frameworks or infrastructures, but the lack of business semantics added to the variety of technologies involved in its implementation make it difficult to reuse views, controllers, and models outside the context for which they were designed.

The major consequence of this lack of reusability is that it has been more cost effective to develop new systems and integrate them with one another via synchronization or replication mechanisms rather than carefully design each system for potential reuse. This runaway mode can only stop when the cost of the integration and management of all these systems overrun the IT budget. Unfortunately, even today, new technologies appear, such as the AJAX (Garret, 2005) pattern to develop the next generation of Web applications, that do not seem to be concerned by the development of modular and composable assets. Even worse, AJAX specifies an architecture in which the controller can be implemented either on the client (browser) within the AJAX engine or within the service-side systems. It is likely that most applications will operate in mixed mode, creating assets that cannot be reused easily.

Service Orientation

In the last couple of years, the industry has heavily marketed a series of new principles for designing modular enterprise systems under the name of service-oriented architecture and based on a set of platform-neutral open standards commonly referred as the Web-services technology stack (Papazoglou, 2005). The objective of service orientation is to enable autonomous software or hardware agents to be assembled and participate in complex connected systems that perform units of work cooperatively. Agents expose a contract, of which the unit of definition is called a service. A service interface defines the messages exchanged by the agent to perform its work. The message exchanges are grouped in operations. By extension, these agents are called services.

Agents interact as peers by exchanging messages to initiate work, send notifications, synchronize their state, and request and send information. Service interfaces differ from object interfaces because they contain also the outbound operations in addition to the inbound ones, explicitly specifying the entire contract they rely on to perform their work. These interfaces may be discovered manually or automatically in a registry. Service interfaces are often qualified by policies that specify the service capabilities. These policies are used to form agreements when two or more services are assembled to perform a unit of work.

As we will see in the following sections, services can be made composable and capable of being reused in several contexts of utilization. For instance, a *CalculateSalesTax* service should be designed to be used in the context of an order, invoice, quote, and so forth, as well as across different legislation contexts. From an operations perspective, it would be more efficient for the enterprise to outsource such a service from existing business applications since the maintenance and management becomes shared across a large number of companies that otherwise would have to upgrade their systems each time sales-tax regulations would change in their geographical area of operations. Services may encapsulate systems of records, computations, or user-driven activities. Service granularity may be as small as a database insert or update, or as large as managing the life cycle of a business entity such as a purchase order, as we will see in an example below.

The financial-services sector (*Microsoft Case Study*, 2003) offers early examples of implementations of service-oriented architectures. The adoption of an enterprise-wide SOA is often motivated by requirements such as the following.

- Deliver cross-functional systems capable of supporting end-to-end activities
- Exploit existing assets to as great a degree as possible
- Adapt to changing business conditions or resolve new problems
- Develop new business opportunities by exposing existing assets to customers or partners
- Reduce the number of unstructured activities that cross departmental boundaries

While service-oriented architectures have already shown clear returns on investments, their benefits could be extended with a service-oriented application model that would leverage a series of internal or external services to build new information systems.

The Web has already opened up new ways to think about application models by taking advantage of its ubiquitous connectivity to aggregate functionality from different back-end systems as portals in a seemingly unified user interface. But this is not enough; this type of user-interface composition presents some limitations as portlets or views are often directed to one system of record, forcing users to interact with several portlets to perform a single activity.

The industry is looking for a new application model offering activity, process, and information composability (Gilpin, 2005). This model is commonly referred to as the composite application model. Combined with Web-services technologies, composite applications would be capable of delivering functionality crossing technology or even organization boundaries. Composite applications would provide a unified user experience, integrating information and user tasks from different systems into a single user interface and enabling the vision of a process-driven IT set forth almost a decade ago with the business-process engine concept (Rymer, 1998).

In the following sections, we will explore the foundations of this composite application model, first by looking at the principles of composability in service-oriented technologies.

Principles of Composability in Service Orientation

Service orientation is not just about architecture, design principles, or service interfaces. Service orientation is introducing a new computing model (Papazoglou, 2003) designed to master connectivity, where the message is becoming a first-class citizen along with data and code to enable autonomous services assembled as peers to work together collaboratively to perform activities. In SOC (service-oriented computing), service interactions do not just follow a synchronous remote-procedure

Figure 4. Operation composition via a WS-BPEL definition

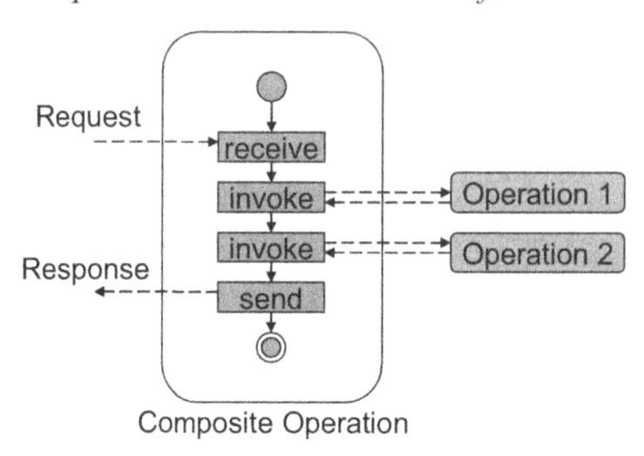

call model, but they also shift to a peer-to-peer, long–running, and asynchronous model. These types of interactions combined with a platform-neutral interchange infrastructure enable new levels of composition.

Service Composition

SOC offers two models of service composition. The most common is referred to here as operation composition (Srivastava & Koehler, 2004). In this model, a service operation is implemented by the invocation of other operations via some programming logic (Figure 4). This is similar to all composition models found in structured programming. The programming logic is often expressed using WS-BPEL (Web services business process execution language; Brown & Szefler, 2003; IBM, 2002), which is currently under development at OASIS. WS-BPEL is well supported by several products and is often used for enterprise application-integration purposes. Some vendors have also elected to use WS-BPEL solely as a standard interchange format rather than an execution language because of some of its limitations, such as the lack of scripting or message-transformation capabilities.

However, we suggest that WS-BPEL represents the first programming language that is specifically designed for SOC with the notion of message interchange at its core and providing a run time capable of managing the context of long-running interactions in a standard way. This use of WS-BPEL is actually supported by a new specification, the service-component architecture (SCA), published by a consortium of vendors including BEA, IBM, Oracle, and SAP (Beisiegel & Blohm, 2005).

The second model of service composition is referred to here as service-interface composition. This model does not rely on explicit business logic but rather on the interaction assemblies of services. Figure 5 represents a unit of work (procurement) performed as the interactions of five services under the control of different business parties. As we see in this example, there is no explicit business logic controlling the advancement of the whole unit of work; it is the mere autonomous execution

Figure 5. Service-interface composition model

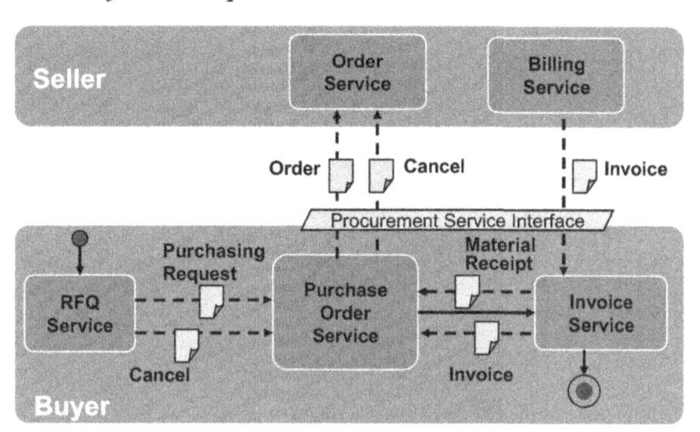

of the services, as peers, that advance its state. In this model, services that were not necessarily designed to work together are assembled and configured to work with other peers to perform a specific unit of work. The same services may support multiple configurations to perform the same or a different role in several other units of work. This unit of work exposes a buyer's composite procurement service interface, combining operations from the internal order and invoice services.

This approach is not yet fully supported by Web-services technologies. In particular, the *WS-I Basic Profile* (Ballinger & Ehnebuske, 2004) recommends not using outbound operations (i.e., the *sendOrder* and *sendCancel* operations of the buyer's order service are forbidden per WS-I rules, though they can be specified using WSDL [Web service description language]). The reason is that when outbound operations are used, it is not possible to just discover the service interfaces for this type of composition to work since the binding of outbound operations can only be known when all the participants are defined, therefore breaking the discovery paradigm introduced by UDDI (universal description, discovery, and integration), the Web-service interface registry.

Even though a service implementation may use any technology, WS-BPEL represents an ideal programming language for implementing services participating in a complex assembly of services performing a unit of work. Figure 6 shows the WS-BPEL elements that receive and send the messages exchanged with other services. Internally, WS-BPEL may invoke subprocesses for creating the purchase order in back-end systems (not represented in Figure 5).

Choreography standards such as WS-CDL (Kavantzas, 2005) and OASIS ebBP (Dubray & Saint-Amand, 2005) provide the foundation for modeling this composition model.

Even though this type of composition itself is often self-coordinated, some units of work may require technical services such as shared context management between participants, activation of the units of work, and different forms of coordination

Figure 6. Partial representation of the WS-BPEL implementation of the buyer's order service

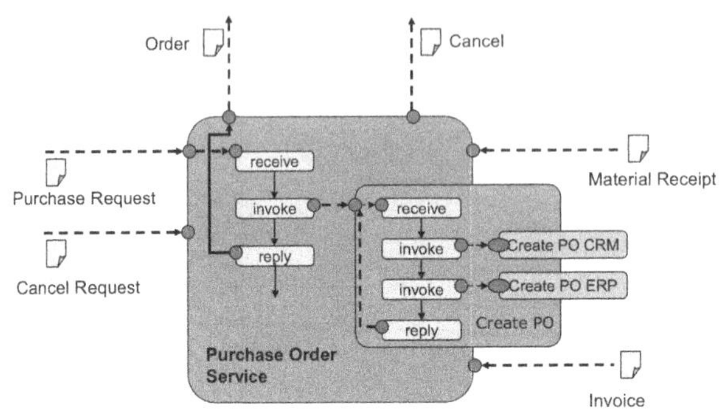

(routing, transaction, transformations, publish, subscribe, etc.). The WS-TX (IBM, 2005) specification developed at OASIS provides a coordination architecture and a specification for some of these technical services.

WS-Policy (IBM, 2003) is another specification from the Web-services technology stack that could be used for automating the composition of services as it specifies service properties that might be computed to be incompatible (such as service-level agreements).

The key innovation of service-interface composition is that it enables a clear separation for the control and management of IT assets, leading to a granularity of service definitions that is meaningful. Topologically, we could implement the buyer role with a single BPEL definition instead of three, invoking all the atomic operations within the three buyers' services implementations. However, this BPEL definition would couple the procurement business process with the business logic that controls the life cycle of the business objects (RFQ, purchase order, and invoice). This would be undesirable as some of the business rules of the internal implementation of services may change without changing the performance of the unit of work provided that the composition contract is respected.

Information-Entity Composition

XML (extensible markup language; and SGML before it) introduced a new concept as a data container that associates the data structure (metadata) and the instance data in a single document. In older technologies, the data structure is generally separate from the instance data itself (object orientation, relational databases, etc.). That capability combined with validation technologies such as XML schema enables flexible information composition. For instance, XML data sets that contain product descriptions from different suppliers, often with differing schemas, may be captured within the same purchase order request. This order request would then be processed to create the individual purchase orders sent back to each supplier with their specific product information and associated schema.

In addition, the fact that metadata and data are stored jointly enables semantic access to XML data structures. In other words, it is possible to access information in an XML document without fully being knowledgeable about the entire structure of the document. Technologies like XPath, XSLT, or XQuery provide relative access to any node in the XML document, compared to other data-structure technologies that offer only absolute access to data structures, requiring the knowledge of the original schema used to specify the corresponding data structure. XSLT technologies represent the composition engine enabling the declarative, yet relative, composition or transformation of multiple documents. XML technologies reinforce the composition capabilities of not just information entities, but also services, which are able to consume document parts of an incoming message regardless of the incoming message structure. This capability is best illustrated with the composability of the specifications of the Web-services stack (Ferguson et al., 2003).

More formal forms of composition have been developed in the *Core Component Technical Specification* (UN/CEFACT, 2003), which is based on the concept that all information entities can be composed of generic elements, sometimes dynamically, based on their context of utilization (e.g., a purchase order, in France, in the automotive industry). Basic types can be composed into complex components via aggregation or association. The *Tax* component, for example, can be used whenever taxes and duties are involved. However, the *Tax* component might be different based on the context of its information-entity containers (industries, countries, etc.).

In a service-oriented architecture, an information entity is composed (Figure 7) of an identity (typically globally unique), content, states, and information related to its life cycle in a connected system: the location of where representations were sent to (associated to instances of work in progress), the replication requirements

Figure 7. Information-entity elements in service orientation

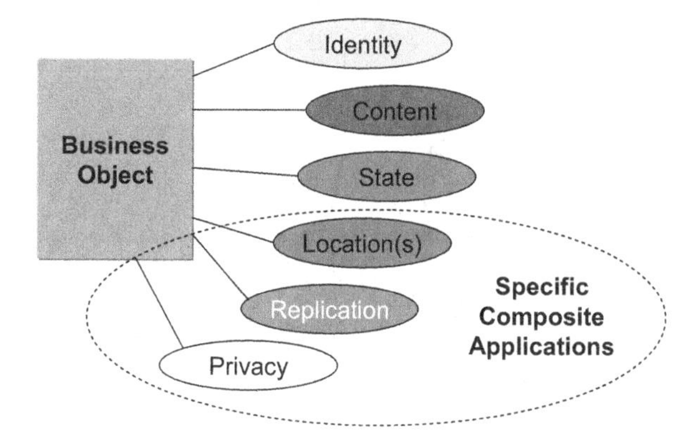

that specify who would need to be notified should the content or state change, and finally privacy policies that specify how representations of this information entity should be handled if and when passed to other parties.

All these principles of composability represent an opportunity to build assets that can be reused in contexts unknown at the time the asset was designed and implemented. In the next section, we are going to explore how an application model can be built on top of a service-oriented architecture.

Composite Application Models

Enterprise Services

Many attempts have been made to create a common information model at the enterprise level as the foundation to build new solutions. For instance, in the early '90s, many companies developed the concept of a common database where all enterprise data would be stored in one common schema. This approach failed mainly because some business logic (the model-oriented business logic) was required to be migrated along with the data. In addition, performance problems made it very difficult to tune these databases for all scenarios and it was often impossible to keep up with the changes in the individual systems, to move data to the common database as needed, and to update the corresponding solutions to use the common database. By the mid-'90s, most companies had preferred to adopt an enterprise application-integration solution that synchronizes data between islands of information. A decade later and after a steady adoption of integration platforms, companies find that the sheer number of systems synchronizing their state makes it difficult to

Figure 8. Composite services as logical information model

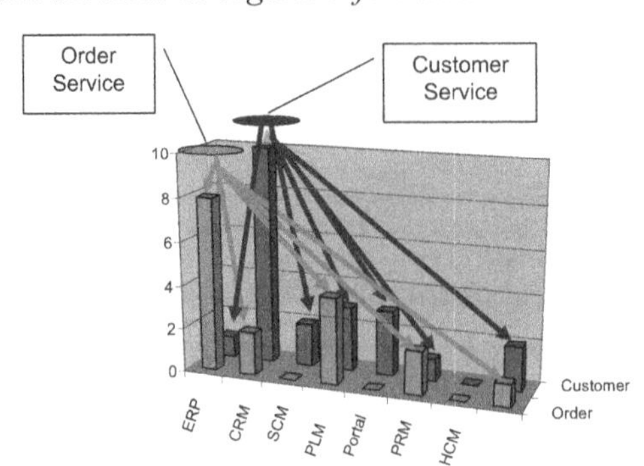

keep up with integration projects while the cost of integration itself is consuming too much of the budget of new projects.

Composite applications represent a new approach to the creation of normalized information models as it relies on services to provide a composite view of the data model and the business logic associated to it. This approach actually represents a unification of the integration and application models. Services represent prepackaged integration elements maintained somewhat independently of the activities that consume them. For instance, if a new system is installed and requires to be updated each time a customer record is updated, all the integration work happens behind the service interface once, without impacting the applications that are dependent on this particular service. Of course, this approach works if all solutions consuming customer data access it via the customer service.

This approach in itself is not revolutionary; it is rather a best-practice pattern of application integration. The difficulty in implementing this pattern was that services must be built in a heterogeneous environment as they may be consumed by and consume processes that run on all types of platforms. The industry has developed collaboratively a series of open standards that govern the interoperable message exchanges between services regardless of their implementation platform. This set of standards (Figure 9) is often referenced to as the WS-* stack (STAR stands for secure, transacted, reliable). It is divided into three broad categories of standards:

Figure 9. Service-oriented computing architecture stack (Dashed lines represent the specification that work is still in progress)

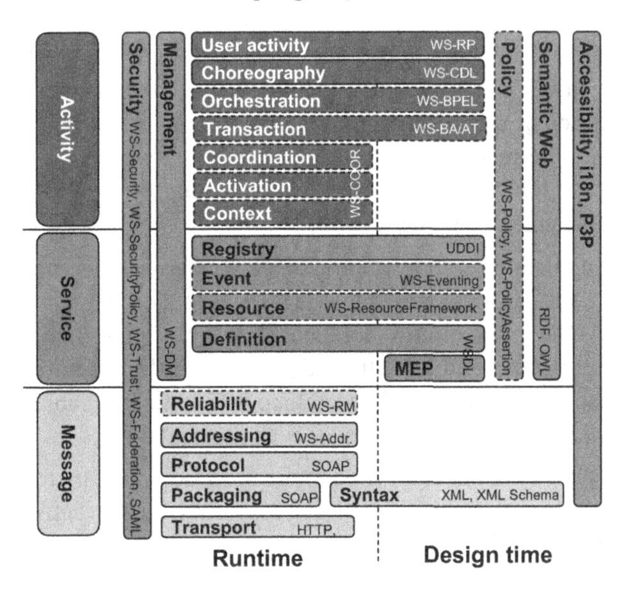

message based, service based, and service-oriented based. A thorough review has been written by Cabrera and Curt (2004) and Papazoglou and Dubray (2004). These technologies altogether enable us to create autonomous components capable of exchanging messages securely and reliably.

The three specifications (WS-BPEL, WS-CDL, and WS-TX) constitute the foundation of service-oriented computing, providing a new programming language, a service-assembly model, and an infrastructure to manage the units of work performed by services. This is also the foundation of a new application model: a service-oriented application model.

Service-Oriented Application Model

Enterprise services tend to expose data interactions with a coarse granularity. This is partly because in a connected system, it is still expensive to exchange messages, especially in a secure, reliable, and transacted way. This is also due to the fact that it is easier to agree on and enforce a document format rather than a sequence of operation invocations. These coarse data interactions, which are the keystone of service orientation, create an increasing mismatch between traditional user-interface application models and technologies such as Java Server Faces (Burns & Horwat, 2004), which encourages fine-grained interactions triggered by GUI control events.

The era of the user interface "bolted" on a single system of record has ended. The complexity of the tasks that any given user performs today requires a federated, composite, and collaborative point of usage. User productivity can be increased by avoiding redundant data entry, so the user interface has to be federated across all systems of records. This is achieved by consuming enterprise services. Users also want to perform all their tasks in the same composite environment, just like the browser. Finally, users want to collaborate with other users as they perform their activities for training or escalation purposes. In addition, the security model of any user activity has to be federated. In a connected world, we cannot expect that all services will be available at all times. Hence, this composite client should be able to operate in a disconnected mode as well.

In addition, the application model should also support process composition with no technical or physical boundaries, just legal and logical boundaries (company, department, etc.). The application model should be easy to deploy and evolve without breaking existing systems and activities, requiring as little integration as possible. Finally, the application model should be able to deal with unreliable environments and offer the scalability and availability of Web applications.

To support this type of application model, we propose a new pattern (Figure 10) based on service-oriented concepts involving resources, events, activities, services, and coordinators (REASC).

Figure 10. Pattern of REASC

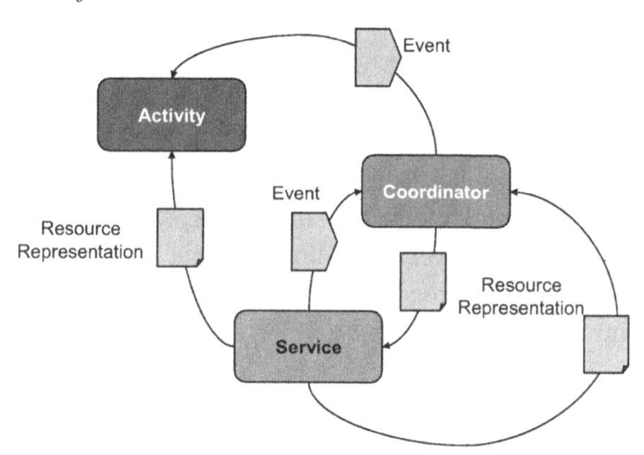

Activities, services, and coordinators are peers, exchanging messages that contain resource representations and/or events. All elements of the pattern are composable. Service composition can be achieved via technologies such as orchestration (WS-BPEL) or choreography (WS-CDL, ebXML/ebBP). Resource composition or at a minimum resource-representation composition is possible via the underlying XML foundation as XML documents can represent composite data sets coming from different data sources, and an XML schema enables further composition of existing XML documents via the open-content model. Coordinators are also composable by design in the Web-service composite application framework or the WS-Coordination specifications. For instance, a transaction coordinator can be registered as a participant with respect to another transaction coordinator. The coordinator role is to ensure the existence of the supporting (technical) services necessary for the performance of a unit of work by multiple services. At a minimum, there is a need for managing the context and the life cycle of the unit of work such that all services may be able to refer to its context and its state (started, ..., completed). Different types of coordinators can provide services such as transaction management, transformation, message routing, and so forth.

Events represent requests for action and may or may not contain resource representations.

An activity is a unit of work in its broadest sense (while a coordination or a service represents a more specialized or atomic unit of work). Examples of activities include business processes, user tasks (such as creating or approving a purchase order), transactions, and so on. These types of activities must rely on a model that is also composable.

The pattern does not make any assumption on boundaries or technologies between the different actors of the pattern. The pattern does not mandate a model-driven (Frankel, 2001) implementation for the different elements; however, for agility purposes, it is recommended that activities, enterprise services, and coordination definitions be model driven. There are two general approaches to building model-driven systems: the OMG's MDA (Object Management Group's model-driven architecture) and domain-specific languages (Greenfield & Short, 2004); both are equally applicable when implementing the pattern.

Figure 11 represents an example of the application of this pattern for building a composite application that implements the "quote-to-cash" scenario. The scenario shows two business parties (Buyer and Supplier) that exchange three business transactions: request for quote, process order, and process invoice. The supplier's business process shows three components.

- A sales-force automation component that contains a *Quote* service that manages the life cycle of quotes
- A salesperson, who performs manual operations such reviewing the quote, creating new accounts, and so forth
- An ERP component that contains services such as *Accounts*, *Orders*, and *Billing*

Some of these components rely on external services to perform their work: *Credit Check* and *Sales Tax Calculation.*

The supplier's coordinator acts as a B2B (business-to-business) integration server and provides routing and transformation services in this particular scenario. The

Figure 11. Supplier's quote-to-cash composite application

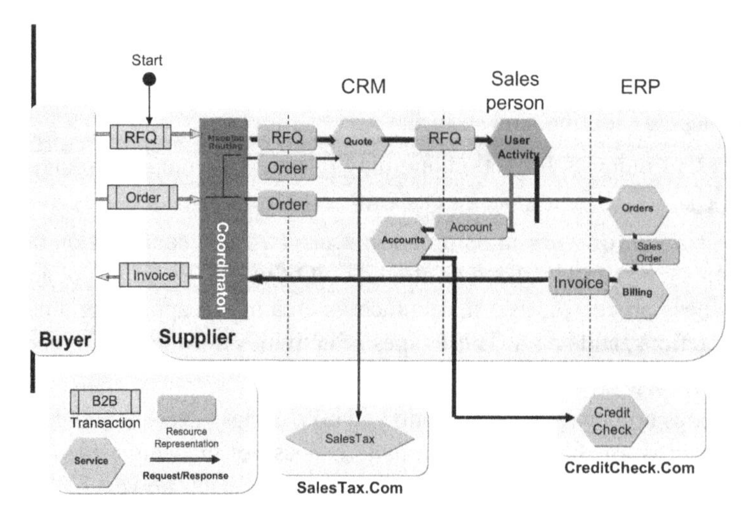

services exchange resource representations such as RFQ, *Order*, *Invoice*, *Account*, and *Sales Order*.

Conclusion

Information systems have grown to a point of extreme complexity, making them difficult to manage, evolve, or replace whilst the demand for access and connectivity is increasing at a rapid pace. After having been the engine enabling record productivity gains and innovative business models, this complexity has now become the bottleneck for innovation and change as most of the IT budget is tied into operations or integration projects.

A major evolution of enterprise architectures is under way fueled by technologies such as Web services and service-component architecture, guided by principles of the autonomy and composition of IT assets, and driven by the need to restore IT's leadership to help the enterprise innovate, adapt, and optimize. This evolution will be as significant as the one introduced by object orientation and will lead to the emergence of a service-oriented computing model.

Note

Jean-Jacques is a standards architect at SAP Laboratories. He has been a pioneer of the business-process-management field since 1997. He has been involved in developing composite application models since 2001. He has contributed to several standards (OAGIS, STAR-XML, ebXML BPSS and ebBP, WS-CDL, and WS-CAF), and he is the editor of the OASIS ebBP specifications (v1.1 and v2.0). He enjoys spending time with his kids, Marie and Matthieu, and when he gets the chance, to travel to Corsica, to the village of Aquadilici where his family is from.

References

Ballinger, K., & Ehnebuske, D. (2004). *WS-I basic profile v1.0.* Web Services Interoperability Organization. Retrieved from http://www.ws-i.org/Profiles/BasicProfile-1.0-2004-04-16.html#refinement16612720

Beisiegel, M., & Blohm, H. (2005). *SCA service component architecture version 0.9.* BEA, IBM, Oracle, & SAP.

Brown, P., & Szefler, M. (2003). BPEL for architects. *FiveSight.* Retrieved from http://www.fivesight.com/downloads/BPEL4ProgArchies.pdf

Burns, E., & Horwat, J. (2004). *An introduction to Java server faces.* Sun Microsystems. Retrieved from http://java.sun.com/j2ee/javaserverfaces/jsfintro.html

Cabrera, L. F., & Curt, K. (2004). *An introduction to the Web services architecture and its specifications version 1.0.* Microsoft.

Dubray, J. J., & Saint-Amand, S. (2005). *OASIS ebXML business process specification.* OASIS. Retrieved from http://www.oasis-open.org/committees/download.php/13906/ebxmlbp-v2.0.1-Spec-cd-en-html.zip

Ferguson, D. F., et al. (2003). *Secure, reliable, transacted Web services: Architecture and composition.* MSDN. Retrieved from http://msdn.microsoft.com/webservices/default.aspx?pull=/library/en-us/dnwebsrv/html/wsoverview.asp

Frankel, D. (2001). *Model driven architecture reality and implementation.* OMG. Retrieved from http://www.omg.org/mda/mda_files/DFrankel_MDA_v01-00_PDF.pdf

Garret, J. J. (2005). *AJAX: A new approach to Web applications.* Retrieved from http://www.adaptivepath.com/publications/essays/archives/000385.php

Garritano, T. (2003). On the grid. *ClusterWorld, 1*(1), 30-31, 50.

Gilpin, M. (2005). *How will the next generation of applications be built?* Forrester Research. Retrieved from http://www.ebizq.net/webinars/5869.html

Greenfield, J., & Short, K. (2004). *Software factories: Assembling applications with patterns, models, frameworks, and tools.* Wiley.

IBM. (2002). *Business process execution language for Web services v1.1.* Retrieved from http://www-128.ibm.com/developerworks/library/specification/ws-bpel/

IBM. (2003). *Web service policy framework.* Retrieved from http://www-128.ibm.com/developerworks/library/specification/ws-polfram/

IBM. (2005). *Web services transaction specifications.* Retrieved from http://www-128.ibm.com/developerworks/library/specification/ws-tx/

Kavantzas, N., & Burdett, D. (2004). *Web service choreography definition language v1.0, working draft.* W3C. Retrieved from http://www.w3.org/TR/ws-cdl-10/

McClanahan, C., & Husted, T. (2005). *Struts.* Apache Software Foundation. Retrieved from http://struts.apache.org/

Microsoft case study. (2003). *Allstate Insurance Co.* Retrieved from http://www.microsoft.com/resources/casestudies/CaseStudy.asp?CaseStudyID=13648

Papazoglou, M. P. (2003). Service-oriented computing: Concepts, characteristics and directions. *Keynote for the 4th International Conference on Web Information Systems Engineering (WISE 2003)*. IEEE CS.

Papazoglou, M. P., & Dubray, J. J. (2004). *A survey of Web service technologies* (Tech. Rep. No. DIT-04-058). University of Trento, Informatica e Telecomunicazioni.

Reenskaug, T. (2003). *The model-view-controller (MVC): Its past and present.* JavaZONE. Retrieved from http://heim.ifi.uio.no/~trygver/2003/javazone-jaoo/MVC_pattern.pdf

Rymer, J. (1998). *Business process engines, a new category of server software, will burst the barriers in distributed application performance engines.* Emeryville, CA: Upstream Consulting.

Srivastava, B., & Koehler, J. (2004). *Web service composition: Current solutions and open problems.* IBM. Retrieved from http://www.zurich.ibm.com/pdf/ebizz/icaps-ws.pdf

UN/CEFACT. (2003). *Core component technical specification.* Retrieved from http://www.untmg.org/artifacts/CCTS_v2.01_2003-11-15.pdf

Chapter VIII

Three-Point Service-Oriented Design and Modeling Methodology for Web Services Composition

Xiang Gao, Supercom Canada, Canada

Jen-Yao Chung, IBM T.J. Watson Research, USA

Abstract

In this chapter we present a three-point service-oriented conceptual design and modeling methodology for Web-service composition based on the object-oriented research results in model driven, existence dependency relation, object-oriented development (MERODE) method. Thus, on the concept level, we first clarify the research issue on the semantic consistency of Web-services composition, and on the logic level, we precisely define and demonstrate the semantic consistency of Web-services composition using some standard schemes of an order-handling system. Afterward, we establish a service model using formal techniques, and define the formal semantics of services interactions and formal historical semantic conform-

ances within Web-services interactions and composition on the conceptual system level. These aspects of research results can effectively and significantly tackle one big inhibitor of Web-services adoption: the lack of semantic consistency in business processes within Web-services interactions and composition.

Introduction

Web services have emerged as a new paradigm for loosely coupled distributed computing on the World Wide Web. This new paradigm aims at spanning organizational boundaries and interacts over the open Internet through the use of standard protocols. These standards include messaging formats (SOAP, simple object access protocol); interaction definitions (WSDL, Web service description language); and universal description, discovery, and integration (UDDI), which provide rudimentary mechanisms for defining interactions amongst services that may be located in different organizations. In order for Web services to realize their full potential and become widely accepted, researchers must tackle a number of important challenges and solve some major research issues, which include various aspects of Web-services composition. Web-services composition allows us to combine some existing Web services into a new, value-added Web service. During the Web-services composition, if we have two Web services that do not know each other in the open Web environment, but we prefer or need them to be combined into an advanced value-added Web service, then we will meet at least two apparently simple but profound research issues: The first one is how to combine these two Web services, and second one is how to deal with the semantic consistency between the two Web services, which means how to keep two Web services consistent when the involved Web service has changed. These two research issues are interesting and important in the field of Web-services composition. These aspects of research results can effectively and significantly tackle one of the three main inhibitors of Web-services adoption and acceptance: the lack of semantic consistency in business processes such as ordering, shipping, or billing within Web-services interactions (Shirky, 2002). Correspondingly, three research questions may be raised.

- **Question 1:** Does there exist a kind of semantic consistent relationship (or semantic constraints) within Web-services interactions in addition to the semantic consistency in terms of uniform data interpretation (we call this uniform data interpretation the first layer of semantic consistency within Web-services interactions)?
- **Question 2:** What is this semantic consistency in the context of Web-services interactions?

- **Question 3:** How do we recognize and precisely define this semantic consistency and tackle its complexity?

In this chapter we clarify and answer these questions, thus making major contributions to the hard point of Web-services composition: the semantic consistency of Web-services composition. We propose and create a novel and sound analysis, design, and modeling methodology for Web-services composition: a three-point service-oriented conceptual design and modeling methodology. This methodology captures the semantic-consistent relationship between Web services. Its conceptual design and modeling method is based on some object-oriented modeling research results in MERODE. In order to cater to the needs of the Web-services computing environment, we analyze and redefine the concept of semantic consistency in terms of Web-services interactions for business transactions. We claim that the information dependency rules are the key to semantic-consistency checking within Web-services interactions and composition. In the second part of this chapter, we formally define semantic consistency and semantic conformance using formal techniques in the context of Web-services interactions on the conceptual system level. This part of the work provides a formal theoretical framework for the semantic consistency of Web-services interactions. Our major contributions in this chapter can be summarized as follows.

1. Creating a novel and sound analysis, design, and modeling methodology for Web-services composition: a three-point service-oriented conceptual design and modeling methodology.
2. Clarifying on the concept level the semantic consistency in the research area of Web-services interactions and composition.
3. Precisely defining on the logic and system level the semantic consistency of Web-services composition using formal techniques.
4. Enhancing the semantic-consistency concept of object-oriented theory in MERODE to the service-oriented layer, which meets the needs of this new distributed Web computing environment.

The outline of our chapter is as follows. First we present our three-point service-oriented conceptual analysis, design, and modeling methodology for Web-services interactions based on some object-oriented research results in MERODE (Snoeck, Dedene, Verhelst, & Depuydt, 1999). Then we define some basic concepts in the methodology, and then point out and clarify the semantic consistency rules within Web-services interactions in terms of the principles of the three-point methodology, and end the section by formally defining and demonstrating the semantic consistency of Web-services composition on the conceptual logical level through demonstrating some standard schemes of an order-handling system. Afterward, we formally establish a service model using the formal techniques, and define the formal semantics of services interactions and formal history semantic conformances within Web-services interactions and composition on the conceptual system level. We analyze the related works, and then conclude the chapter and propose the further research directions.

Three-Point Service-Oriented Design and Modeling Methodology for Web Services

Interactions

As we know, in the standard business transaction there are two basic participants or roles that are the buyer and the supplier. In the open Internet environments, if we want to realize the same functionality of business transaction, we also need these same two participants or roles. In terms of Web services, the buyer and the supplier have the exposed input and output interfaces, which means they can be expressed as a WSDL description. If they are combined as a composite ordering system on the Web, which we call the Web-services composition, the key point is how to deal with their consistency. That means if the buyer changes its request, how does the supplier correspondingly adjust its supply in order to make the whole composite system consistent?

There is no doubt that there should be a coordinated part that acts as an intermediary. If we point out this part, it can also be described as an independent Web service. This intermediary handles the relationship between the other two Web services, and makes the composite ordering system work smoothly and consistently. (Afterward, you will know this is our novel three-point service-oriented design and modeling methodology.) Here we call this intermediary role the agent (seller).

Also, the relationship between the two Web services is expressed through the messages transferring between them that are visible within the Web-services interactions. In the domain of object-oriented enterprise modeling, the object types and event types are the first-class entities, and the semantics of existence dependency are the keys to semantic-integrity checking that enables one to check the semantic integrity and consistency between object types and event types. In the domain of Web-services composition, the Web services are the first-class entities; when we combine the services, we can also point out and define the similar semantic relationship (existence dependency) between the Web services in order to handle the semantic consistency between the Web services. The semantic relationship between the Web services is expressed through the existence dependency of message holders in which the message transferring is visible, and can be checked and tested. Our design and modeling methodology can be summarized as follows.

The Service-Oriented Analysis and Design Method

In light of the needs of Web-services composition (the functionality of the composite system, for example, for an order-handling system may provide ordering, shipping, and billing services), we recognize the participants or roles. For example, for the ordering service of this order-handling system, they are the buyer and the supplier. For the shipping service, they are the buyer and the shipper, and for the billing service, they are the buyer and the billing service provider.

Three-Point Design and Modeling Methodology for the Participants

In order to clarify and handle the semantic consistency between the Web services, we point out and give prominence to an intermediary role between the requesting service and the answering service, and the existence of this intermediary role is dependent on that of the other two parent services. For example, the existence of the agent (seller) is dependent on that of the buyer and the supplier, which is illustrated in Figure 1.

Three-Point Design and Modeling Methodology for Message Holders

Because the semantic relationship between the Web services is built up through message transferring, the existence dependency of the participants or roles is expressed in message holders. These message holders belong to the corresponding participants or roles. For example, because of the above existence dependency of the buyer, the

Figure 1. Existence dependency among the Web services

Figure 2. Existence dependency among the message holders

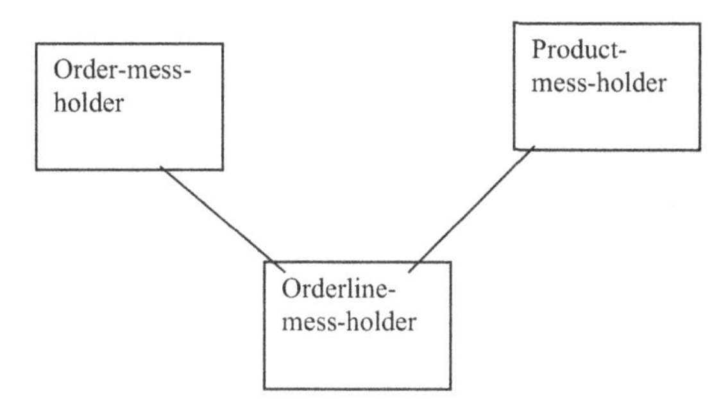

supplier, and the agent (seller), correspondingly, the existence of Orderline-mess-holder is dependent on that of Order-mess-holder and Product-mess-holder. The existence dependency of the message holders is illustrated in Figure 2.

Some researchers may argue that using this methodology raises the service numbers, but we believe that although the service numbers are raised, this method significantly reduces the logic complexity between the services and explicitly addresses the semantic consistency between the services. It can effectively tackle the challenges of the semantic consistency of Web-services composition and keep the service interactions in order. The key advantage is the semantic consistency of Web-services composition can be tested on the concept level and on the system level.

Example for an Order-Handling System Using Our Design and Modeling Methodology

By using the service-oriented analysis, and in light of the functionality analysis of Web-services composition for the order-handling composite system, we recognize the different participants or roles for the different services. For the ordering service of this order-handling system, we recognize the buyer and the supplier. For the shipping service, we recognize the buyer and the shipper, and for the billing service, we recognize the buyer and the billing service provider.

By using the three-point design and modeling methodology for the above-mentioned participants, in order to clarify and handle the semantic consistency between the Web services, we give prominence to an intermediary role between the requesting service and the answering service, and the existence of this intermediary role is dependent on that of the other two parent services. For example, the existence of the agent (seller) is dependent on that of the buyer and the supplier, which is illustrated

Figure 3. Existence dependency of the shipping service for the order-handling system

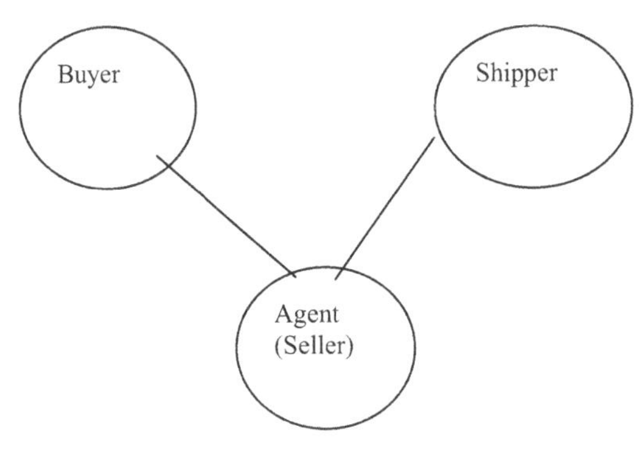

Figure 4. Existence dependency of the billing service for the order-handling system

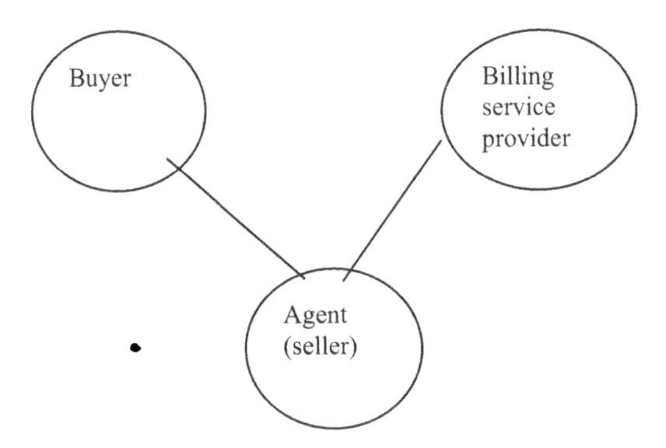

Figure 5. Existence dependency of the message holders for the shipping service of the order-handling system

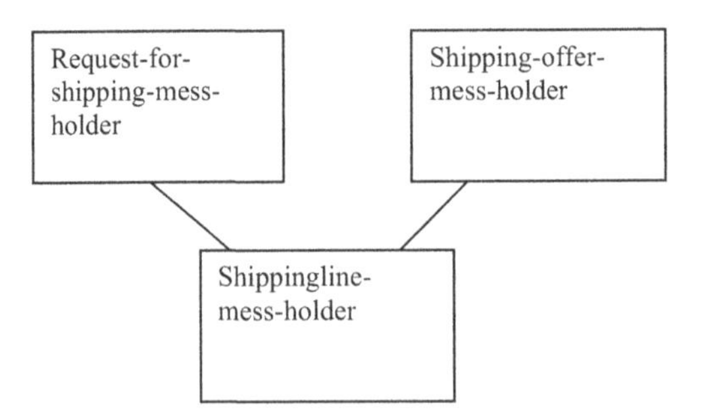

Figure 6. Existence dependency of the message holders for the billing service of the order-handling system

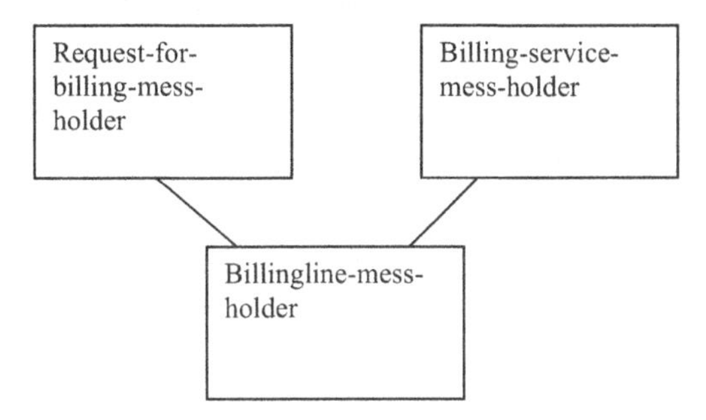

in Figure 1. The existence of the agent (seller) is also dependent on that of the buyer and the shipper, which is illustrated in Figure 3, and additionally dependent on that of the buyer and the billing service provider, as in Figure 4.

By using our three-point design and modeling methodology for message holders, we know the semantic relationship between the Web services is built through message transferring. The existence dependency of the participants or roles (Web services) is external to that of message holders. Therefore, the existence of Orderline-mess-holder is dependent on that of Order-mess-holder and Product-mess-holder in Figure 2. The existence of Shippingline-mess-holder is dependent on that of Request-for-shipping-mess-holder and Shipping-offer-mess-holder, which is illustrated in Figure 5, and the existence of Billingline-mess-holder is dependent on that of Request-for-billing-mess-holder and Billing-service-mess-holder, as illustrated in Figure 6.

Semantic Consistency on the Conceptual Logical Level

Definition 1: Three-Point Principle

When two independent participants or roles (normally one is requesting the service and the other is answering with the service) have to be combined into an advanced system and the semantic relationship in terms of Web-services interactions need to be established in order to fulfill the certain task or achieve a goal, we point out and give prominence to an intermediary role between the requesting service and answering service, which is responsible for handling the semantic consistency among the services. The existence of this intermediary role is dependent on that of the requesting service and answering service.

From the three-point principle, we know the existence of the intermediary role is dependent on that of its parent roles. The semantic relationship among the roles is expressed through message transferring because the content of transferred messages is visible and external in the context of Web-services interactions. Correspondingly, the existence of the intermediary role's message holder is dependent on that of its parent roles' message holders.

Definition 2: Message Dependency

PM is message dependent on OM, and PM is message dependent on QM if and only if the following is true.

1. Let O, P, and Q be three participants (Web services) in the context of Web-services composition that hold three message holders OM, PM, and QM.
2. O, P, and Q are the results when using the three-point service-oriented design and modeling methodology. P is the intermediary role, and O and Q are the parent roles. The existence of P is dependent on that of O and Q.

We call PM a dependent message holder, and OM and QM parent message holders. We use the formal notations PM $\Leftarrow$ OM; PM $\Leftarrow$ QM. (The concept of message dependency is based on the content of the message holders, and it does not refer to the type, structure, or template of the corresponding message.)

Definition 3: The Message-Dependency Rules within Web-Services Interactions

According to Definition 1 (three-point principle) and Definition 2 (message dependency), the existence of the intermediary role is dependent on its parent participants or roles, and the existence of the intermediary role's message holder is dependent on its parent roles' message holders. Therefore, the semantic relationship can be summarized and expressed as follows in terms of message-dependency rules.

1. Within a transaction, if the intermediary role's message holder, which is dependent on the message holders of the other two parent roles, contains a C (create), then on the same row a C must appear in the column of each parent message holder.
2. Within a transaction, if the intermediary role's message holder, which is the dependent on the message holders of the other two parent roles, contains an M (modify), then on the same row an M must appear in the column of each parent message holder.

3. Within a transaction, if the intermediary role's message holder, which is the dependent on the message holders of the other two parent roles, contains an F (fulfill or end), then on the same row an F must appear in the column of each parent message holder.

So far, we have answered Question 1 and Question 2 from the beginning of the chapter. Indeed, there exists a kind of consistent semantic relationship within Web-services interactions. We call this kind of semantic consistency the second layer of semantic consistency within Web-services interactions. In addition, there is semantic consistency in terms of uniform data interpretation. We call uniform data interpretation the first layer of semantic consistency within Web-services interactions. The content of the second layer is defined as message-dependency rules in the context of Web-services interactions by the three-point service-oriented design and modeling methodology.

In the following section, we will address and answer Question 3. First, we would like to address the semantic consistency within Web-services interactions on the logic level.

Definition 4: Holder Function, Holder (T2, T1, and Value)

We define the kind of message (T2), the type of transferring document (T1), and the value in a message holder as the holder function, which has three kinds of messages (C, M, F) and a type and value. This holder function reflects the state of a message holder. For example, Holder(Order-mess-holder) = M: Order(a to b), which means the sate of Order-mess-holder is modified, and the state of the order is changed from a to b.

The Standard Schemes for the Ordering Service of an Order-Handling System (One Supplier)

Scheme A: One buyer wants to book goods and sends an order message. This piece of information is passed on to the supplier by the agent (seller). If it is accepted by the supplier, then this message is sent back to the agent and is confirmed. Here, the order message can be created by message-dependency rules.

Conditions: Orderline-mess-holder $\Leftarrow$ Order-mess-holder and

Orderline-mess-holder $\Leftarrow$ Product-mess-holder and

Match: Holder(Product-mess-holder) = C(create) and

Holder(Online-mess-holder) = C(create)

Under the message-dependency rule 1,

Holder(Order-mess-holder) = C(create).

Scheme B: This buyer would like to change the content of the booking and sends the modified message. This piece of information is passed to the supplier by the agent. If it is accepted by the supplier, then this message is sent back to the gent and is confirmed. Here, the order message can be modified by information-dependency rules.

Conditions: Orderline-mess-holder ⇐ Order-mess-holder and

Orderline-mess-holder ⇐ Product-mess-holder and

Match: Holder(Product-mess-holder) = M(modify) and

Holder(Orderline-mess-holder)= M(modify)

Under the message-dependency rule 2,

Holder(Order-mess-holder) = M(modify).

Scheme C: The supplier fulfills the contract and this piece of information is passed to the buyer by the agent. If it is accepted the buyer, then this message is sent back to the agent and is confirmed. Here, the product message can be fulfilled by the supplier.

Conditions: Orderline-mess-holder ⇐ Order-mess-holder and

Orderline-mess-holder ⇐ Product-mess-holder and

Match: Holder(Order-mess-holder) = F(fulfill) and

Holder(Orderline-mess-holder) = F(fulfill)

Under the message-dependency rule 3,

Holder(Order-mess-holder) = F(fulfill).

Lemmas of the Message-Dependency Rules

Lemma 1

Within a transaction, if the intermediary role's message holder, which is dependent on the message holders of the other two parent roles, does not contain a C, then although there is one parent message holder that contains a C, the other parent message holder does not contain a C.

Lemma 2

Within a transaction, if the intermediary role's message holder, which is dependent on the message holders of the other two parent roles, does not contain an M, then although there is one parent message holder that contains an M, the other parent message holder does not contain an M.

Lemma 3

Within a transaction, if the intermediary role's message holder, which is dependent on the message holders of the other two parent roles, does not contain an F, then although there is one parent message holder that contains an F, the other parent message holder does not contain an F.

Proof Sketch

- **Approach 1:** The proof relies on the definition of the three-point principle and the definition of message dependency.
- **Approach 2:** It can apply the reduction to absurdity.

More Schemes for the Ordering Service of an Order-Handling System of One Supplier

Scheme D: One buyer would like to book goods and sends the order message. This piece of information is passed to the supplier by the agent. If it is not accepted by the supplier (due to type mismatch, limited quantity, etc.), then this message is sent back to the agent and is confirmed by the agent. Accordingly, the order message is not created in Order-mess-holder by the buyer.

Conditions: Orderline-mess-holder $\Leftarrow$ Order-mess-holder and

Orderline-mess-holder $\Leftarrow$ Product-mess-holder and

Match: Holder(Product-mess-holder) $\neq$ C(create) and

Holder(Online-mess-holder) $\neq$ C(create)

Under the message-dependency lemma 1,

Holder(Order-mess-holder) ≠ C(create).

Scheme E: This buyer would like to change the content of the booking and sends the modified message. This piece of information is passed to the supplier by the agent. If it is not accepted by the supplier, then this message is sent back to the agent and is confirmed. Accordingly, the order message is not modified in Order-mess-holder by the buyer.

Conditions: Orderline-mess-holder ⇐ Order-mess-holder and

Orderline-mess-holder ⇐Product-mess-holder and

Match: Holder(Product-mess-holder) ≠ M(modify) and

Holder(Orderline-mess-holder) ≠ M(modify)

Under the message-dependency lemma 2,

Holder(Order-mess-holder) ≠ M(modify).

Scheme F: The supplier thinks that the contract has been fulfilled. This piece of information is passed to the buyer by the agent. If it is not accepted by the buyer, then this message is sent back to the agent and is confirmed. Then the product message is not ended by the supplier.

Conditions: Orderline-mess-holder ⇐ Order-mess-holder and

Orderline-mess-holder ⇐ Product-mess-holder and

Match: Holder(Order-mess-holder) ≠ F(fulfill) and

Holder(Orderline-mess-holder) ≠ F(fulfill)

Under the message-dependency lemma 3,

Holder(Product-mess-holder) ≠ F(fulfill).

In summary, if one parent role with one parent message holder would like to take one of three activities (C, M, F), its activity needs to get the agreement of the other parent role with another parent message holder, which then needs to be confirmed by the intermediary role with the dependent message holder. Here we define this principle as the three-point cooperative principle.

Definition 5: Three-Point Cooperative Principle

According to Lemmas 1, 2, and 3 of the message-dependency rules, we recognize the three-point cooperative principle.

If one parent role wants to take one of the three activities (C, M, F), its activity needs to obtain the agreement of the other parent role and be confirmed by the intermediary role.

So far, we have clarified the semantic consistency within Web-services interactions on the conceptual logic level. In the remainder of the chapter, we would like to address the semantic consistency within Web-services interactions on the conceptual system level.

The Semantic Consistency on the Conceptual System Level

Previously we demonstrated the semantic consistency of Web-services interactions and composition on the conceptual logic level. In the remainder of chapter, we would like to build up a formal model using the formal techniques, and then formally define the semantic consistency of Web-services composition on the system level. In this way, the correctness of the Web-services composite system can be checked and tested. Also, by providing a formal theoretical framework, it will improve the ability of reasoning within the formal system.

Service Model

Basically, we can abstractly model a Web service as an entity that has two pieces: an interface and back-end application implementation. The interface takes interservice messaging and dispatches incoming messages to the back-end application implementation. The back-end application implementation contains the application-specific behavior of a service. The boundary between the back end and the interface in a given service allows the application implementation to send and receive messages through the interface, and it allows the back end to instantiate interactions and composition definitions to specify the legal sequences of such send and receive actions.

A send action may either commit or abort. That is, when a back end tries to send a message as part of a conversation, the interface may either abort or commit the sending. The sending of a message and the signaling of the outcome of the message are asynchronous actions. That is, to send a message, the back end simply deposits the message in the interface. The send action then remains tentative until the interface subsequently signals that the sending has either been committed or aborted. This asynchronous semantic for message-based communication preserves the loose coupling of services.

We model the behavior of a service in terms of the send, receive, abort, and commit events that occur between the interface and back end of the service.

Since we are interested in providing a theoretical foundation of observable conversation between services, we first define the notion of observable actions.

Definition 6

A basic action is a member of the set {A-s(v), A-r(v), A-c(v), A-a(v)}, where

A-s(v) represents the action that service A sends the value v,

A-r(v) represents the action that service A receives the value v,

A-c(v) represents the action that service A commits the send of value v, and

A-a(v) represents the action that service A aborts the send of value v.

Note that the abort and commit actions are status messages pertaining to send actions. Here the values that are communicated between services are unconstrained. They could be XML (extensible markup language) documents, integers, and so forth.

Definition 7

The set of valid interaction histories H is defined as follows.

1. $\lambda \in$ H (λ is the history with no actions in it).
2. For any basic action e and the interaction history h $\in$ H, e.h $\in$ H.

An action e is said to occur in an interaction history h if h = h_1.e. h_2 for some h_1 and h_2. We write e$\in$h if action e occurs in interaction history h. Interaction histories with at least one action in them are said to be nonempty.

Definition 8

A basic action e_1 is said to occur before an action e_2 in an interaction history h if there exist interaction histories h_1, h_2, and h_3 such that h = $h_1 e_1 h_2 h_3$. We write $e_1 \langle e_2$ if e_1 occurs before e_2 in h.

Definition 9

Given a basic action of the form A-x(v), where $x \in \{s, r, a, c\}$, the service A is said to be the principal role of the action.

Definition 10

Given two services A and B, we define the relevant interaction history of actions between them as a string of basic actions in which the principal roles of the actions are the service A and service B.

Formal Expression of Service Interactions

We introduce a formal expression to specify two-party service interactions.

Abstract Syntax

Definition 11: The Formal Expression to Specify Two-Party Service Interactions

$$I ::= A \rightarrow B: T2 \quad (1)$$

$$T2 ::=(C: T1(v)) \cup (M: T1(v1 \text{ to } v2)) \cup (F: T1(v)) \quad (2)$$

A conversation is an exchange of typed documents over time between two services. A conversation consists of a number of interactions, including validating the value and type of the message holder. Each interaction has a sender and receiver. An interaction between the services A and B is specified as $A \rightarrow B$: T2, where T2 is one of three interactions, which includes C, M, and F. T1 is the type of the document being passed from A to B, and v is the value of a document being passed from A to B. For M, we define one value (v1) as modified to another value (v2).

Definition 11-A

$$I ::= A \rightarrow B: QT2 \quad (3)$$

$$QT2 ::=(QC: T1(v)) \cup (QM: T1(V1 \text{ to } V2)) \cup (QF:T1(V))$$

Within one life cycle of one transaction, if one canceling occurs at one of the three phrases (C, M, F), it can be expressed by equation 3.

Here we call the semantic consistency of T1 the first layer of semantic consistency within the Web-services conversation, and the semantic consistency of T2 the second layer of semantic consistency.

To define such a conversation, we rely on a set of services (A, B $\in$ Service) and a set of document types (T1 $\in$ Type). We do not explicitly define the type of the system associated with the messages exchanged between documents, but it encompasses the type of the system for XML.

Formal Semantics on Services Interactions

We can define the semantics of a conversation as the set of typed interaction histories that can arise when executing the conversation between two Web services. We map a conversation to a set of interaction histories as a two-step process: We map a conversation to an intermediate representation, called a typed interaction history, and we then define a conformance relation between interaction histories and typed interaction histories. A typed interaction history is a typed trace. That is, the elements of a typed interaction history include the types of information that may be exchanged in a conversation. We map a given conversation to a set of typed interaction histories, and the value and type of the message container within the conversation. We then define a notion of conformance between interaction histories and typed interaction histories.

Definition 12

A typed interaction history is a trace with the type of information for the observable actions, where each typed action has the following format.

- A typed action E is a member of the set {A-S(T2:T1(v)), A-R(T2:T1(v)), A-A(T2:T1(v), A-C(T2:T1(v))}, where
- A-S(T2:T1(v)) represents the action that service A sends a value v that matches or validates against the templates of T2 and T1,
- A-R(T2:T1(v)) represents the action that service A receives a value v that matches or validates against the templates of T2 and T1,
- A-A(T2:T1(v)) represents the action that service A aborts a send of a value v that matches or validates against the template of T2 and T1, and
- A-C(T2:T1(v)) represents the action that service A commits a send of a value v that matches or validates against the template of T2 and T1.

Definition 13

The set of the valid typed interaction history T is defined as follows.

1. $\lambda \in T$
2. For any typed action E and typed interaction history $t \in T$, $E.t \in T$.

A typed action E is said to occur in a typed interaction history t if $t = t_1.E.t_2$. We write $E \in t$ if typed action E occurs in typed interaction history t.

In essence, a typed interaction history is like an interaction history except for the fact that it has templates of documents or types of the values associated with the actions as opposed to the actual values themselves.

Definition 14

Given an interaction i of the form A → B: T2: T1(v), the interaction semantics of i is the set {A-S(T2:T1(v)).A-C(T2:T1(v)).B-R(T2: T1(v))}. We write $i \Rightarrow S$ if S is the semantics of the interaction i.

This typed interaction history captures the fact that one of the end points of the interaction did a successful send followed by a commit, while the other end of the interaction did a successful receive.

Definition 15

Given two sets of typed interaction histories T_1 and T_2, the concatenation T of T_1 and T_2 (written as $T = T_1.T_2$) is defined as a set of sequences whose elements are made by concatenating any elements of the first set with any element of the second set.

Definition 16

Given a conversation definition of the form $c = i_1;c'$, where i_1 is a simple interaction of the form A → B: T2: T1(v) and c' is a conversation fragment, c' can be the interaction fragment or the evaluation of a holder function, or it can be both the interaction fragment and the evaluation of a holder function. Suppose further that $i_1 \Rightarrow S_{i1}$ and $c' \Rightarrow S_{c'}$. The set S_c such that $c \Rightarrow S_c$ is defined as $S_c = S_{i1} \cdot S_{c'}$.

The Formal Semantics of the Standard Schemes

In terms of the standard schemes for the ordering service of an order-handing system (one supplier) and Definitions 13 and 15, we would like to address the formal semantics of the standard schemes on the conceptual system level by taking the example of the ordering service of an order-handing system. In this chapter, the definitions of schemes have general meanings and are not limited to a concrete example.

The Formal Semantics of Scheme A

For Scheme A, the conversation fragment including records of matching message holders and received messages is:

Holder(Order-mess-holder) = C: Order(a);
Buyer → Agent: C: Order(a);
Holder(Orderline-mess-holder) = C: Order(a);
Agent → Supplier: C: Order(a).
Evaluating Holder(Product-Mess-Holder)
Holder(Product-mess-holder) = C: Order(a);
Supplier → Agent: C: Order(a);
Holder(Orderline-mess-holder) = C: Order(a);
Agent → Buyer: C: Order(a);
Holder(Order-mess-holder) = C: Order(a).

Correspondingly, the semantics of the conversation with the interaction semantics are the following.

Holder(Order-mess-holder) = C: Order(a).
{Buyer-S(C: Order(a)).Buyer-C(C: Order(a)).Agent-R(C: Order(a))}.
Holder(Orderline-mess-holder) = C: Order(a).
{Agent-S(C: Order(a)).Agent-C(C: Order(a)).Supplier-R(C: Order(a))}.
Evaluating Holder(Product-Mess-Holder)
Holder(Product-mess-holder) = C: Order(a).
{Supplier-S(C: Order(a)).Supplier-C(C: Order(a)).Agent-R(C: Order(a))}.

Holder(Orderline-mess-holder) = C: Order(a).

{Agent-S(C: Order(a)).Agent-C(C: Order(a)).Buyer-R(C: Order(a))}.

Holder(Order-mess-holder) = C: Order(a).

The semantics of Scheme B and Scheme C can be represented as those of Scheme A.

Definition 17

We can define Scheme A, Scheme B, and Scheme C as positive (denoted with *) schemes since they are all successful. Correspondingly, their formal semantics can be expressed in the following:

$$\text{st (Scheme A)} = (* \, s_C)$$
$$\text{st (Scheme B)} = (* \, s_M)$$
$$\text{st (Scheme C)} = (* \, s_F).$$

The Formal Semantics of Scheme D

For Scheme D, the conversation fragment including records of matching message holders and received messages is:

Holder(Order-mess-holder) = C: Order(a);

Buyer → Agent: C: Order(a);

Holder(Orderline-mess-holder) = C: Order(a);

Agent → Supplier: C: Order(a).

Evaluating Holder(Product-Mess-Holder)

Holder(Product-mess-holder) ≠ C: Order(a);

Due to Definition 5,

Holder(Orderline-mess-holder) ≠ C: Order(a);

Holder(Order-mess-holder) ≠ C: Order(a).

In the above conversation fragment, the three-point cooperative principle, which is deducted from the lemmas of the message-dependency rules, can act as the reasoning engine within the conversation between the services; that is, if one parent role wants to take one of three activities (C, M, F), its activity needs to obtain the agreement of the other parent role and be confirmed by the intermediary role.

Correspondingly, the semantics of the conversation with the interaction semantics are the following.

Holder(Order-mess-holder) = C: Order(a).

{Buyer-S(C: Order(a)).Buyer-C(C: Order(a)).Agent-R(C: Order(a))}.

Holder(Orderline-mess-holder) = C: Order(a).

{Agent-S(C: Order(a)).Agent-C(C: Order(a)).Supplier-R(C: Order(a))}.

Evaluating Holder(Product-Mess-Holder)

Holder(Product-mess-holder) ≠ C: Order(a).

Due to Definition 5,

Holder(Orderline-mess-holder) ≠ C: Order(a);

Holder(Order-mess-holder) ≠ C: Order(a).

The semantics of Scheme E and Scheme F can be represented as those of Scheme D.

Definition 18:

We can define Scheme D, Scheme E, and Scheme F as negative (denoted with !) schemes since they are not successful. Correspondingly, their formal semantics can be expressed in the following:

st (Scheme D) = $(!s_C)$

st (Scheme E) = $(!s_M)$

st (Scheme F) = $(!s_F)$.

The Schemes of Canceling an Order in the Order-Handling System of One Supplier.

In the following, we would like to address the situation of canceling.

Scheme G: One buyer would like to book goods and sends the order message. This piece of information is passed to the supplier by the agent. At the same time the agent is preparing to send this message to supplier (denoted as ||), the buyer wants to cancel the order and sends the canceling information to the agent. We assume the buyer has priority over the supplier, denoted as Buyer ∠ Supplier.

The Formal Semantics of Scheme G

The conversation fragment including records of matching message holders and received messages is

(Buyer ∠ Supplier)

Holder(Order-mess-holder) = C: Order(a);

Buyer → Agent: C: Order(a);

Holder(Orderline-mess-holder) = C: Order (a);

Agent → Supplier: C: Order(a) || Buyer → Agent: QC: Order(a).

When evaluating Holder(Product-mess-holder),

Holder(Product-mess-holder) ≠ C: Order(a).

Due to Definition 5,

Holder(Orderline-mess-holder) ≠ C: Order(a);

Holder(Order-mess-holder) ≠ C: Order(a).

Correspondingly, the semantics of the conversation with the interaction semantics are the following.

Holder(Order-mess-holder) = C: Order(a).

{Buyer-S(C: Order(a)).Buyer-C(C: Order(a)).Agent-R(C: Order(a))}.

Holder(Orderline-mess-holder) = C: Order(a).

{Agent-S(C: Order(a)).[Buyer-S(QC: Order(a)).Buyer-C(QC:Order(a)).

Agent-R(QC:Order(a))].Agent-A(C:Order(a))}.

Evaluating Holder(Product-Mess-Holder)

Holder(Product-mess-holder) ≠ C: Order(a).

Due to Definition 5,

Holder(Orderline-mess-holder) ≠ C: Order(a);

Holder(Order-mess-holder) ≠ C: Order(a).

Scheme H: The buyer would like to change the content of the booking and sends the modified message. This piece of information is passed to the supplier by the agent. At the same time the agent is preparing to send this message to supplier, the buyer wants to cancel this change and sends the canceling information to the agent. We assume the buyer has priority over the supplier.

Scheme I: The Supplier fulfills the contract. This piece of information is passed to the buyer by the agent. At the same time, the buyer cancels the order (we assume the buyer has priority over the supplier). Therefore, the order is not fulfilled by the supplier. (Within one transaction, this canceling may occur in one of the three phases C, M, or F.)

The semantics of Scheme H and Scheme I can be represented as those of Scheme G.

Definition 19

We can define Scheme G, Scheme H, and Scheme I as negative schemes with canceling (denoted with #). Correspondingly, their formal semantics can be expressed in the following:

$$\text{st (Scheme G)} = (\# S_C)$$
$$\text{st (Scheme H)} = (\# S_M)$$
$$\text{st (Scheme I)} = (\# S_F)\,.$$

Table 1. Formal semantics of the ordering of an order-handling system of one supplier

Phase	Scheme	Positive or Negative	Semantics Expression
Creating	A	Positive	$(* S_C)$
Creating	D	Negative	$(! S_C)$
Creating	G	Negative with Canceling	$(\# S_C)$
Modifying	B	Positive	$(* S_M)$
Modifying	E	Negative	$(! S_M)$
Modifying	H	Negative with Canceling	$(\# S_M)$
Fulfilling	C	Positive	$(* S_F)$
Fulfilling	F	Negative	$(! S_F)$
Fulfilling	I	Negative with Canceling	$(\# S_F)$

Definition 20

Because the semantics of the canceling means there is no possibility of success, the parts of Scheme G, Scheme H, and Scheme I are the same as those of Scheme D, Scheme E, and Scheme F.

Correspondingly, the semantics of Scheme D from evaluating Holder(Product-mess-holder) are the same as those of Scheme G. The semantics of Scheme E from evaluating Holder(Product-mess-holder) are the same as those of Scheme H, and the semantics of Scheme F from evaluating Holder(Order-mess-holder) are the same as those of Scheme I.

In summary, the formal semantics of the ordering of an order-handling system of one supplier can be summarized in the following table.

Definition 21

For any set of semantics of conversation S and finite integer i, S^i = S.S.S… *i* times.

Definition 22

In terms of the positive schemes, the successful standard semantics specification (denoted as SP) of the system within one life cycle of one transaction can be expressed as follows:

- $SP = (*S_c)^1 \cdot (*S_M) i \cdot (*S_F)^1$,

where i is a finite integer.

Formal History Conformance

Local History Conformance to Schemes

In Definition 6 and Definition 7, we formally defined the valid interaction history. In Definition 12 and Definition 13, we formally defined the valid typed interaction history. We also formally defined local typed traces, which we call the formal semantics of schemes (in terms of the ordering service of an order-handling system of one supplier). In the following, we first define the local history, and then we define the local history conformance to the formal semantics of schemes.

Definition 23: Local History

In terms of our three-point service-oriented design and modeling methodology for Web-services composition, on the conceptual system level, given two services A and B and their message holders, we define the local history (denoted as lh) to be a set of history scripts, which includes all the values of the message holder function, the valid interaction history, the evaluating message holder, and the reasoning scripts from Service A to Service B.

Definition 24

For any set of local history lh and the finite integer i, $(lh)^i = (lh).(lh).(lh)\ldots$ *i* times.

Definition 25: The Conformance Relation Between the Local History and the Formal Semantics of Schemes (Scheme A...Scheme I)

In the following, we define a conformance relation between the local history and the formal semantics of the schemes (Scheme A…Scheme I). Given a local history and a scheme, if lh $\approx$ st(Scheme A…Scheme I), we say lh conforms to st (Scheme A...Scheme I). Next we would like to define them in detail in terms of the individual schemes.

The conformance relation has two aspects. There is a structure aspect, requiring that the sequences agree between a local history and the formal semantics of a scheme, which includes the sequence of interactions, the matching or validating of the holder value, the evaluation of the holder, and the reasoning results. There is also a value aspect, which requires that the data exchanged as part of the history match or validate against the templates in the formal semantics of the schemes.

Definition 26: Well-Formed Local History

A local history is said to be well formed if it conforms to the formal semantics of one of the schemes (Scheme A…Scheme I).

Successful History of One Life Cycle of One Transaction

Definition 27: Successful History of One Life Cycle of One Transaction

In Definition 22, we defined the successful standard semantics specification of a system within one life cycle of one transaction. Here we define the successful history of one life cycle of one transaction. A history (denoted as His) of one life cycle of one transaction is said to be successful if one of the following conditions holds.

A: If there is no modification within one life cycle of one transaction, then

$$[lh_1 \approx (*\, S_C)]^1 . [lh_2 \approx (*\, S_F)]^1 . \quad (4)$$

B: If there are i ($i \geq 1$) modifications within one life cycle of one transaction, then

$$[lh_1 \approx (*\, S_C)]^1 . [lh_{2 \ldots (1+i)} \approx (*\, S_M)]^i . [lh_{(2+i)} \approx (*\, S_F)]^1 , \quad (5)$$

where i ($i \geq 1$) is a finite integer.

According to Definition 17, Definition 18, and Definition 19, a successful history of one life cycle of one transaction can be expressed in one of the following expressions as well.

A: If there is no modification within one life cycle of one transaction, then

$$[lh_1 \approx \text{st (Scheme A)}]^1 . [lh_2 \approx \text{st (Scheme C)}]^1 . \quad (6)$$

B: If there are i ($i \geq 1$) modifications within one life cycle of one transaction, then

$$[lh_1 \approx \text{st (Scheme A)}]^1 . [lh_{2 \ldots (1+i)} \approx \text{st (Scheme B)}]^i . [lh_{(2+i)} \approx \text{st (Scheme C)}]^1 , \quad (7)$$

where i ($i \geq 1$) is a finite integer.

System History Conformance

Definition 28: Well-Formed History of One Life Cycle of One Transaction on the System Level

We are interested in defining a well-formed history of one life cycle of one transaction on the system level because if we have this definition, we can avert the non-well-formed system behaviors that can be tested and found on the system level for Web-services composition.

A history of one life cycle of one transaction is said to be well formed if one of the following conditions holds.

a: $[lh_1 \approx (!s_C)]^1$ (8)

b: $[lh_1 \approx (\#s_C)]^1$ (9)

c: $[lh_1 \approx (*s_C)]^1.[lh_2 \approx (!s_M)]^1$ (10)

d: $[lh_1 \approx (*s_C)]^1.[lh_2 \approx (\#s_M)]^1$ (11)

e: $[lh_1 \approx (*s_C)]^1.[lh_{2...(1+i)} \approx (*s_M)]^i.[lh_{(2+i)} \approx (!s_F)]^1$, (12)

where i ($i \geq 1$) is a finite integer.

f: $[lh_1 \approx (*s_C)]^1.[lh_{2...(1+i)} \approx (*s_M)]^i.[lh_{(2+i)} \approx (\#s_F)]^1$, (13)

where i ($i \geq 1$) is a finite integer.

g: $[lh_1 \approx (*s_C)]^1.[lh_2 \approx (*s_F)]^1$ (14)

h: $[lh_1 \approx (*s_C)]^1.[lh_{2...(1+i)} \approx (*s_M)]^i.[lh_{(2+i)} \approx (*s_F)]^1$, (15)

where i ($i \geq 1$) is a finite integer.

In terms of our three-point service-oriented design and modeling methodology for Web-services composition, given the fact that there is a service buyer and service supplier, the intermediary service is the service agent.

When equation 8 happens within one life cycle of one transaction, the service buyer does not successfully create an order.

When equation 9 happens within one life cycle of one transaction, the service buyer first successfully creates an order, but later cancels the order.

When equation 10 happens within one life cycle of one transaction, the service buyer first successfully creates an order, but later does not successfully modify the order.

When equation 11 happens within one life cycle of one transaction, the service buyer first successfully creates an order, but later cancels the Order during the modifying phase.

When equation 12 happens within one life cycle of one transaction, the service buyer successfully creates and modifies an order i ($i \geq 1$) times, but the service supplier does not fulfill the order in the end.

When equation 13 happens within one life cycle of one transaction, the service buyer successfully creates and modifies an order i ($i \geq 1$) times, but later cancels the order during the fulfillment phase under the condition that the service buyer has priority over the service supplier.

When equation 14 happens within one life cycle of one transaction, the service buyer and service supplier successfully complete one life cycle of one transaction without modifying the order.

When equation 15 happens within one life cycle of one transaction, the service buyer and service supplier successfully complete one life cycle of one transaction and modify the order i ($i \geq 1$) times.

According to Definition 17, Definition 18, and Definition 19, a well-formed history of one life cycle of one transaction can be expressed as well in one of the following expressions:

a: $$[lh_1 \approx \text{st (Scheme D)}]^1 \quad (16)$$

b: $$[lh_1 \approx \text{st (Scheme G)}]^1 \quad (17)$$

c: $$[lh_1 \approx \text{st (Scheme A)}]^1.[lh_2 \approx \text{st (Scheme E)}]^1 \quad (18)$$

d: $$[lh_1 \approx \text{st (Scheme A)}]^1.[lh_2 \approx \text{st (Scheme H)}]^1 \quad (19)$$

e: $$[lh_1 \approx \text{st (Scheme A)}]^1.\ [lh_{2...(1+i)} \approx \text{st (Scheme B)}]_i.[lh_{(2+i)} \approx \text{st (Scheme F)}]^1, \quad (20)$$

where i ($i \geq 1$) is a finite integer,

f: $$[lh_1 \approx \text{st (Scheme A)}]^1.\ [lh_{2...(1+i)} \approx \text{st (Scheme B)}]^i.[lh_{(2+i)} \approx \text{st (Scheme I)}]^1, \quad (21)$$

where i ($i \geq 1$) is a finite integer,

g: $[lh_1 \approx \text{st (Scheme A)}]^1 . [lh_2 \approx \text{st (Scheme C)}]^1$ (22)

h: $[lh_1 \approx \text{st (Scheme A)}]^1 . [lh_{2...(1+i)}\ \text{st (Scheme B)}]^i . [lh_{(2+i)} \approx \text{st (Scheme C)}]^1$, (23)

where i ($i \geq 1$) is a finite integer.

The Key Results of Formal History Conformance

Theorem 1

There exists a decision procedure *pro* that, given a local history *lh* and the formal semantics of one of the schemes *st* (Scheme A…Scheme I), decides whether $lh \approx st$ (Scheme A…Scheme I). (Our theory is on the conceptual system level and not on the concrete system level between services, which means we are not considering concrete system problems such as delays of date exchanging, delays of matching and validating against the templates of schemes, delays of evaluating the holder, delays of reasoning, etc.)

Proof Sketch

This can be proven by a straightforward induction on the length of *lh* or by constructing the decision procedure.

Theorem 2

Given the formal semantics of one of the schemes *st* (Scheme A…Scheme I) and a local history *lh* such that $\neq [lh \approx st$(Scheme A…Scheme I], there is a deterministic procedure to determine the first mismatch between them that causes $\neq [lh \approx st$(Scheme A…Scheme I].

This property essentially ensures that there is a way to determine the cause of the deviation of a local history from the formal semantics of one of the schemes.

Essentially, these two theorems tell us that it is possible to decide whether a local history conforms to the formal semantics of one of the schemes. In addition, if a local history does not conform to the formal semantics of one of the schemes, we can determine the first cause of nonconformance. These properties therefore provide a basis of managing and testing Web-services interactions.

Related Works

There are many emerging Web-services standards that are relevant for our work. We enumerate some of major work and briefly describe the relationship with our work.

- BPEL4WS: BPEL4WS (*Business Process Execution Language for Web Services Version 1.1*, 2003) has emerged as a proposal for describing business-process execution. BPEL4WS is built on top of the Web-services architecture, and it is a richer, expressive services-composition language in which business protocols can be expressed. Because it is a machine-executable language and is implementation oriented, it does not explicitly address the composition logic and lacks a clear analysis of the relationships between services and the relationships between messages. Our work makes up for its insufficiency from the high level, providing a set of sound analysis, design, and modeling methodologics for Web-services composition and making clear the semantic consistency within Web-services interactions on the concept level and system level. Its results are significant and profound, and they can be the solid theoretical foundation of the BPEL4WS' further evolution.
- WSCL: The work on WSCL (*Web Services Conversation Language (WSCL) 1.0*, 2002) proposes a conversation language for Web-service protocols and frameworks. It focuses on modeling the sequencing of the interactions or operations of one interface and fills the gap between mere interface definition languages that do not specify any choreography and more complex process or flow languages that describe complex, global, multiparty conversations and processes. Also, its conversation definitions are themselves XML documents and can be interpreted by Web-service development tools. However, its work lacks the semantic analysis, design, and modeling on the semantic consistency of Web-services conversation on the logic level and on the system level, and clear and accurate formal constructs. Therefore, our work can provide a good complement for its insufficiency in semantic consistency within Web-services conversations, and at the same time, provides the formal underpinnings for Web-services conversation and Web-services composition.
- MERODE (Lemahieu, Snoeck, Michiels, & Goethals, 2003; Lemahieu, Snoeck, Michiels, Goethals, Guido, & Vandenbulcke, 2003; Snoeck & Dedene, 1998;

Snoeck, Dedene, et al., 1999; Snoeck et al., 1999): Our work has some connection with the research results in MERODE. MERODE is the object-oriented analysis and design methodology for enterprise modeling. Its first-class entities are object types and event types. It requires that all relationships express existence dependency in object types, thus this methodology provides the key to semantic integrity among object types through presenting the existence dependency in object types. Our work caters to the need of new Web-services computing and regards Web services as first-class entities. Our work enhances the concepts of semantic consistency in MERODE in the Web-services world, whose improvement is profound and significant.

- Other research: Bultan, Fu, Hull, and Su (2003) and Hull, Benedikt, Christophides, and Su (2003) have proposed a mealy-machine-based formalism to express the behavior of Web services. A mealy machine can express the behavior of a Web service as sequences of interactions, where the sequences are constructed by sequential composition and nondeterministic choice. The research results in their works provide a good reference for the formal model of Web-services composition. Although this mealy-machine formalism provides the description about the internal behavior of a service, it lacks the direct analysis, discussion, and expression on the semantic consistency of Web-services interactions and composition.

Conclusion

In order to tackle one of the big inhibitors of Web-services adoption and acceptance—the lack of semantic consistency in business processes within Web-services interactions and composition—in this chapter we proposed a novel analysis, design, and modeling methodology for Web-services composition: a three-point service-oriented methodology based on some object-oriented research results in MERODE. On the concept level, we first clarified the research issue of semantic consistency within Web-services interactions and Web-services composition, and at the same time, we precisely defined and demonstrated the semantic consistency of Web-services composition on the logic level and system level through demonstrating the standard schemes of an order-handling system. Therefore, we made major contributions to the research area of Web-services composition.

In the future, there are some possible directions for our further research. For instance, we can extend our work both in practical and theoretical directions. On one side, we can develop a prototype system that implements the methodology for Web-services composition presented in the chapter. Such a system will enable us to implement the composition techniques for Web-services composition and test whether the composition and interactions for Web services are reasonable and correct. On the theoretical side, we can extend and address some open issues such as how to handle the situation in which there are several buyers and suppliers at the same time. Also, we can further address the coordination of more than two Web services. This issue can be considered through message transferring, and concrete rules (for example, if-then statements) can be defined by the system composer.

References

B-business XML (ebXML). (2004). Retrieved from http://www.ebxml.org

Benatallah, B., Dumas, M., Sheng, Q. Z., & Ngu, A. H. H. (2002). Declarative composition and peer-to-peer provisioning of dynamic Web services. In *Proceedings of the International Conference on Data Engineering*, San Jose, CA.

Business process execution language for Web services version 1.1. (2003). Retrieved from http://www-106.ibm.com/developerworks/library/wsbpel/

Bultan, T., Fu, X., Hull, R., & Su, J. (2003). Conversation specification: A new approach to design and analysis of e-service composition. In *Proceedings of the World Wide Web Conference.*

Bussler, C. (2001). The role of B2B protocols in inter-enterprise process execution. In *Proceedings of the Second VLDS-TES Workshop*, Rome.

Casati, F., Sayal, M., & Shan, M.-C. (n.d.). *Developing e-services for composing e-services* (Tech. Rep.). HP Lab.

Casati, F., & Shan, M.-C. (2001). Dynamic and adaptive composition of e-service. *Information Systems, 26*, 143-163.

Curbera, F., et al. (2003). *Business process execution language for Web services* (Version 1.1). Retrieved from http://www-106.ibm.com/developerworks/webservices/library/wsbpel/

Florescu, D., & Kossmann, D. (2001). An XML programming language for Web services specification and composition. *Bulletin of the IEEE Computer Society Technical Committee on Data Engineering.*

Frolund, S., & Govindarajon, K. (2003). *CL: A language for formally defining Web services interactions.* Hewlett-Packard Company.

Gao, X. (2003). *Web services package reuse and dynamic composition.* Proceedings of the Seventh World Multiconference on Systemises, Cybernetics and Informatics, Orlando, FL.

Gao, X., Yang, J., & Papazoglou, M. P. (2002). The capability matching of Web services. In *Proceedings of the Fourth IEEE International Symposium on Multimedia Software Engineering*, Newport Beach, CA.

Heuvel, W. J. Van, Yang, J., & Papazoglou, M. P. (2001). Service representation, discovery, and composition for e-marketplaces.In *Proceedings of the International Conference on Cooperative Information Systems (cooPIS01).*

Hoare, C. A. R. (1978). Communicating sequential processes. *Communications of the ACM, 21*(8), 666-677.

Hull, R., Benedikt, M., Christophides, V., & Su, J. (2003). E-services: A look behind the curtain. In *Proceedings of the ACM Symposium on Principles of Database Systems.*

IBM Alphaworks: Web services outsourcing manager. (2004). Retrieved from http://www.alphaworks.ibm.com/aw.nsf/FAQs/wsom

Kuno, H., Lemon, M., Karp, A., & Beringer, D. (2001). Conversations + interface = business logic. In *Proceedings of the Second VLDS-TES Workshop*, Rome.

Lemahieu, W., Snoeck, M., Michiels, C., & Goethals, F. (2003). An event based approach to Web service design and interaction. In *Proceedings of APWeb'03.*

Lemahieu, W., Snoeck, M., Michiels, C., Goethals, F., Guido, D., & Vandenbulcke, J. (2003). Event based Web service description and coordination. In *Proceedings of WES'03: Web Services, E-Business, and the Semantic Web: Workshop Proceedings of CAiSE 2003*, Klagenfurt, Austria.

Leymann, F. (2004). *Web services flow language (WSFL 1.0).* IBM Software Group. Retrieved from http://www-4.ibm.com/software/solutions/webservices/pdf/WSFL.pdf

Milner, R. (1980). A calculus of communicating systems (LNCS 92). Springer Verlag.

Service framework specification. (2001). Retrieved from http://www.hpl.hp.com/techreports/2001/HPL-2001-138.HTML

Shirky, C. (2002). Web services and context horizons. *Computer*, 93-94.

Snell, J. (2004). *The Web services insider: Part 5. Getting into the flow: Business process modelling with WSFL.* Retrieved from http://www106.ibm.com/developerworks/webservices/

Snoeck, M. (1995). *On a process algebra approach to the construction and analysis of MERODE-based conceptual models.* Unpublished doctoral dissertation, Leuven, Belgium: Katholieke Universiteit Leuven, Faculty of Science & Department of Computer Science.

Snoeck, M. (1999). Separating business process aspects from business object behaviour. In J. Vandenbulcke & M. Snoeck (Eds.), *New directions in software engineering: Liber amicorum maurice verhelst.* Leuven, Belgium: Leuven University Press.

Snoeck, M. (2002). Sequence constraints in business modeling and business progress modeling. In *Proceedings of the Fourth International Conference on Enterprise Information Systems* (pp. 683-690).

Snoeck, M., & Dedene, G. (1998). Existence dependency: The key to semantic integrity between structural and behavioural aspects of object types. *IEEE Transactions on Software Engineering, 24*(24), 233-251.

Snoeck, M., Dedene, G., Verhelst, M., & Depuydt, A. M. (1999). *Object-oriented enterprise modeling with MERODE.* Leuven, Belgium: Leuven University Press.

Snoeck, M., Lemahieu, W., Michiels, C., & Guido, D. (2003). Event-based software architectures. In *Proceedings of the OOIS'03 Conference.* Retrieved from http://www.springer.de/comp/lncs/index.html

Snoeck, M., Poelmans, S., & Dedene, G. (2000). A layered software specification architecture. In A. H. F. Laendler, S. W. Liddle, & V. C. Storey (Eds.), *Conceptual Modeling: ER2000, 19th International Conference on Conceptual Modeling* (LNCS 1920, pp. 454-469). Salt Lake City, UT: Springer.

SOAP version 1.2, part 0: Primer. (2003). Retrieved from http://www.w3.org/TR/soap12-part0/

Tartanoglu, F., Issarny, V., Romanovsky, A., & Levy, N. (2004). *Dependability in the Web service architecture.* Retrieved from http://www.citeseer.com

Thatte, S. (2001). *XLANG: Web services for business process design.* Microsoft Corporation. Retrieved from http://www.gotdotnet.com/team/xml_wsspecs/xlang-c/default.htm

Universal description, discovery and integration (UDDI). (2003). Retrieved from http://www.uddi.org/

Web services conversation language (WSCL) 1.0. (2002). Retrieved from http://www.w3.org/TR/2002/NOTE-wscl10-20020314/

Web services description language (WSDL) version 1.2 part 1: Core language. (2003). Retrieved from http://www.w3.org/TR/2003/WD-wsdl11220030611/

Yang, J., Papazoglou, M. P., & Heuvel, W. J. Van. (2002). *Tackling the challenges of service composition.* ICDE-RIDE Workshop on Engineering E-Commerce/E-Business, San Jose, CA.

Zhang, L.J., Chang, H., & Chao, T. (2002). Web services relationships binding for dynamic e-business integration. *International Conference on Internet Computing (IC'02).*

Section IV

Enterprise Service Computing: Technologies

Chapter IX

Data Replication Strategies in Wide-Area Distributed Systems

Sushant Goel, University of Melbourne, Australia

Rajkumar Buyya, University of Melbourne, Australia

Abstract

Effective data management in today's competitive enterprise environment is an important issue. Data is information, and information is knowledge. Hence, fast and effective access to data is very important. Replication is one widely accepted phenomenon in distributed environments, where data is stored at more than one site for performance and reliability reasons. Applications and architectures of distributed computing have changed drastically during the last decade, and so have replication protocols. Different replication protocols may be suitable for different applications. In this chapter, we present a survey of replication algorithms for different distributed storage and content-management systems including distributed database-management systems, service-oriented data grids, peer-to-peer (P2P) systems, and storage area networks. We discuss the replication algorithms of more recent architectures, data grids and P2P systems, in detail. We briefly discuss replication in storage area networks and on the Internet.

Introduction

Computing infrastructure and network-application technologies have come a long way over the past 20 years and have become more and more detached from the underlying hardware platform on which they run. At the same time, computing technologies have evolved from monolithic to open and then to distributed systems (Foster & Kesselman, 2004). Both scientific and business applications today are generating large amounts of data; typical applications, such as high-energy physics and bioinformatics, will produce petabytes of data per year. In many cases, data may be produced or are required to be accessed or shared at geographically distributed sites. The sharing of data in a distributed environment gives rise to many design issues, for example, access permissions, consistency issues, and security. Thus, effective measures for easy storage and access of such distributed data are necessary (Venugopal, Buyya, & Ramamohanarao, 2005). One of the effective measures to access data effectively in a geographically distributed environment is replication.

Replication is one of the most widely studied phenomena in a distributed environment. Replication is a strategy in which multiple copies of some data are stored at multiple sites (Bernstein, Hadzilacos, & Goodman, 1987). The reason for such a widespread interest is due to the following facts.

1. Increased availability
2. Increased performance
3. Enhanced reliability

By storing the data at more than one site, if a data site fails, a system can operate using replicated data, thus increasing availability and fault tolerance. At the same time, as the data are stored at multiple sites, a request can find the needed data close to the site where the request originated, thus increasing the performance of the system. But the benefits of replication, of course, do not come without overheads of creating, maintaining, and updating the replicas. If the application has a read-only nature, replication can greatly improve the performance. But, if the application needs to process update requests, the benefits of replication can be neutralised to some extent by the overhead of maintaining consistency among multiple replicas, as will be seen in the following sections of the chapter.

A simple example of a replicated environment is shown in Figure 1. Site 1, Site 2, Site 3, ..., and Site *n* are distributed site locations and are connected through a middleware infrastructure (for the time being, it does not matter what the middleware consists of). Data stored in a file, File X, is stored at Site 2 and is replicated at all other sites. Suppose User 1 tries to access File X in Figure 1. For pedagogical simplicity, let the distance shown in the figure be proportional to the access cost

Figure 1. A simple scenario of File X being replicated at all sites

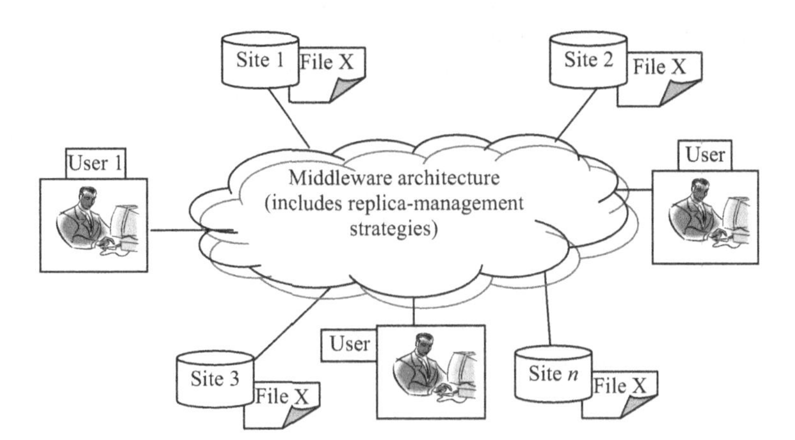

of the file. The above-mentioned benefits of replication are clear in this scenario (as Sites 1 and 3 are close to the user in comparison to Site 2, where the file was originally stored). The files can be accessed at a cheaper cost (thus increasing the performance), and the file can still be accessed even if three out of four sites are down (thus increasing availability).

At the same time, the continuously changing nature of computing has always created new and challenging problems in the replicated environment. The purpose of this chapter is to review the motivation of various replication schemes in different distributed architectures. In particular, we will be covering the following architectures.

1. Distributed database-management systems (DBMSs)
2. Peer-to-peer (P2P) systems
3. Data grids
4. World Wide Web (WWW)

Replication Scenario and Challenges in Replicated Environment

Various combinations of events and access scenarios of data are possible in a distributed replicated environment. For example, an application may want to download chunks of data from different replicated servers for speedy access to data; replicated data may be required to consolidate at a central server on a periodic basis; data may be distributed on a network of servers, where some of the servers may be mobile or frequently connected (http://www.dbmsmag.com/9705d15.html); or data stored at multiple sites may need to be accessed and updated. Based on these requirements, three types of replication scenarios can be identified.

1. Read-only queries
2. Update transactions
3. Mobile-client management

For read-only queries, the data can be accessed by a query without one worrying about the correctness of the data. As is typical, the data may be generated at some site and can be read by other sites. The data can be conveniently stored at different replicated servers.

Contrary to read-only queries, update transactions need special consideration during design time. The replica-management protocol may be simple if only a single site is to update the data. But, as the data can be modified by multiple sites, the consistency of the data may be compromised. To maintain the consistency of data, the order in which the transactions are executed must be maintained. One of the widely accepted correctness criterions in replicated environments is one-copy serializability (1SR; Bernstein et al., 1987). Conflicts can also be resolved with other requirements such as priority-based requirements (a server with a higher priority update is given preference over those with lower priority), timestamp-based requirements (the sequence of conflicting operations must be maintained throughout scheduling), and data partitioning (the data is partitioned and specific sites are given update rights to the partition).

Mobile computing has changed the face of computing in recent times, as well as introduced new and challenging problems in data management. In today's scenario, many employees work away from the office, interacting with clients and collecting data. Sometimes mobile devices do not have enough space to store the data, while at other times employees need to access real-time data from the office. In these cases, data is downloaded on demand from the local server (http://www.dbmsmag.com/9705d15.html).

Challenges in replicated environments can be summarised as follows:

1. **Data consistency:** Maintaining data integrity and consistency in a replicated environment is of prime importance. High-precision applications may require strict consistency (e.g., 1SR, as discussed above) of the updates made by transactions.

2. **Downtime during new-replica creation:** If strict data consistency is to be maintained, performance is severely affected if a new replica is to be created. Sites will not be able to fulfill requests due to consistency requirements.
3. **Maintenance overhead:** If the files are replicated at more then one site, they occupy storage space and have to be administered. Thus, there are overheads in storing multiple files.
4. **Lower write performance:** The performance of write operations can be dramatically lower in applications requiring frequent updates in a replicated environment because transactions may need to update multiple copies.

Classification of Distributed Storage and Data- Distribution Systems

Considering the vast architectural differences in distributed data-storage systems, we classify the data-storage systems as shown in Figure 2. The classification is based on the architectural and data-management polices used in different systems. Distributed DBMSs are tightly synchronised and homogeneous in nature, while data grids are asynchronous and heterogenous in nature. The ownership of data is also a major issue in different systems; for example, in peer-to-peer systems, data are meant to be shared over the network, while on data grids, the data can be application specific and can be easily shared among a particular group of people. Tightly synchronised DBMSs may store organisation-specific, proprietary, and extremely sensitive data. Storage area networks (SANs) also store organisation-specific data, which organisations may not want to share with other organisations. Databases in the World Wide Web environment are mostly meant to serve client requests being generated throughout the globe. Thus, technologies such as Akamai (http://www.akamai.com) and disk mirroring may be a viable option in the WWW environment as the data-access requests are widely distributed.

Figure 2. Classification of data-storage and content-management systems

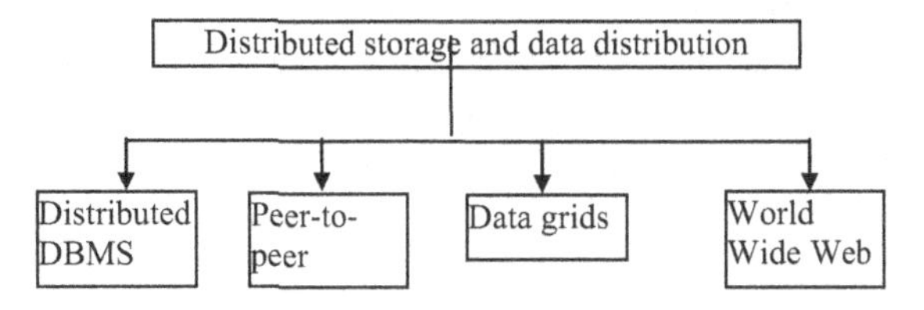

In the following sections, we will discuss the first three technologies—distributed DBMSs, peer-to-peer systems, and data grids—in detail, and we will briefly touch on SANs and the WWW.

Distributed DBMS

A replicated database is a distributed database in which multiple copies of some data items are stored at multiple sites. By storing multiple copies, the system can operate even though some sites have failed. Maintaining the correctness and consistency of data is of prime importance in a distributed DBMS. In a distributed DBMS, it is assumed that a replicated database should behave like a database system managing a single copy of the data. As replication is transparent from users' point of view, users may want to execute interleaved executions on a replicated database so it is equivalent to a one-copy database, the criterion commonly known as one-copy serializability (Bernstein et al., 1987).

Replication Protocols

ROWA and ROWA-Available

In most cases, the system is aware of which data items have replicas and where they are located. A replica-control protocol is required to read and write replicated data items. The most simple replica-control protocol is the read-one-write-all (ROWA) protocol. In the ROWA protocol, a transaction requests to read an item and the system fetches the value from the most convenient location. If a write operation is requested, the system must update all the replicas. It is clearly evident that the read operation benefits from data replication as it can find a replica near the site of request. But, write operations may adversely affect the performance of the system. A very obvious alternative of the ROWA protocol is ROWA-Available (ROWA-A). ROWA-A was proposed to provide more flexibility to the ROWA algorithm in the presence of failures. Read operations of ROWA-A can be performed similar to ROWA, that is, on any replicated copy. But to provide more flexibility, write operations are performed only on the available copies and any failed replicas are ignored. ROWA-A solves the availability problem, but the correctness of the data may be compromised. After the failed site has recovered, it stores the stale value of the data. Any transaction reading that replica reads an out-of-date copy of the replica and thus the resulting execution is not 1SR.

Table 1. Lock compatibility matrix

Lock held / Lock requested	*Read*	*Write*
Read	No conflict	Conflict
Write	Conflict	Conflict

Quorum Based

An interesting proposal to update only a subset of replicas and still not compromise correctness and consistency is based on quorums (Bernstein et al., 1987). Every copy of the replica is assigned a nonnegative vote (quorum). Read and write thresholds are defined for each data item. The sum of read and write thresholds as well as twice the write threshold must be greater than the total vote assigned to the data. These two conditions ensure that there is always a nonnull intersection between any two quorum sets. The nonnull set between the read quorum and write quorum guarantees to have at least one latest copy of the data item in any set of sites. This avoids the read-write and write-write conflicts. The conflict table is shown in Table 1.

All transactions must collect a read-write quorum to read or write any data item. A read-write quorum of a data item is any set of copies of the data with a weight of at least the read-write threshold. Quorum-based protocols maintain the consistency of data in spite of operating only on a subset of the replicated database.

Details of the majority-consensus quorum protocol are shown below.

Q = Total number of votes (maximum quorum) = number of sites in the replicated system (assuming each site has equal weight)

Q_R and Q_W = Read and write quorum, respectively

In order to read an item, a transaction must collect a quorum of at least Q_R votes, and in order to write, it must collect a quorum of Q_W votes. The overlapping between read and write quorums makes sure that a reading transaction will at least get one up-to-date copy of the replica. The quorums must satisfy the following two threshold constraints:

(i) $Q_R + Q_W > Q$ and

(ii) $Q_W + Q_W > Q$.

A quorum-based replicated system may continue to operate even in the case of site or communication failure if it is successful in obtaining the quorum for the data item. Thus, we see that the main research focus in distributed DBMS is in maintaining the consistency of replicated data.

Types of Replication Protocols

For performance reasons, the system may either implement (a) synchronous replication or (b) asynchronous replication. A synchronous system updates all the replicas before the transaction commits. Updates to all replicas are treated in the same way as any other data item. Synchronous systems produce globally serializable schedules.

In asynchronous systems, only a subset of the replicas is updated. Other replicas are brought up-to-date lazily after the transaction commits. This operation can be triggered by the commit operation of the executing transaction or another periodically executing transaction.

The synchronous strategy is also known as eager replication, while the asynchronous strategy is known as lazy replication. Another important aspect on which the replication strategies can be classified is based on the concept of primary copy. It involves the concepts of (a) group and (b) master (Gray, Helland, O'Neil, & Shasha, 1996).

(i) Group: Any site having a replica of the data item can update it. This is also referred to as update anywhere.

(ii) Master: This approach delegates a primary copy of the replica. All other replicas are used for read-only queries. If any transaction wants to update a data item, it must do so on the master or primary copy.

A classification used in Gray et al. (1996) is shown in Table 2.

Table 2. Classification of replication schemes

	Lazy/Asynchronous	**Eager/Synchronous**
Group	*N* transactions *N* owners	One transaction *N* owners
Master/Primary	*N* transactions One owner (primary site)	One transaction One owner (primary site)

Commit Protocol in Distributed DBMS

The two-phase commit (2PC) protocol is the most widely accepted commit protocol in the distributed DBMS environment that helps in achieving replica synchronisation. A 2PC protocol is defined as follows.

A coordinator is typically the site where the transaction is submitted or any other site that keeps all the global information regarding the distributed transaction. Participants are all other sites where the subtransaction of the distributed transaction is executing. The following steps are taken in a 2PC.

The coordinator sends a *vote_request* to all the participating sites.

After receiving the request to vote, the site responds by sending its vote, either *yes* or *no*. If the participant voted *yes*, it enters a *prepared* or *ready* state and waits for the final decision from the coordinator. If the vote was *no*, the participant can abort its part of the transaction.

The coordinator collects all votes from the participants. If all votes including the coordinator's vote are *yes*, then the coordinator decides to *commit* and sends the message accordingly to all the sites. Even if just one of the votes is *no*, the coordinator decides to abort the distributed transaction.

After receiving the *commit* or *abort* decision from the coordinator, the participant commits or aborts accordingly from the *prepared* state.

There are two phases (hence the name two-phase commit) in the commit procedure: the voting phase (Step 1 and Step 2) and the decision phase (Step 3 and Step 4). The state diagram of 2PC is shown below.

Storage Area Network

A SAN is an interconnected network of storage devices, typically connected by a high-speed network (gigabits/second). The storage resources may be connected to

Figure 3. State diagram of coordinator and participants in a two-phase commit

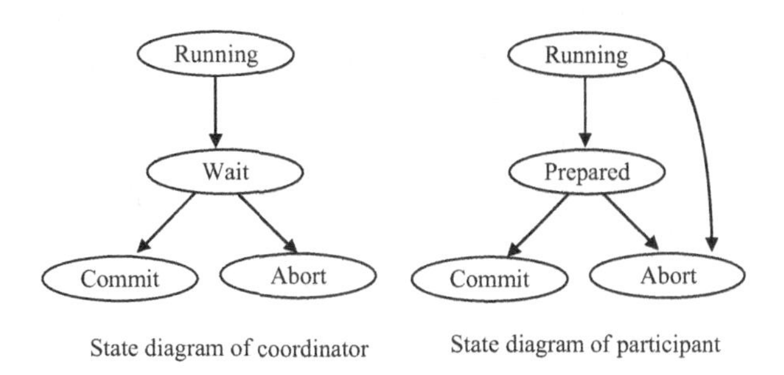

Figure 4. Schematic representation of SAN and normal distributed storage

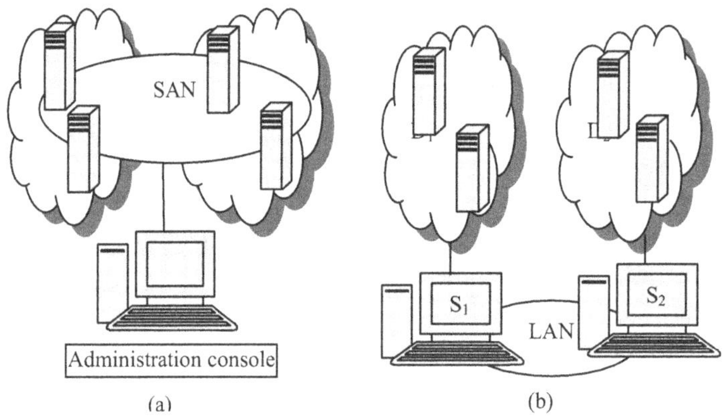

(a)

(b)

one or more servers, unlike direct attached storage (DAS), where each server has dedicated storage devices. One of the main reasons for continued interest in SAN is the flexibility of managing the distributed data. As the amount of data being stored in scientific and business applications is increasing exponentially, the management and accessing of data are also getting complex. A SAN makes administering all the storage resources in such an environment manageable and less complex by providing centralised data-storage operations. This should not be confused with centralised storage, as is clear from Figure 4a.

Figure 4b shows a distributed data-management architecture where the administration control and servers are connected via a LAN (local area network). If a request is made at a server S_1 to access data from data cluster D_2, then the request must be routed through the LAN to access the data. The transfer speed of LAN is typically much slower than that of SAN. SANs are connected by high-speed fibre-optic cables, and gigabits-per-second speed can be achieved as opposed to typical LAN speeds of megabits per second.

SAN increases the storage performance, reliability, and scalability of high-end data centers. Additional storage capacity can also be added in SAN without the need to shut down the servers. A SAN consists of various hardware components such as hubs, switches, backup devices, an interconnection network (typically fibre-optic cables), and RAIDs (redundant array of independent disks).

SAN provides improved data-availability features through data replication. Two main types of replication are discussed in the literature (http://www.dothill.com/tutorial/tutorial.swf).

Storage replication: This strategy focuses on bulk data transfers. It replicates the bulk data that are independent of the application. Thus, more than one application may be running on a server while the replication is being carried out to multiple servers.

Application-specific replication: Application-specific replication is done by the application itself and is performed by the transaction. If multiple transactions are running on the same server, then the application-specific replication must be used for each application.

Table 3. Survey of replication methods in different storage systems

Attributes / Systems	Type of system	Replication method
Arjuna (Parrington, Shrivastava, Wheater, & Little, 1995)	Object-oriented programming system	Default is primary copy, but simultaneous update also possible
Coda (Kistler & Satyanarayanan, 1992)	Distributed file system	Variants of ROWA
Deceit (Siegel, Biman, & Marzullo, 1990)	Distributed file system	File replication with concurrent read and writes. Updates use a write token. Thus, write update is synchronous as the site having the token can update the data. Also provides multiversion control.
Harp (Liskov, Ghemawat, Gruber, Johnson, Shrira, & Williams, 1991)	Distributed file system	Primary-copy replication. Uses two-phase protocol for update. Communicates with backup servers before replying to the request.
Mariposa (Sidell et al., 1996)	Distributed DBMS	Asynchronous replication. Updates are propagated within a fixed time limit (thus having stale data among replicas).
Oracle (Baumgartel, 2002)	(Distributed) DBMS	Provides basic and advanced replication. Basic: Replicated copies are read only. Advanced: Replica copies are updateable. Various conflict-resolution options can be specified such as latest time stamp, site priority, etc. Provides synchronous and asynchronous replication.
Pegasus (Ahmed et al., 1991)	Distributed object-oriented DBMS (supports heterogeneous data)	Supports global consistency (synchronous)
Sybase (*Sybase FAQ*, n.d.)	Distributed DBMS (supports heterogeneity)	Asynchronous replications

Replication can be done either on the storage-array level or host level. In array-level replication, data is copied from one disk array to another. Thus, array-level replication is mostly homogeneous. The arrays are linked by a dedicated channel. Host-level replication is independent of the disk array used. Since arrays used in different hosts can be different, host-level replication has to deal with heterogeneity. Host-level replication uses the TCP/IP (transmission-control protocol/Internet protocol) for data transfer. The replication in SAN also can be divided in two main categories based on the mode of replication: (a) synchronous and (b) asynchronous, as discussed earlier.

Survey of Distributed Data-Storage Systems and Replication Strategies Used

A brief explanation of systems in Table 3 follows. Arjuna (Parrington et al., 1995) supports both active and passive replication. Passive replication is like primary-copy replication, and all updates are redirected to the primary copy. The updates can be propagated after the transaction has committed. In active replication, mutual consistency is maintained and the replicated object can be accessed at any site.

Coda (Kistler & Satyanarayanan, 1992) is a network-distributed file system. A group of servers can fulfill the client's read request. Updates are generally applied to all participating servers. Thus, it uses a ROWA protocol. The motivation behind using this concept was to increase availability so that if one server fails, other servers can take over and the request can be satisfied without the client's knowledge.

The Deceit (Siegel et al., 1990) distributed file system is implemented on top of the Isis (Birman & Joseph, 1987) distributed system. It provides full network-file-system (NFS) capability with concurrent read and writes. It uses write tokens and stability notification to control file replicas (Siegel et al.). Deceit provides variable file semantics that offer a range of consistency guarantees (from no consistency to semantics consistency). However, the main focus of Deceit is not on consistency, but on providing variable file semantics in a replicated NFS server (Triantafillou, 1997).

Harp (Liskov, 1991) uses a primary-copy replica protocol. Harp is a server protocol and there is no support for client caching (Triantafillou & Nelson, 1997). In Harp, file systems are divided into groups, and each group has its own primary site and secondary sites. For each group, a primary site, a set of secondary sites, and a set of sites as witnesses are designated. If the primary site is unavailable, a primary site is chosen from the secondary sites. If enough sites are not available from the primary and secondary sites, a witness is promoted to act as a secondary site. The data from such a witness are backed up in tapes so that if it is the only surviving site, then the data can be retrieved. Read and write operations follow typical ROWA protocol.

Mariposa (Sidell et al., 1996) was designed at the University of California (Berkley) in 1993 and 1994. Basic design principles behind the design of Mariposa were the scalability of distributed data servers (up to 10,000) and the local autonomy of sites. Mariposa implements an asynchronous replica-control protocol, thus distributed data may be stale at certain sites. The updates are propagated to other replicas within a time limit. Therefore it could be implemented in systems where applications can afford stale data within a specified time window. Mariposa uses an economic approach in replica management, where a site buys a copy from another site and negotiates to pay for update streams (Sidell et al.).

Oracle (Baumgartel, 2002) is a successful commercial company that provides data-management solutions. Oracle provides a wide range of replication solutions. It supports basic and advanced replication. Basic replication supports read-only queries, while advanced replication supports update operations. Advanced replication supports synchronous and asynchronous replication for update requests. It uses 2PC for synchronous replication. 2PC ensures that all cohorts of the distributed transaction completes successfully, or rolls back the completed part of the transaction.

Pegasus (Ahmed et al., 1991) is an object-oriented DBMS designed to support multiple heterogeneous data sources. It supports Object Structured Query Language (SQL). Pegasus maps a heterogeneous object model to a common Pegasus object model. Pegasus supports global consistency in replicated environments as well as it respects integrity constraints. Thus, Pegasus supports synchronous replication.

Sybase (*Sybase FAQ*, 2003) implements a Sybase replication server to implement replication. Sybase supports the replication of stored procedure calls. It implements replication at the transaction level and not at the table level (Helal, Hedaya, & Bhargava, 1996). Only the rows affected by a transaction at the primary site are replicated to remote sites. The log-transfer manager (LTM) passes the changed records to the local replication server. The local replication server then communicates the changes to the appropriate distributed replication servers. Changes can then be applied to the replicated rows. The replication server ensures that all transactions are executed in correct order to maintain the consistency of data. Sybase mainly implements asynchronous replication. To implement synchronous replication, the user should add his or her own code and a 2PC protocol (http://www.dbmsmag.com/9705d15.html).

Peer-to-Peer Systems

P2P networks are a type of overlay network that uses the computing power and bandwidth of the participants in the network rather than concentrating it in a relatively few servers (Oram, 2001). The word peer-to-peer reflects the fact that all participants have equal capability and are treated equally, unlike in the client-server model where clients and servers have different capabilities. Some P2P networks use the client-server model for certain functions (e.g., Napster uses the client-server model for searching; Oram). Those networks that use the P2P model for all functions, for example, Gnutella (Oram), are referred to as pure P2P systems. A brief classification of P2P systems is shown below.

Types of Peer-to-Peer Systems

Today P2P systems produce a large share of Internet traffic. A P2P system relies on the computing power and bandwidth of participants rather than relying on central servers. Each host has a set of neighbours.

P2P systems are classified into two categories.

1. **Centralised P2P systems:** Centralised P2P systems have a central directory server where the users submit requests, for example, as is the case for Napster (Oram, 2001). Centralised P2P systems store a central directory, which keeps information regarding file location at different peers. After the files are located, the peers communicate among themselves. Clearly centralised systems have the problem of a single point of failure, and they scale poorly when the number of clients ranges in the millions.
2. **Decentralised P2P systems:** Decentralised P2P systems do not have any central servers. Hosts form an ad hoc network among themselves on top of the existing Internet infrastructure, which is known as the overlay network. Based on two factors—(a) the network topology and (b) the file location—decentralised P2P systems are classified into the following two categories.
   (i) Structured decentralised: In a structured architecture, the network topology is tightly controlled and the file locations are such that they are easier to find (i.e., not at random locations). The structured architecture can also be classified into two categories: (a) loosely structured and (b) highly structured. Loosely structured systems place the file based on some hints, for example, as with Freenet (Oram, 2001). In highly structured systems, the file locations are precisely determined with the help of techniques such as hash tables.
   (ii) Unstructured: Unstructured systems do not have any control over the network topology or placement of the files over the network. Examples of such systems include Gnutella, KaZaA, and so forth (Oram, 2001). Since there is no structure, to locate a file, a node queries its neighbours.

Table 4. Examples of different types of P2P systems

Type	Example
Centralised	Napster
Decentralised structured	Freenet (loosely structured) Distribute hash table (DHT) (highly structured) FatTrack eDonkey
Decentralised unstructured	Gnutella

Flooding is the most common query method used in such an unstructured environment. Gnutella uses the flooding method to query.

In unstructured systems, since the P2P network topology is unrelated to the location of data, the set of nodes receiving a particular query is unrelated to the content of the query. The most general P2P architecture is the decentralised, unstructured architecture.

Main research in P2P systems have focused on architectural issues, search techniques, legal issues, and so forth. Very limited literature is available for unstructured P2P systems. Replication in unstructured P2P systems can improve the performance of the system as the desired data can be found near the requested node. Especially in flooding algorithms, reducing the search even by one hop can drastically reduce the number of messages in the system. Table 4 shows different P2P systems.

A challenging problem in unstructured P2P systems is that the network topology is independent of the data location. Thus, the nodes receiving queries can be completely unrelated to the content of the query. Consequently, the receiving nodes also do not have any idea of where to forward the request for quickly locating the data. To minimise the number of hops before the data are found, data can be proactively replicated at more than one site.

Replication Strategies in P2P Systems

Based on Size of Files (Granularity)

1. **Full-file replication:** Full files are replicated at multiple peers based upon which node downloads the file. This strategy is used in Gnutella. This strategy is simple to implement. However, replicating larger files at one single file can

Figure 5. Classification of replication schemes in P2P systems

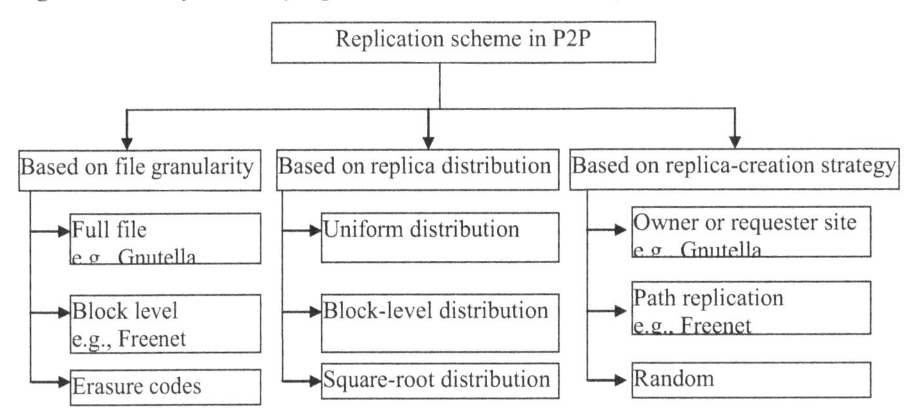

be cumbersome in terms of space and time (Bhagwan, Moore, Savage, & Voelker, 2002).

2. **Block-level replication:** This replication divides each file into an ordered sequence of fixed-size blocks. This is also advantageous if a single peer cannot store a whole file. Block-level replication is used by eDonkey. A limitation of block-level replication is that during file downloading, it is required that enough peers are available to assemble and reconstruct the whole file. Even if a single block is unavailable, the file cannot be reconstructed. To overcome this problem, erasure codes (ECs), such as Reed-Solomon (Pless, 1998), are used.
3. **Erasure-code replication:** This provides the capability for original files to be constructed from less available blocks. For example, k original blocks can be reconstructed from l (l is close to k) coded blocks taken from a set of ek (e is a small constant) coded blocks (Bhagwan et al., 2002). In Reed-Solomon codes, the source data are passed through a data encoder, which adds redundant bits (parity) to the pieces of data. After the pieces are retrieved later, they are sent through a decoder process. The decoder attempts to recover the original data even if some blocks are missing. Adding EC in block-level replication can improve the availability of the files because it can tolerate the unavailability of certain blocks.

Based on Replica Distribution

The following need to be defined.

Consider that each file is replicated on r_i nodes.

Let the total number of files (including replicas) in the network be denoted as R (Cohen & Shenker, 2002).

$R = \sum_{i=1}^{m} r_i$, where m is the number of individual files or objects.

(i) **Uniform:** The uniform replication strategy replicates everything equally. Thus, from the above equation, replication distribution for the uniform strategy can be defined as follows:

$r_i = R / m.$

(ii) **Proportional:** The number of replicas is proportional to their popularity. Thus, if a data item is popular, it has more chances of finding the data close to the site where the query was submitted.

$r_i \propto q_i$,

where, q_i = the relative popularity of the file or object (in terms of the number of queries issued for the ith file).

$$\sum_{i=1}^{m} q_i = 1$$

If all objects were equally popular, then

$q_i = 1/m.$

However, results have shown that object popularity show a Zipf-like distribution in systems such Napster and Gnutella. Thus, the query distribution is as follows:

$q_i \propto 1/ i^{\alpha}$, where α is close to unity.

(iii) **Square root:** The number of replicas of a file i is proportional to the square root of query distribution q_i.

$r_i \propto \sqrt{q_i}$

The necessity of square-root replication is clear from the following discussion. The uniform and proportional strategies have been shown to have the same search space, as follows.

m: number of files

n: number of sites

r_i: number of replicas for the ith file

R = total number of files

The average search size for file i is $A_i = \frac{n}{r_i}$.

Hence, the overall average search size is $A = \sum_i q_i A_i$.

The assumed average number of files per site is $\mathrm{m} = \frac{R}{n}$.

Following the above equations, the average search size for the uniform replication strategy is as follows.

Since $r_i = R / m$, the following equations are true.

$$A = \sum q_i \frac{n}{r_i} \text{ (replacing the value of } A_i\text{)}$$

$$A = \sum q_i \frac{n\, m}{R}$$

$$A = \frac{m}{\mu} \text{ (as, } \sum_{i=1}^{m} q_i = 1\text{)} \qquad (1)$$

The average search size for the proportional replication strategy is as follows.

Since $r_i = R\, q_i$ (as, $r_i \propto q_i$, and $q_i = 1$), the following are true.

$$A = \sum q_i \frac{n}{r_i} \text{ (replacing the value of } A_i\text{)}$$

$$A = \sum q_i \frac{n}{R q_i}$$

$$A = \frac{m}{\mu} \text{ (as, } \sum_{i=1}^{m} q_i = 1, \frac{n}{R} = \frac{1}{\mu} \text{ and } \frac{1}{q_i} = m \text{ for proportional replication} \qquad (2)$$

It is clear from Equations 1 and 2 that the average search size is the same in the uniform and proportional replication strategies.

It has also been shown in the literature (Cohen & Shenker, 2002) that the average search size is the minimum under the following condition:

$$A_{optimal} = \frac{1}{\mu}(\sum\sqrt{q_i})^2.$$

This is known as square-root replication.

Based on Replica-Creation Strategy

1. **Owner replication:** The object is replicated only at the requester node once the file is found. For example, Gnutella (Oram, 2001) uses owner replication.
2. **Path replication:** The file is replicated at all the nodes along the path through which the request is satisfied. For example, Freenet uses path replication.
3. **Random replication:** The random-replication algorithm creates the same number of replicas as path replication. However, it distributes the replicas in a random order rather than following the topological order. It has been shown in Lv, Cao, Cohen, Li, and Shenker (2002) that the factor of improvement in path replication is close to 3, and in random replication, the improvement factor is approximately 4. The following tree summarises the classification of replication schemes in P2P systems, as discussed above.

Replication Strategy for Read-Only Requests

Replica Selection Based on Replica Location and User Preference

The replicas are selected based on users' preferences and the replica location. Vazhkudai, Tuecke, and Foster (2001) propose a strategy that uses Condor's ClassAds (classified advertisements; Raman, Livny, & Solomon, 1998) to rank the sites' suitability in the storage context. The application requiring access to a file presents its requirement to the broker in the form of ClassAds. The broker then does the *search*, *match*, and *access* of the file that matches the requirements published in the ClassAds.

Dynamic replica-creation strategies discussed in Ranganathan and Foster (2001) are as follows:

1. **Best client:** Each node maintains a record of the access history for each replica, that is, which data item is being accessed by which site. If the access frequency of a replica exceeds a threshold, a replica is created at the requester site.

2. **Cascading replication:** This strategy can be used in the tired architecture discussed above. Instead of replicating the data at the best client, the replica is created at the next level on the path of the best client. This strategy evenly distributes the storage space, and other lower level sites have close proximity to the replica.
3. **Fast spread:** Fast spread replicates the file in each node along the path of the best client. This is similar to path replication in P2P systems.

Since the storage space is limited, there must be an efficient method to delete the files from the sites. The replacement strategy proposed in Ranganathan and Foster (2001) deletes the most unpopular files once the storage space of the node is exhausted. The age of the file at the node is also considered to decide the unpopularity of the file.

Economy-Based Replication Policies

The basic principle behind economy-based polices are to use the socioeconomic concepts of emergent marketplace behaviour, where local optimisation leads to global optimisation. This could be thought of as an auction, where each site tries to buy a data item to create the replica at its own node and generate revenue in the future by selling the replica to other interested nodes. Various economy-based protocols such as those in Carman, Zini, Serafini, and Stockinger (2002) and Bell, Cameron, Carvajal-Schiaffino, Millar, Stockinger, and Zini (2003) have been proposed, which dynamically replicate and delete the files based on the future return on the investment. Bell et al. use a reverse-auction protocol to determine where the replica should be created.

For example, following rule is used in Carman et al. (2002). A file request (FR) is considered to be an n-tuple of the form

$$FR_i = \langle t_i, o_i, g_i, n_i, r_i, s_i, p_i \rangle,$$

where the following are true.

t_i: time stamp at which the file was requested

o_i, g_i, and n_i: together represent the logical file being requested (o_i is the virtual organisation to which the file belongs, g_i is the group, and n_i is the file identification number)

Figure 6. A tiered or hierarchical architecture of a data grid for the particle physics accelerator at the European Organization for Nuclear Research (CERN)

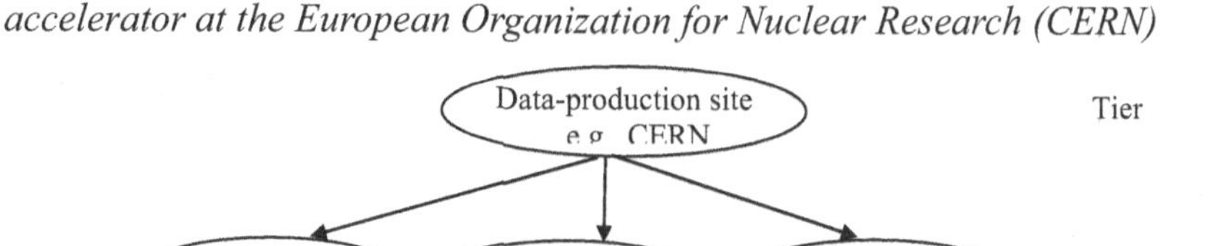
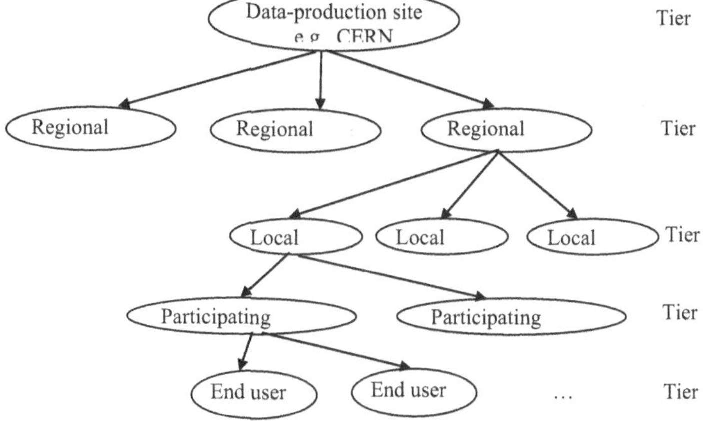

r_i and s_i: represent the element requesting and supplying the file, respectively

p_i: represents the price paid for the file (price could be virtual money)

To maximise the profit, the future value of the file is defined over the average life time of the file storage T_{av}.

$$V(F, T_k) = \sum_{i=k+1}^{k+n} p_i \partial(F, F_i) \partial(s, s_i),$$

where V represents the value of the file, p_i represents the price paid for the file, s is the local storage element, and F represents the triple (o, g, n). ∂ is a function that returns 1 if the arguments are equal and 0 if they differ. The investment cost is determined by the difference in cost between the price paid and the expected price if the file is sold immediately.

As the storage space of the site is limited, the choice of whether it is worth deleting an existing file must be made before replicating a file. Thus, the investment decision between purchasing a new file and keeping an old file depends on the change in profit between the two strategies.

Cost-Estimation Based

The cost-estimation model (Lamehamedi, Shentu, Szymanski, & Deelman, 2003) is very similar to the economic model. The cost-estimation model is driven by the estimation of the data-access gains and the maintenance cost of the replica. While the investment measured in economic models (Bell et al., 2003; Carman et al., 2002) are only based on data access, it is more elaborate in the cost-estimation model. The cost calculations are based on network latency, bandwidth, replica size, run-time-accumulated read and write statistics (Lamehamedi et al.), and so forth.

Replication Strategy for Update Request

Synchronous

In the synchronous model, a replica is modified locally. The replica-propagation protocol then synchronises all other replicas. However, it is possible that other nodes may work on their local replicas. If such a conflict occurs, the job must be redone with the latest replica. This is very similar to the synchronous approach discussed in the distributed-DBMS section.

Figure 7. Classification of replication scheme in data grids

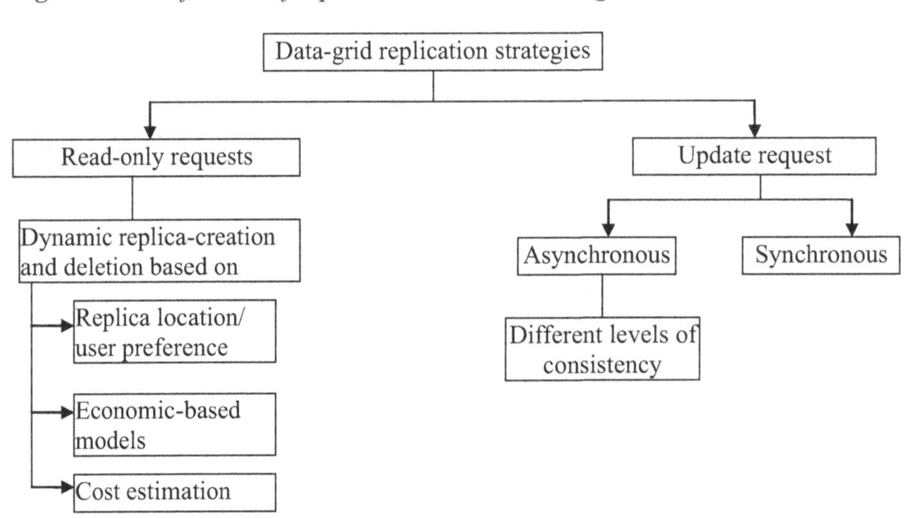

Asynchronous

Various consistency levels are proposed for asynchronous replication. Asynchronous replication approaches are discussed as follows (Dullmann et al., 2001):

1. **Possible inconsistent copy (consistency level: -1):** The content of the file is not consistent with two different users. For example, one user is updating the file while the other is copying it: a typical case of the "dirty read problem."
2. **Consistent file copy (consistency level: 0):** At this consistency level, the data within a given file correspond to a snapshot of the original file at some point in time.
3. **Consistent transactional copy (consistency level: 1):** A replica can be used by clients without internal consistency problems. However, if the job needs to access more than one file, then the job may have an inconsistent view.

Figure 7 shows the classification of the replication scheme discussed above. The major classification criterion is the update characteristics of the transaction.

Data-Grid Replication Strategies

Data-Grid Replication Schemes

An overview of replication studies in data grids follows along with a brief explanation of each strategy.

Vazhkudai et al. (2001) propose a replica-selection scheme in the globus data grid. The method optimises the selection of replica in the dynamic grid environment. A high-level replica-selection service is proposed. Information such as replica location and user preferences is considered to select the suitable replica from multiple replicas.

Lamehamedi et al. (2003) propose a method for dynamically creating replicas based on the cost-estimation model. Replication decision is based on gains of creating a replica against the creation and maintenance cost of the replica.

Regarding economy-based replica protocols, Carman et al. (2002) aim to achieve global optimisation through local optimisation with the help of emerging marketplace behaviour. The paper proposes a technique to maximise the profit and minimise the cost of data-resource management. The value of the file is defined as the sum of the future payments that will be received by the site.

Another economy-based approach for file replication proposed by Bell et al. (2003) dynamically creates and deletes replicas of files. The model is based on the reverse Vickery auction where the cheapest bid from participating replica sites is accepted to replicate the file. It is similar to the work in Carman et al. (2002) with the difference in predicting the cost and benefits.

Consistency issues have received limited attention in data grids. Dullmann et al. (2001) propose a grid-consistency service (GCS). GCS uses data-grid services and supports replica-update synchronisation and consistency maintenance. Different levels of consistency are proposed, starting from level -1 to level 3 in increasing order of strictness.

Lin and Buyya (2005) propose various policies for selecting a server for data transfer. The least-cost policy chooses the server with the minimum cost from the server list. The minimise-cost-and-delay policy considers the delay in transferring the file in addition to the cost of transferring it. A scoring function is calculated from the time and delay in replicating files. The file is replicated at the site with the highest score. The policy of minimising cost and delay with service migration considers the variation in service quality. If the site is incapable of maintaining the promised service quality, the request can be migrated to other sites.

World Wide Web

The WWW has become a ubiquitous media for content sharing and distribution. Applications using the Web spans from small-business applications to large scientific calculations. Download delay is one of the major factors that affect the client base of the application. Hence, reducing latency is one of the major research foci in WWW. Caching and replication are two major techniques used in WWW to reduce request latencies. Caching is typically on the client side to reduce the access latency, whereas replication is implemented on the server side so that the request can access the data located in a server close to the request. Caching targets, reducing download delays, and replication improve end-to-end responsiveness. Every caching technique has an equivalent in replica systems, but the reverse is not true.

Large volumes of requests at popular sites may be required to serve thousands of queries per second. Hence, Web servers are replicated at different geographical locations to serve requests for services in a timely manner. From the users' perspective, these replicated Web servers act as a single powerful server. Initially, servers were manually mirrored at different locations. But the continuously increasing demand for hosts has motivated the research of the dynamic replication strategy in WWW. The following major challenges can be easily identified in replicated systems on the Internet (Loukopoulos, Ahmad, & Papadias, 2002).

1. How to assign a request to a server based on a performance criterion
2. The number of placements of the replica
3. Consistency issues in the presence of update requests

Here we would briefly like to mention Akamai Technologies (http://www.akamai.com). Akamai Technologies has more than 16,000 servers located across the globe. When a user requests a page from the Web server, it sends some text with additional information for getting pages from one of the Akamai servers. The user's browser then requests the page from Akamai's server, which delivers the page to the user.

Most of the replication strategies on the Internet use a primary-copy approach (Baentsch, Baum, Molter, Rothkugel, & Sturm, 1997; Baentsch, Molter, & Sturm, 1996; Khan & Ahmad, 2004). Replication techniques in Baentsch et al. (1997) and Baentsch et al. (1996) use a primary server (PS) and replicated servers (RSs). In Baentsch et al. (1997), the main focus is on maintaining up-to-date copies of documents on the WWW. A PS enables the distribution of most often requested documents by forwarding the updates to the RS as soon as the pages are modified. An RS can act as a replica server for more than one PS. An RS can also act as a cache for nonreplicated data. RSs also reduce the load on the Web servers as they can successfully answer requests.

Replica management on the Internet is not as widely studied and understood as in other distributed environments. We believe that due to changed architectural challenges on the Internet, it needs special attention. Good replication placement and management algorithms can greatly reduce the access latency.

Discussion and Analysis

In this section, we discuss different data-storage technologies such as distributed DBMSs, P2P systems, and data grids for different data-management attributes.

Data Control

In distributed DBMSs, the data are owned mostly by a single organisation and hence can be maintained with central-management policies. In P2P systems, control of the data is distributed across sites. The site where the data are stored is thought of as owning the data, and there is no obligation to follow a central policy for data control. Considering the most widely used data-grid environment (the large hadron collider [LHC] experiment), the data are produced at a central location but are hierarchically distributed to processing sites.

Table 5. Comparison of different storage and content-management systems

Attributes \ Systems	Distributed DBMSs	P2P systems	Data grid	WWW
Data control	Mostly central	Distributed	Hierarchical	Mostly central
Autonomy among sites	Tightly coupled	Autonomous	Autonomous, but in a trusted environment	Tightly coupled
Load distribution	Central and easy	Decentralised	Hierarchical	Central
Update performance	Well understood and can be controlled	Difficult to monitor	Not well studied yet (most studies are in read-only environments)	Mostly read content
Reliability	Can be considered during designing and has a direct relation with performance (in replication scenario)	Difficult to account for during system design (as a peer can disconnect at any time from the system)	Intermediate	Central management, hence it can be considered at design time
Heterogeneity	Mostly homogeneous environment	Heterogeneous environment	Intermediate as the environment is mostly trusted	Mostly homogeneous
Status of replication strategies	Read and update scenarios are almost equivalent	Mostly read environment	Mostly read but does need to update depending on the application requirement	Mostly read environment with lazy replication

Autonomy

Distributed DBMSs are usually tightly coupled, mainly because they belong to a single organisation. Hence, the design choices depend on one another, and the complete system is tightly integrated and coupled. P2P systems are autonomous as there is no dependency among any distributed sites. Each site is designed according to independent design choices and evolves without any interference from each other. In data grids, sites are autonomous in relation to each other, but the typical characteristic is that they mostly operate in a trusted environment.

Load Distribution

The load distribution directly depends on the data-control attribute. If the data are centrally managed, it is easy to manage the load distribution among distributed servers as compared to distributed management. It is easy to manage the distributed data in a DBMS environment as compared to P2P systems because central policies can be implemented in DBMSs for data management while it is virtually impossible to implement a central management policy in P2P systems.

Update Performance

Update performance in databases is easy to monitor and analyse during the database design (again, due to the fact that it is centrally designed). Databases, in general, have well-defined data-access interfaces and access patterns. Due to the decentralised management and asynchronous behaviour of P2P systems, it may be difficult to monitor update performance in such systems. In data grids, applications are mainly driven by read-only queries, and hence update performance is not well studied and understood. But, with advancements in technology, applications will need to update stored data in data grids as well. Hence, there is a need to study update performance in greater detail.

Reliability

As distributed DBMSs work under a central policy, the downtime of a particular site can be scheduled and the load of that site can be delegated to other sites. Thus, DBMS systems can be designed for a guaranteed quality of service (QoS). A P2P system's architecture is dynamic. Sites participating in P2P systems can join and leave the network according to their convenience and hence cannot be scheduled. In grids, though the architecture is dynamic, research has focussed on providing a QoS guarantee and some degree of commitment toward the common good.

Heterogeneity

Distributed DBMSs typically work in homogeneous environments as they are built bottom-up by the designer. P2P systems can be highly heterogenous in nature since sites are autonomous and are managed independently. As shown in Figure 6, data grids have hierarchical architectures, and individual organisations and institutes may choose homogeneous environments. However, different participants may opt for heterogenous components or policies.

Replication strategies

Database designers pay attention to update requests as well as the performance of read-only queries. At the same time, applications also demand update transactions and read-only queries almost equally. P2P systems are designed for applications requiring only file sharing. Thus, P2P systems mostly focus on read-only queries. Data grids, so far, have mainly focused on read-only queries, but the importance of write queries is also being realised and is attracting research interest.

As we are discussing data-management systems, we would briefly want to mention the preservation work done by Stanford Peers Group (http://www-db.stanford.edu/peers/) for the sake of completeness. Data preservation mainly focuses on archiving the data for the long term, for example, in digital libraries. Data replication in such an environment can improve the reliability of the data. Such systems should be able to sustain long-term failures. Replication can help in preserving online journal archives, white papers, manuals, and so forth against single-system failures, natural disasters, theft, and so on. Data trading is one such technique proposed in Cooper & Garcia-Molina (2002) to increase the reliability of preservation systems in P2P environments.

The purpose of this chapter is to gather and present the replication strategies present in different architectural domains. This chapter will help researchers working in different distributed data domains to identify and analyse replication theories present in other distributed environments, and borrow some of the existing theories that best suit them. Replication theories have been studied and developed for many years in different domains, but there has been a lack of comparative studies. In this chapter, we presented the state of the art and the research directions of replication strategies in different distributed architectural domains, which can be used by researchers working in different architectural areas.

Conclusion

We presented different replication strategies in distributed storage and content-management systems. With changing architectural requirements, replication protocols have also changed and evolved. A replication strategy suitable for a certain application or architecture may not be suitable for another. The most important difference in replication protocols is due to consistency requirements. If an application requires strict consistency and has lots of update transactions, replication may reduce the performance due to synchronisation requirements. But, if the application requires read-only queries, the replication protocol need not worry for synchronisation and performance can be increased. We would like to conclude by mentioning that though there are continuously evolving architectures, replication is now a widely studied area and new architectures can use the lessons learned by researchers in other architectural domains.

References

Ahmed, R., DeSmedt, P., Du, W., Kent, W., Ketabchi, M., Litwin, W., et al. (1991, April). *Using an object model in Pegasus to integrate heterogeneous data.* Retrieved from http://www.bkent.net/Doc/usobpeg.htm

Baentsch, M., Baum, L., Molter, G., Rothkugel, S., & Sturm, P. (1997, April). *Caching and replication in the World-Wide Web.* Retrieved from http://www.newcastle.research.ec.org/cabernet/workshops/plenary/3rd-plenary-papers/13-baentsch.html

Baentsch, M., Molter, G., & Sturm, P. (1996). Introducing application-level replication and naming into today's Web. *International Journal of Computer Networks and ISDN Systems, 28*(7), 921-930.

Baumgartel, P. (2002, September). Oracle replication: An introduction. Retrieved from http://www.nyoug.org/200212baumbartel.pdf

Bell, W. H., Cameron, D. G., Carvajal-Schiaffino, R., Millar, A. P., Stockinger, K., & Zini, F. (2003). Evaluation of an economy-based file replication strategy for a data grid. In *Proceedings of the Third IEEE International Symposium on Cluster Computing and the Grid (CCGrid)*, Tokyo.

Bernstein, P. A., Hadzilacos, V., & Goodman, N. (1987). *Concurrency control and recovery in database systems.* New York: Addison-Wesley Publishers.

Bhagwan, R., Moore, D., Savage, S., & Voelker, G. M. (2002). Replication strategies for highly available peer-to-peer storage. In *Proceedings of the International Workshop on Future Directions in Distributed Computing.* Retrieved from http://www.cs.unibo.it/fudico/

Birman, K. P., & Joseph, T. A. (1987). Reliable communication in the presence of failures. *ACM Transactions on Computer Systems, 5*(1), 47-76.

Carman, M., Zini, F., Serafini, L., & Stockinger, K. (2002). Towards an economy-based optimisation of file access and replication on a data grid. In *Proceedings of the 1st IEEE/ACM International Conference on Cluster Computing and the Grid (CCGrid)* (pp. 340-345).

Cohen, E., & Shenker, S. (2002). Replication strategies in unstructured peer-to-peer networks. In *Proceedings of the Special Interest Group on Data Communications (SIGCOMM)*, 177-190.

Cooper, B., & Garcia-Molina, H. (2002). Peer-to-peer data trading to preserve information. *ACM Transactions on Information Systems, 20*(2), 133-170.

Domenici, A., Donno, F., Pucciani, G., Stockinger, H., & Stockinger, K. (2004). Replica consistency in a data grid. *Nuclear Instruments and Methods in Physics Research: Section A. Accelerators, Spectrometers, Detectors and Associated Equipment, 534*(1-2), 24-28.

Dullmann, D., Hosckek, W., Jaen-Martinez, J., Segal, B., Samar, A., Stockinger, H., et al. (2001). Models for replica synchronisation and consistency in a data grid. In *Proceedings of the 10th IEEE International Symposium on High Performance and Distributed Computing (HPDC)* (pp. 67-75).

Foster, I., & Kesselman, C. (Eds.). (2004). *The grid: Blueprint for a new computing infrastructure* (2nd ed.). San Francisco: Morgan Kaufmann Publishers.

Gray, J., Helland, P., O'Neil, P., & Shasha, D. (1996). The dangers of replication and a solution. *Proceedings of the International Conference on Management of Data (ACM SIGMOD)* (pp. 173-182).

Helal, A. A., Hedaya, A. A., & Bhargava, B. B. (1996). *Replication techniques in distributed systems.* Boston: Kluwer Academic Publishers.

Khan, S. U., & Ahmad, I. (2004). *Internet content replication: A solution from game theory* (Tech. Rep. No. CSE-2004-5). Arlington: University of Texas at Arlington, Department of Computer Science and Engineering.

Kistler, J. J., & Satyanarayanan, M. (1992). Disconnected operation in the Coda file system. *ACM Transactions on Computer Systems, 10*(*1*), 3-25.

Lamehamedi, H., Shentu, Z., Szymanski, B., & Deelman, E. (2003). Simulation of dynamic data replication strategies in data grids. In *Proceedings of the 17th International Parallel and Distributed Processing Symposium (PDPS)*, Nice, France.

Lin, H., & Buyya, R. (2005). *Economy-based data replication broker policies in data grids.* Unpublished bachelor's honours thesis, University of Melbourne, Department of Computer Systems and Software Engineering, Melbourne, Australia.

Liskov, B., Ghemawat, S., Gruber, R., Johnson, P., Shrira, L., & Williams, M. (1991). Replication in the Harp file system. In *Proceedings of 13th ACM Symposium on Operating Systems Principles* (pp. 226-238).

Loukopoulos, T., Ahmad, I., & Papadias, D. (2002). An overview of data replication on the Internet. In *Proceedings of the International Symposium on Parallel Architectures, Algorithms and Networks (ISPAN)*, 694-711.

Lv, Q., Cao, P., Cohen, E., Li, K., & Shenker, S. (2002). Search and replication in unstructured peer-to-peer networks. In *Proceeding of the 16th Annual ACM International Conference on Supercomputing* (pp. 84-95).

Oram, A. (Ed.). (2001). *Peer-to-peer: Harnessing the power of disruptive technologies.* Sebastopol, CA: O'Reilly Publishers.

Paris, J.-F. (1986). Voting with witnesses: A consistency scheme for replicated files. In *Proceedings of the Sixth International Conference on Distributed Computing Systems* (pp. 606-612).

Parrington, G. D., Shrivastava, S. K., Wheater, S. M., & Little, M. C. (1995). The design and implementation of Arjuna. *USENIX Computing Systems Journal, 8*(2), 255-308.

Pless, V. (1998). *Introduction to the theory of error-correcting codes* (3rd ed.). New York: John Wiley and Sons.

Raman, P., Livny, M., & Solomon, M. (1998). Matchmaking: Distributed resource management for high throughput computing. In *Proceedings of the 7th IEEE Symposium on High Performance Distributed Computing (HPDC)* (pp. 140-146).

Ranganathan, K., & Foster, I. (2001). Identifying dynamic replication strategies for a high-performance data grid. In *Proceedings of the International Workshop on Grid Computing* (pp. 75-86).

Sidell, J., Aoki, P. M., Barr, S., Sah, A., Staelin, C., Stonebraker, M., et al. (1996). Data replication in Mariposa. In *Proceedings of the 17th International Conference on Data Engineering* (pp. 485-494).

Siegel, A., Birman, K., & Marzullo, K. (1990). *Deceit: A flexible distributed file system* (Tech. Rep. No. 89-1042). Ithaca, NY: Cornell University, Department of Computer Science.

Sybase FAQ. (2003). Retrieved from http://www.faqs.org/faqs/databases/sybase-faq/part3/

Triantafillou, P., & Neilson, C. (1997). Achieving strong consistency in a distributed file system. In *Proceedings of IEEE Transactions on Software Engineering, 23*(1), 35-55.

Vazhkudai, S., Tuecke, S., & Foster, I. (2001). Replica selection in the Globus data grid. In *Proceedings of the International Workshop on Data Models and Databases on Clusters and the Grid (DataGrid 2001)*, 106-113.

Venugopal, S., Buyya, R., & Ramamohanarao, K. (2005). *A taxonomy of data grids for distributed data sharing, management and processing* (Tech. Rep. No. GRIDS-TR-2005-3). Melbourne, Australia: University of Melbourne, Grid Computing and Distributed Systems Laboratory.

URLs

http://www.akamai.com

http://www.dbmsag.com/970d15.html

http://www.dothill.com/tutorial/tutorial.swf

http://www-db.stanford.edu/peers/

Chapter X

Web Services vs. ebXML: An Evaluation of Web Services and ebXML for E-Business Applications

Yuhong Yan, Canada National Research Council, Canada

Matthias Klein, University of New Brunswick, Canada

Abstract

Web services and ebXML are modern integration technologies that represent the latest developments in the line of middleware technologies and business-related integration paradigms, respectively. In this chapter, we discuss relevant aspects of the two technologies and compare their capabilities from an e-business point of view.

Introduction

For companies operating in an increasingly globalized business environment, e-business means online transactions, automated business collaborations, and system integration. This means not only the provision of products through supply chains, but also the delivery of services and information through networks. The e-business tools and standards come from two domains known as Web services and e-business XML (ebXML; electronic business using extensible markup language).

Web services are a technology-oriented approach. Its ancestors include CORBA (common object request broker architecture) and other middleware technologies such as TPM (transaction processing monitor) and RPC (remote procedure call). The W3C (World Wide Web Consortium) is a big sponsor of Web-service technologies. Many Web-services standards, such as SOAP (simple object access protocol), WSDL (Web service description language), UDDI (universal description, discovery, and integration), and so forth, are W3C standards or recommendations. Many world-level IT companies currently support Web-service technology. Web services are moving from a middleware solution to a tool of business-process integration (BPI) by adding more functions for business-entity descriptions and business-process management.

In comparison, ebXML is the successor of EDI (electronic data interchange). ebXML is sponsored by UN/CEFACT (United Nations Centre for Trade Facilitation and Electronic Business) and OASIS (Organization for Advancement of Structured Information Standards). It is the latest achievement in a long line of business-integration paradigms that include EDI, ANSI X12 (American National Standards Institute X12; X12 stands for the originator of this standard, the Accredited Standards Committee X12 [ASC X12]), EDIFACT (electronic data interchange for administration, commerce, and transport), EAI (enterprise application integration), XML-EDI, B2Bi (business-to-business integration), and BPI. Compared to Web services, ebXML is more at the executive business level (Alonso, Casati, Kuno, & Machiraju, 2003). Although currently there is a lack of software tools implementing ebXML specifications, existing Web-service software can be modified as an implementation of ebXML specifications through binding.

In this chapter, we discuss relevant aspects of the two technologies and compare their capabilities from an e-business point of view. We see a B2B process as following. Before doing business with someone, a business needs to find its partner. While negotiating with this potential partner, documents and messages must be processed via reliable and secure channels, such as post or courier services. Those documents must be designed in a semantic fashion in forms that both partners understand. In order to ensure smooth business operation, the companies will have to agree upon the processes the resulting transactions are to follow. Ultimately, a contract or trading-partner agreement (TPA) must be signed to establish this new business relationship. Therefore, we compare the two technologies from the above aspects. We point out the capabilities and the limitations of both and discuss trends in the future development of both technologies. This also helps the readers to make right decisions about choosing the specifications and implementation software when facing a new B2B integration project.

Overall Functionality

Both Web services and ebXML put their service entities on a network and have means for service description, service discovery, and service invocation. A Web service adopts a service-oriented architecture (SOA) with three kinds of parties: service providers, service requesters, and service registries (as shown in Figure 1). The service providers register their service descriptions in the service registries for service-discovery purposes. The service requesters search the service registries for services that meet their requirements. The service requesters then can communicate with the service providers directly and use their services. Similar to Web services, ebXML also has a service register to collect the service descriptions. Different from Web services, however, the business partners are not distinguished as service providers or requesters, but are treated as the same role of business partners. The service discovery and invocation are similar to Web services (details in this section). For some people in the ebXML community, ebXML is not an SOA solution. If we consider SOA as a kind of architecture in computing technology, the argument is true that SOA is a solution to software-component reuse, analogous to object-oriented architectures. However, we can expect that the implementation of ebXML should be a kind of SOA that comprises loosely joined, highly interoperable application services. In fact, the current practice shows that ebXML adopts some SOA technology such as SOAP.

In Web services, the interactions among the parties are implemented in a straightforward manner. The communications between any parties use SOAP, which is based on Internet protocols and XML technology. It is exactly SOAP that makes Web services interoperable across the platforms and programming languages. UDDI is the protocol used by a service registry to describe the information of the services. One important piece of information in the business descriptions is the URI (uniform resource indicator) for the WSDL file. WSDL is an XML file describing how the service can be invoked from a software-engineering point of view.

Web-services invocation is similar to an RPC (Figure 2). The client side wraps the parameters for a remote function call into a SOAP message using the encoding

Figure 1. Both Web services and ebXML have means for service description, service discovery, and service invocation

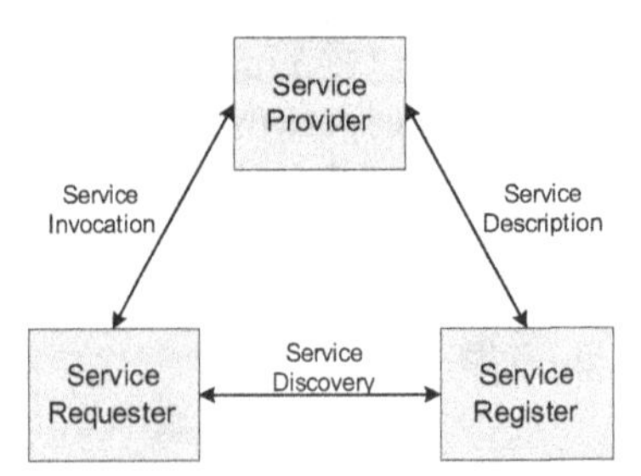

Figure 2. A Web service follows a simple RPC-like communication pattern.

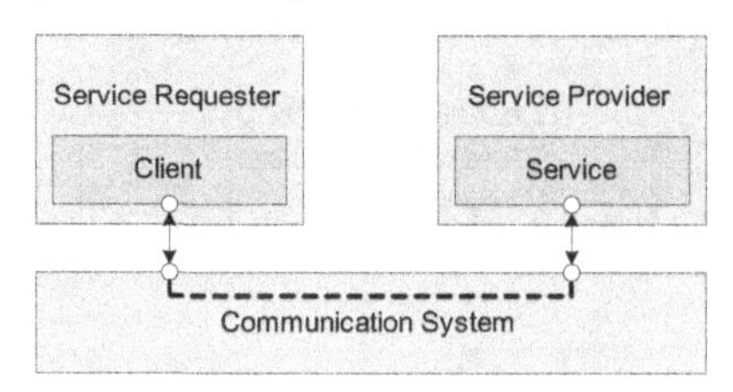

convention (marshaling). The SOAP message is transported to the server end and unwrapped. The parameter information is used to invoke the service. The same method is used to send information back to the client side. Many companies and W3C are working to move Web services beyond the function of RPC. For example, standards are being suggested for business-process modeling (see the section about business-process modeling).

The interactions among the parties are far more complicated in an ebXML-enabled system than for Web services. ebXML is geared toward the business-oriented collaboration of arbitrary partners. It works in two phases.

Implementation Phase

A company that wishes to enter a new business sector queries the ebXML registry to determine if third parties, such as existing vertical standardization organizations (e.g., ODETTE, Organization for Data Exchange by Tele Transmission in Europe, for the European automotive industry), have already placed an industry profile there. This profile contains business processes, conventions of this sector, the specific documents and forms used, and the rules on how to do business in this industry. If such a profile already exists, the new company downloads it and adapts its own system to comply with these rules and processes. This is a manual step that is needed only once when a business enters a new business sector as opposed to once per business partner when using Web services. One can reasonably assume that a company changes its business sectors far less often than its business partners. Nevertheless, this manual adaptation can be further reduced if the provider of the company's business system (e.g., SAP, Systeme, Anwendungen, Produkte in der Datenverarbeitung; it is the third largest independent software supplier in the world and is known for its enterprise software products; see http://www.sap.com) provides templates for all existing business sectors. Even then, however, the newcomer has to decide which of the many processes described in the industry profile it wishes to support. The technical parameters of the message-exchange capabilities are described by the collaboration-protocol profile (CPP). This CPP is uploaded to the ebXML registry so that other companies can find it.

Figure 3. The implementation phase of ebXML prepares the system of a joining company for business collaboration within a vertical industry branch

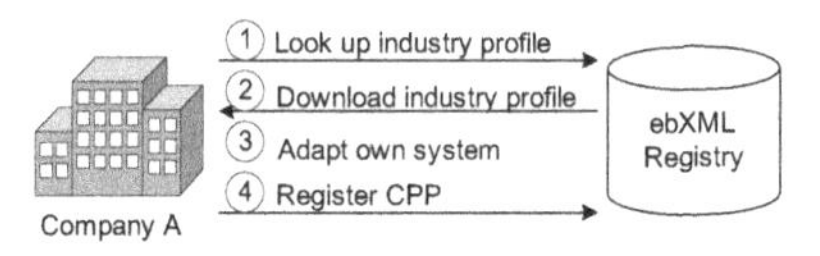

Run-Time Phase

Any other company can now download this CPP from the registry. Assume Company B downloads the CPP of Company A. Company B can compare those constraints to its own guidelines and rules and propose a collaboration-protocol agreement (CPA), which is the agreed technical parameters for message exchange. If the proposal complies with the rules defined in the CPP, Company A will agree to it and business transactions can begin. A CPA does not cover all aspects the companies may want to agree on. It is just the technical part of a trading-partner agreement. ebXML currently defines only CPP and CPA. The business-related agreements seem to involve paperwork and human beings.

We point out that we do not include the design phase of ebXML in this chapter. It is because there is no explicit equivalent process in Web-service standards. In short, the design phase in ebXML standards defines the work flows and worksheets that are used for business-process acquisition and modeling. Any organization can describe ebXML-compliant business processes. One can check Chappell et al. (2001) for more information on the design phase.

From the implementation phase and run-time phase, one can see that ebXML is a much more complex system than Web services. Indeed, it is true that Web-services-enabled systems can be implemented very quickly if developers use the existing powerful libraries. Those libraries allow developers to accomplish technology-centric

Figure 4. The steps of the run-time phase of ebXML can be carried out automatically; this, however, does not mean that manual intervention is not possible

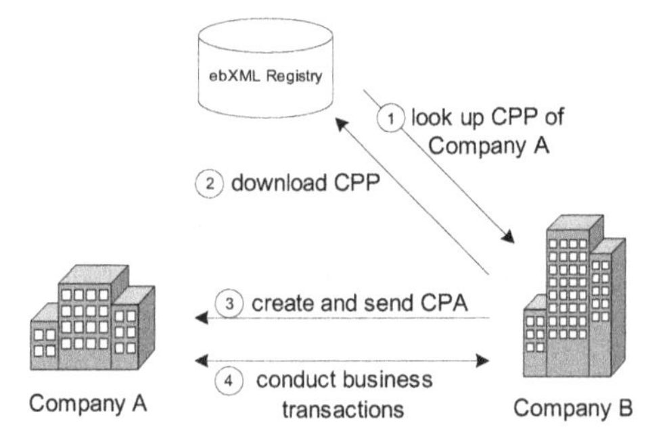

integration assignments quickly. But from a business-integration point of view, Web services have two major drawbacks.

1. **Web services rely on stubs.** A client must implement one stub for each service with which it is to interact. In a rapidly changing business environment where companies maintain collaborations with hundreds or thousands of arbitrary international partners, such technology-oriented bottom-up architecture might prove to be too limited.
2. **Web services describe systems, not businesses.** More precisely, Web services describe the parameter types for service invocation from the software-engineering point of view, but not the semantics of the parameters from the business point of view. While this is valid for simple Web services, the limitations when dealing with more complex business scenarios are quite evident. New standards and research efforts are trying to change this.

On the other hand, the major drawbacks of ebXML are its complexity and the fact that implementations are still rare. While it is possible to have a simple Web service up and running within minutes, a simple ebXML system will require much more effort and time. Accordingly, small technology-oriented integration scenarios remain the strength of Web services.

Message Transport

In order to convey messages between service providers and requestors, both ebXML and Web services use SOAP. SOAP can be transferred via arbitrary protocols, yet the only binding defined in the SOAP specification 1.2 (Gudgin, Hadley, Mendelsohn, & Moreau, 2003) is of SOAP to HTTP (hypertext transfer protocol).

All SOAP messages are XML documents, but the structures of the SOAP messages are different between Web services and ebXML (see Figure 5; Barton, Thatte, & Nielsen, 2000).

SOAP messages for Web services always contain a SOAP envelope and a SOAP body inside the envelope. The payload is in the SOAP body. The SOAP messages may contain an optional SOAP header that provides a mechanism for adding information about the message. For example, information about message routing, authentication, and transaction management can be contained in the SOAP header.

Since Web services are built on top of the XML serialization architecture, the data types are limited to those defined in the XML schema. Though some implementations from companies provide mechanisms to define complex data types, it should not be encouraged because it introduces potential interoperability problems. The transfer of binary data, such as images, can only be performed using something like byte arrays. But the communicating parties must agree on the format before being able to parse the data. Floating-point data can also bring in interoperability problems between different programming languages (Cohen, 2002).

The biggest difference in SOAP messages for ebXML as compared to those for Web services is that ebXML uses multipart MIME (multipurpose Internet mail extensions) attachments for the payload. The ebXML message header is divided into the two parts that are placed into the SOAP header and SOAP body, respectively. Part 1 of the ebXML message header contains mandatory information, such as routine information, and optional information such as error control or signatures. Part 2 of the ebXML message header contains the manifest information that is mainly an index of the payload. The ebXML payload is entirely stored in one or more MIME attachment. This enables ebXML to easily transmit binary data such as picture catalogues. We point out that SOAP messages for Web services also support MIME attachments. This is exactly the reason why ebXML adopts SOAP instead of developing its own messaging mechanism.

Since ebXML itself relies on its core components, interoperability issues are not likely to occur. However, it is possible to use any type of payload within ebXML, so it is necessary to be aware of the common issues associated with the chosen payload format. For migration purposes, it is necessary to be able to process discretionary payloads.

The basic SOAP specifications do not fully satisfy e-business demands and requirements for security and reliability. This is why OASIS (2002b) defines a set of layered extensions to the basic SOAP specifications for ebXML. Those extensions define a

Figure 5. The ebXML message container (right) is more complex than the rather simple Web-services envelope (left)

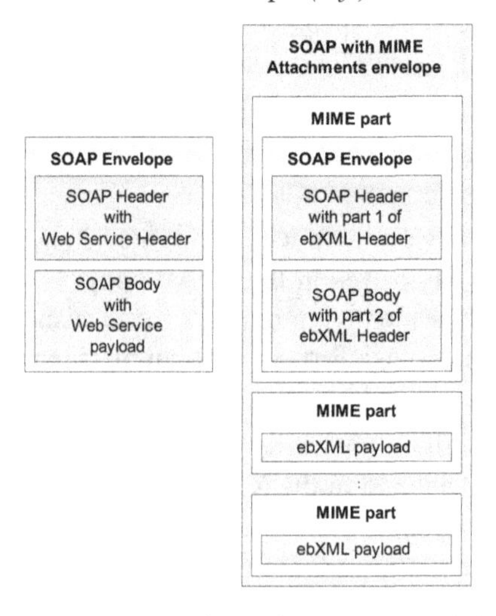

message-reliability layer that handles the delivery and acknowledgement of ebXML messages to ensure QoS (quality of service) and transactions.

Security

The Web-services specification does not define a security framework, but refers to the SOAP extensibility model. The W3C has specified those extensions for digital signatures, encryption, credentials, and authentications. It remains the responsibility of the provider of a Web service to implement security services. This may not always be done.

Since business transactions require integrity, confidentiality, and availability of all participating systems, the ebXML Security Team conducted a risk assessment in UN/CEFACT & OASIS (2001b) and proposed further specifications for ebXML. Even though there is no complete security model defined for the overall ebXML specification yet, available security technologies are integral parts of the subspecifications defining business processes, trading partners, registry and repository, and the messaging layer. For example, the CPP or CPA defines authorization, authentication, and confidentiality. It also provides the means to create tamperproof documents.

Service Discovery

To locate a service, Web services use UDDI, while ebXML relies on a registry with repositories.

The platform-independent, XML-based UDDI is a directory service that allows access to WSDL information using a SOAP interface. A UDDI business registration consists of three main components.

- **White Pages:** Business names, contact information, human-readable descriptions, and identifiers such as tax IDs
- **Yellow Pages:** Services and products index, industry codes, and geographic index
- **Green Pages:** E-business rules, service descriptions, application invocation, and data binding

The contents of this register can be found using API (application programming interface) calls.

In ebXML, the registry contains descriptions of business artefacts and the (distributed) repositories actually store them. Those business artefacts are usually the following (OASIS, 2001).

- Business-process and information metamodels
- Business library
- Core library
- Collaboration-protocol profiles
- List of scenarios
- Messaging constraints
- Security constraints

Apart from those objects, the ebXML registry is designed to handle arbitrary information fragments such as XML schema, documents, process descriptions, ebXML core components, context descriptions, UML (unified modeling language) models, information about parties, and software components (OASIS, 2002a). Compared to UDDI, the ebXML repositories are intended for more general-purpose storage. UDDI is more specialized and geared toward the type of information that can be stored in the white, yellow, and green pages. While UDDI stores mostly flat lists, ebXML is capable of handling classification information and information about relationships between business artefacts.

Generally, the UDDI model focuses on middleware connectivity and describes the systems companies use through XML. In contrast, ebXML standardizes the way XML is used in B2Bi.

Semantics

When carrying out business transactions, business information (e.g., documents) must be processed. The semantic triangle (Figure 6) illustrates that the term associated with a concept must be defined according to the context of the communication. There is an ambiguity of linguistic terms and the objects to which they refer. In a bilateral environment, it might be acceptable for two partners to define proprietary and nonreusable document formats for their transactions. If we want the business process and the business information to be reusable on a global basis, we need to add a layer of core components.

In the example in Figure 7, companies use different terms, supplier or contractor, for the same concept. If both companies map their concepts to an object whose identifier is CC 0815, they may talk to each other without ambiguity. Such an agreed-upon

Figure 6. The semantic triangle describes, among other things, the fact that several linguistic terms might describe the same object

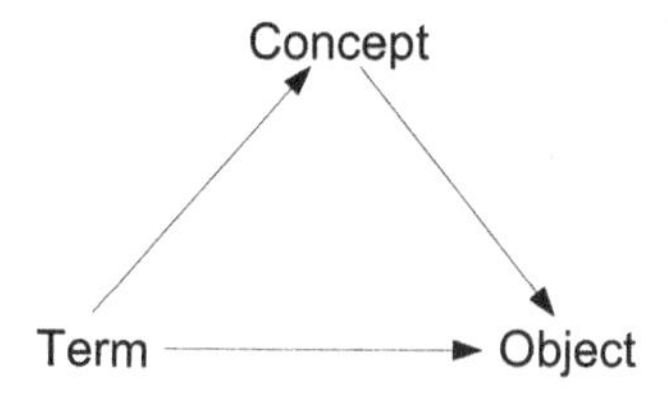

Figure 7. Unique identifiers can solve semantic problems by introducing a common vocabulary so that all business partners refer to the same unique object, even if they use different terms

use of common descriptors is vital to industries or businesses if they want their systems to carry out cross-company processes.

Core components in ebXML are the standardized data elements that are used for constructing electronic business documents. In other words, core components are building blocks that serve as the basis to assemble business documents so that the business document can be mutually understandable in business collaboration. Simply, a core component is the object identified as CC 0815 in figure 7.

Core components are in fact the generic representations of information on UML object classes (UN/CEFACT, 2003). Because UML class diagrams have four categories of elements, there are four categories of core components: aggregate core components (ACCs) that represent object classes; basic core components (BCCs) that represent simple properties of object classes; association core components (ASCCs) that represent relations between object classes, where one object class is the (complex) property of another object class; and core component types (CCTs) that define the type of information that a basic core component may contain, like text, a number, or a date. Each aggregate core component, basic core component, and association core component is given a unique name under which the core component can be found in a registry or dictionary. This name is therefore called a dictionary-entry name. The dictionary-entry name consists in principle of three parts or terms: the object-class term (the name of the object class), the property term (the property the core component is representing), and the representation term (the name of the data type that is derived from the core-component type).

The so-called context driver defines the environment where the business process is engaged. The specific business-information entities that are contained by a business document can be derived contextually from the more generic core components.

- **Example:** When a business process or document contains a *date of order* item, its North American (ISO, International Organization for Standardization) representation will be YYYY-MM-DD (where each Y is a digit in the year, M is a digit in the month, and D is a digit in the day), while the European representation of the same component will be DD-MM-YYYY. A context driver can translate the *date of order* core component into the proper format according to whether the geographical context is Europe or North America.

Being able to use core components to create new documents that are mutually understandable is a very powerful semantic instrument. This flexible tool can help diminish the semantic gap of EDI technologies, but only if it is globally accepted and widely adopted. At the same time, the EDI history suggests that core components alone might not be able to close the semantic gap entirely (Kelz, 2004).

Since the WSDL standard for Web services only defines syntax and does not include any semantic definitions, it is the responsibility of the service provider to deal with the resulting problems. To close this semantic gap, one can use the recent OASIS standard UBL (universal business language), which is based on xCBL (XML common business library) and is harmonized with ebXML core-component specifications (OASIS, 2004). UBL defines a set of standard business documents that build a common business vocabulary. Those documents can be used as a semantic layer for existing technologies such as Web services even though the EDI history suggests that it is unlikely that UBL will be the lingua franca of e-business. Nevertheless, UBL can be used to add interoperability to Web services (Gertner, 2002) or to migrate from Web services to ebXML.

Business-Process Modeling

Business transactions of any kind follow certain processes to ensure smooth business operation with predictable and agreed-upon behaviour of the participating parties. In the past, those processes were usually not formalized. Modern companies use modeling tools such as ARIS (an integrated product of the IDS-Scheer AG for the design, implementation, and controlling of business processes; http://www.ids-scheer.de) to represent, formalize, understand, and ultimately optimize the processes relevant to their own organization.

Figure 8. The goal of business-process integration is to integrate the existing systems of individual companies into a single cooperative operating system

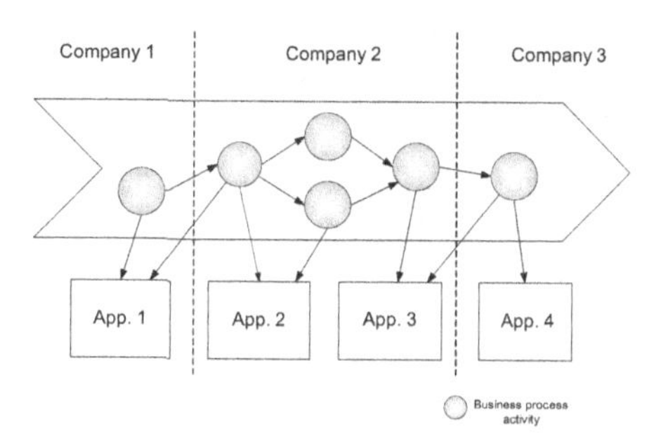

Though it might be possible to develop and enforce a proprietary business model for internal processes (e.g., by using an integrated platform such as SAP), this is not feasible for transactions that go beyond company boundaries. Therefore, the goal of BPI is to integrate the systems of individual companies to carry out business processes smoothly based on changing customer requirements and varying partners. Figure 8 shows how the applications of different companies are integrated to work cooperatively on the same business process.

The great challenge of BPI is to find and introduce a global and cross-industry standard to formalize business processes so that individual companies can interact in this manner. Following the general movement in the e-business community, such a standard should create a machine-readable definition of interactions between business partners to build a declarative system rather than a procedural one (Chappell et al., 2001). In addition, the transactions between partners cannot be repudiated, and have to be legally binding and transmitted in a reliable manner.

The innovative business-process specification schema (BPSS) among ebXML standards promises to solve the above problems. BPSS "provides a standard framework by which business systems may be configured to support the execution of business collaborations consisting of business transactions.... The Specification Schema supports the specification of Business Transactions and the choreography of Business Transactions into Business Collaboration" (UN/CEFACT & OASIS, 2001a).

BPSS provides the semantics, elements, and properties necessary to define business collaborations rather than business processes. BPSS defines the roles that partners may fulfill. It consists of one or more choreographed business transactions and describes the type of business information that needs to be exchanged. BPSS can be used independent of ebXML to capture and communicate business processes that can be understood by all participating parties without ambiguity.

A BPSS instance is composed of the following:

- Business documents
- Business transactions (protocol to exchange the documents)
- Binary collaborations (collaboration of transactions)
- Multiparty collaborations (composition of one or more binary collaborations)
- Substitution sets (replacing existing document definitions for the purpose of specializing collaboration definitions for a specific industry)

In summary, a BPSS instance specifies all business messages and their content, sequence, and timing.

BPSS is designed to accommodate any kind of payload, so it is possible to use the ebXML core-component framework to design machine-readable business documents. In order to ensure message reliability, BPSS provides a message-reliability layer that is distinct from the ebXML messaging-service layer. The aspect of nonrepudiation is based on digital signatures as specified by the W3C XML-DSIG, while legally binding transactions are created by simply using an associated property within a binary collaboration. Substitution sets allow for existing vertical standardization organizations to define reusable process specifications.

The Web-services community also works hard to enable business modeling and work-flow management. Some of those standards are the business process execution language (BPEL) and business process modeling language (BPML): languages that enable Web-service composition and Web-service choreography.

BPEL describes the following:

- The sequence of activities
- The triggering conditions of activities
- The consequences of executing activities
- Partners for external activities
- The composition of Web services
- The binding to WSDL

The abilities and scopes of BPEL and BPML do not differ significantly (Mendling & Müller, 2003). One of the major disadvantages for both is that both can automate a sequence of messages but cannot execute actual transactions. While the ability to automate transactions is essential for a full-scale e-business system, such as one that uses ebXML, even the automation of a few steps leading to a transaction can be a big cost saver. For smaller scale systems, BPEL or BPML might just be the tools to add some aspects of e-business to existing Web-services systems (Fogarty, 2004).

Since they do not provide data transformation, human work flow, trading-partner agreements, or the support of existing business protocols, BPEL and BPML could certainly be seen as inferior when compared to ebXML. But those standards do not promise to provide full-scale e-business over Web services. They aim to compose Web services, which is precisely what they do. There are other standards, such as Web services choreography interface (WSCI), Web services conversation language (WSCL), and Defence Advanced Research Projects Agency (DARPA) agent markup language-service (DAML-S), that aim to solve particular problems in the field of business-process modeling.

The big difference between BPEL and BPSS is the point of view from where the collaboration is described. BPSS describes the collaboration from a neutral view; that is, it describes how party A and party B interact. BPEL describes a collaboration from the point of view of the involved partners, that is, how party A interacts with party B and party C. If B and C interact in the same multiparty collaboration as well, this cannot be seen from the BPEL file of party A. Currently, the W3C conducts the work on Web service choreography description language (WS-CDL), which describes a choreography of Web services from a neutral perspective. From the above, one can see that BPEL supports multiparty definitions. For BPSS, although there is a tag for multiparty collaboration, it is composed by several binary collaborations.

Currently, all the modeling languages in Web services have software implementations. BPSS has no direct implementations. However, it is possible that by binding existing implementations from Web services to BPSS specifications, BPSS can be implemented. Chappell et al. (2001) gives binding between BPML and BPSS, and binding between XLANG and BPSS.

Trading-Partner Agreements

Most operational e-business infrastructures focus on the automation of established (static) business relationships, where the partners already know each other and have made arrangements with which to carry out business. The e-business system simply automates those existing arrangements. However, the e-business community suggests the development of systems that support highly dynamic business relations. Such a system must be able to automate the process of setting up new collaboration agreements on an ad hoc and time-limited basis.

Currently, ebXML defines CPP and CPA, which are the technical parts of a trading-partner agreement. More specifically, CPP and CPA define the technical run-time environment.

Within ebXML, this demand is addressed through the CPPs and CPAs. A CPP defines the technical parameters of the message-exchange capabilities, and a CPA is the agreed technical parameters for message exchange. Previously, we described how they are used when an ebXML forms a process. CPP and CPA define the technical run-time environment of the collaboration.

Web-services specifications do not allow descriptions similar to CPP, and there is no agreement between partners like CPA. The protocol binding is fixed by the service provider. It is a simpler but less flexible solution.

Industrial Support and Compliance

Web services are well accepted and supported by industrial companies and W3C. Many large companies, such as SUN, IBM, Microsoft, HP, and SAP, have their implementations of Web-services specifications, such as SOAP, WSDL, and UDDI. Information about these software packages are not difficult to find from their Web sites. Many other service-providing companies, such as Amazon.com, Google, and eBay, use SOAP as an interface to their platform. Obviously, Web services become a strategic direction in e-business companies. Hogan (2003) reports that International Data Corporation (IDC) predicts global spending for Web services will be $15.2 billion in 2008, up from $3 billion in 2003. Correia & Cantara (2003) report that by 2006, 99% of all new products for application integration will have some level of support for Web services, while the market for Web-services-enabled IT professional services will be worth $29 billion.

Compared to Web services, ebXML is less accepted. UN/CEFACT Techniques and Methodologies Group (TMG) estimates that the acceptance rate of ebXML is only about 3% of that of Web services. ebXML is especially less accepted by small and medium enterprises. However, there are still many implementation projects from various organizations and companies. Here, we list just some of the players.

- Sun Microsystems (http://www.sun.com/software/xml/developers/regrep/)
- Korea Institute of eCommerce (http://www.ebxml.or.kr/)
- Korea Trade Network (http://www.GXMLHub.com/com/english/index.html)
- XML Global (http://www.xmlglobal.com)
- XML.gov registry (http://xml.gov/registries.htm)
- Data Interchange Standards Association (DISA): Open Travel Alliance and Interactive Financial Exchange Forum (http://www.disa.org/drive/)
- Seeburger (http://www.seeburger.com)
- Drummond Group (http://www.drummondgroup.com/)
- Sterling Commerce (http://www.stercomm.com/)

Yet many other companies, such as bTrade, U.S. Centers for Disease Control (CDC), Cyclone Commerce, eXcelon, Fujitsu, GE Global eXchange Services (GXS), IPNet Solutions, and Sybase, have ebXML projects.

While Web services are a well-adopted standard for system integration throughout business sectors, ebXML still lacks industry support. However, it is quite evident that soon ebXML will be the state-of-the-art technology for global cross-company and cross-industry system integration. When a business is planning its overall system-integration strategy or specific integration tasks these days, it is advisable to keep emerging standards such as ebXML in mind. In order to reduce the cost for system integration and interface building, companies might want to aim for a consistent integration strategy that leads to uniformity of system interfaces. Existing strategies might have focused on in-house applications only, treating gateway systems as a whole different world. However, as indicated earlier, it is possible to merge both realms.

Since Web services and ebXML use the same technological foundations, the task of (slowly) migrating from one technology to the other does not require exchanging the underlying infrastructure. At the same time, even a step-by-step migration is possible. Standards such as UBL can add ebXML-compatible semantics to Web services, while the implementation of the ebXML messaging service allows Web services to use secure and reliable message transfer. Since ebXML is modular and uses the same technologies as Web services, businesses can pick individual modules to deal with the integration tasks at hand. At the same time, they protect their investments because they ensure that the modules they implement now for use with existing Web-service interfaces can still be used if the system is switched entirely to ebXML in the future.

However, even if no such full migration is wanted, companies can take advantage of the fact that, if they use Web services for in-house integration and ebXML for cross-company integration, they use compatible technologies. Plus, they can always upgrade individual modules without the need to use different experts for internal and external interfaces.

Conclusion

Web services and ebXML have many things in common and can complement each other. Both technologies provide solutions to integration problems, both use XML over the Internet for message interchange, and both approaches share a common high-level architecture. Observing the e-business world reveals the evolution from tactical systems with limited scope to strategic e-business initiatives. This does not mean, however, that Web services will soon be abolished and replaced by ebXML.

Web services are a well-established and widely adopted standard. A multitude of experienced developers use the numerous available libraries and frameworks to guarantee short time to market for their products. In addition to those strengths, the Web-services domain is much broader than that of ebXML, and its architecture is simpler and easier to handle. As a successor of other middleware technologies, Web services excel in intra-enterprise request-response-type application-integration environments.

At the same time, real-life business, especially in the B2B domain, is far more complicated than a collection of request-response pairs. This is why many initiatives have begun to add layers of powerful business functionality, such as reliable messaging, security, and business-process orchestration, to Web services. But while these aspects were successfully defined within ebXML, the Web-services community could endanger all its efforts through divergence over those technologies.

If Web services want to be more than a middleware standard for intra-enterprise application integration, the Web-services community will have to specify the layers of business standards used to support the complex and collaborative business transactions that organizations demand.

On the other hand, ebXML is a complete solution focused on B2B integration scenarios. It is not surprising that ebXML excels whenever it comes to interenterprise business-process integration. But ebXML is also suitable for intra-enterprise business-process integration, especially when departments of large enterprises are treated as separate companies. Moreover, since ebXML is modular, an enterprise could use single ebXML modules for in-house application-integration projects (e.g., pick the ebXML messaging service to add reliable and secure message transfer to an enterprise application-integration project).

The major drawbacks of ebXML are that the specification is not entirely complete and that industry support is still lacking. If industry fails to provide affordable implementations of ebXML, this standard might follow the destiny of EDIFACT, which was not widely adopted due largely to its cost. Since ebXML is powerful, implementations are likely to be complex and might not be easy to handle. Templates for the most common demands of companies might help to decrease the time to market for system providers that use ebXML implementations.

For the global community, an open ebXML initiative is likely to trigger a whole new industry that could have the potential to change the way we view system integration. So far, several attempts have been made to provide an open-source implementation of ebXML, but none has reached a level of maturity that suggests use in commercial applications.

While ebXML is always intended for e-business, Web services are a bottom-up technology that focuses on the technical aspects of middleware functionality. However, for many integration projects (especially in house), companies do not need full-grown e-business suites. Instead, they need smaller, more reliable, and easier-to-handle technologies that have reached a sufficient level of maturity.

One interesting topic for system architects might be to create migration paths between Web services and ebXML by taking the modules of ebXML and enabling them to be used with Web services, while at the same time suggesting a step-by-step migration path. Companies that already use Web services might be more interested in using certain aspects of ebXML in conjunction with their existing Web-services infrastructure. As their products evolve, they might consider adding more modules until their product is, in fact, a full ebXML framework. If such a migration follows a specified plan, migration issues can be reduced.

References

Alonso, G., Casati, F., Kuno, H., & Machiraju, V. (2003). Web services: Concepts, architectures and applications. Heidelberg, Germany: Springer Verlag.

Barton, J., Thatte, S., & Nielsen, H. S. (2000). *SOAP messages with attachments.* Retrieved January 29, 2005, from http://www.w3.org/TR/2000/NOTE-SOAP-attachments-20001211

Chappell, D. A., Chopra, V., Dubray, J.-J., Evans, C., van der Eijk, P., Harvey, B., et al. (2001). *Professional ebXML foundations.* Birmingham, United Kingdom: Wrox Press Ltd.

Cohen, F. (2002). *Understanding Web service interoperability.* Retrieved December 2004 from http://www-106.ibm.com/developerworks/webservices/library/we-inter.html#4

Correia, J., & Cantara, M. (2003). *Gartner sheds light on developer opps in Web service.* Integration Developers News LLC. Retrieved January 29, 2005, from http://idevnews.com/IntegrationNews.asp?ID=69

Fogarty, K. (2004). *Business process execution language.* Ziff Davis Media. Retrieved January 29, 2005, from http://www.baselinemag.com/print_article2/0,2533,a=123575,00.asp

Gertner, M. (2002). UBL and Web services. *XML-Journal, 3*(6), 16-19.

Gudgin, M. (2003). *SOAP version 1.2 part 2: Adjuncts.* W3C. Retrieved January 29, 2005, from http://www.w3.org/TR/2003/REC-soap12-part2-20030624/

Gudgin, M., Hadley, M., Mendelsohn, N., & Moreau, J. (2003). *SOAP specification 1.2.* Retrieved from http://www.w3.org/TR/soap12-part1/

Hogan, J. (2003). *Gartner: Web services projects riding out budget cuts.* Retrieved January 29, 2005, from http://WebServices.com

Kelz, W. (2004). Allheilmittel? Die universal business language. *XML Magazine & Web Services.* Retrieved January 29, 2005, from http://www.xmlmagazin.de/itr/online_artikel/psecom,id,571,nodeid,69.html

Mendling, J., & Müller, M. (2003). *A comparison of BPML and BPEL4WS.* Retrieved January 29, 2005, from http://wi.wu-wien.ac.at/~mendling/talks/BXML2003.pdf

Organization for Advancement of Structured Information Standards (OASIS). (2001). *ebXML technical architecture specification v. 1.0.4.* ebXML Technical Architecture Project Team. Retrieved January 29, 2005, from http://www.ebxml.org/specs/ebTA.pdf

Organization for Advancement of Structured Information Standards (OASIS). (2002a). *ebXML registry information model.* ebXML Registry Technical Committee. Retrieved January 29, 2005, from http://www.oasis-open.org/committees/ regrep/documents/2.1/specs/ebrim_v2.1.pdf

Organization for Advancement of Structured Information Standards (OASIS). (2002b). *Message service specification version 2.0.* ebXML Messaging Services Technical Committee. Retrieved January 29, 2005, from http://www.oasis-open.org/committees/ ebxml-msg/documents/ebMS_v2_0.pdf

Organization for Advancement of Structured Information Standards (OASIS). (2004). *Universal business language 1.0.* Retrieved January 29, 2005, from http://docs.oasis-open.org/ubl/cd-UBL-1.0/

United Nations Centre for Trade Facilitation and Electronic Business (UN/CEFACT). (2003). *Core components user's guide.* Retrieved January 29, 2005, from http://www.ecp.nl/ebxml/docs/cc_ug_oct03.pdf

United Nations Centre for Trade Facilitation and Electronic Business (UN/CEFACT) & Organization for Advancement of Structured Information Standards (OASIS). (2001a). *ebXML business process specification schema version 1.01.* Retrieved January 29, 2005, from http://www.ebxml.org/specs/ebBPSS.pdf

United Nations Centre for Trade Facilitation and Electronic Business (UN/CEFACT) & Organization for Advancement of Structured Information Standards (OASIS). (2001b). *ebXML technical architecture risk assessment version 1.0.* ebXML Security Team. Retrieved January 29, 2005, from http://lists.oasis-open.org/archives/ security-consider/200103/pdf00000.pdf

Chapter XI

Leveraging Pervasive and Ubiquitous Service Computing

Zhijun Zhang, University of Phoenix, USA

Abstract

The advancement of technologies to connect people and objects anywhere has provided many opportunities for enterprises. This chapter will review the different wireless networking technologies and mobile devices that have been developed, and discuss how they can help organizations better bridge the gap between their employees or customers and the information they need. The chapter will also discuss the promising application areas and human-computer interaction modes in the pervasive computing world, and propose a service-oriented architecture to better support such applications and interactions.

Introduction

With the advancement of computing and communications technologies, people do not have to sit in front of Internet-ready computers to enjoy the benefit of information access and processing. Pervasive computing, or ubiquitous computing, refers to the use of wireless and/or mobile devices to provide users access to information or applications while the users are on the go. These mobile devices can be carried by the users, or embedded in the environment. In either case, these devices are connected, most likely through a wireless network, to the Internet or a local area network (LAN).

Mobile technologies come in a large variety and are ever changing. In order to gain the business value of pervasive computing, and at the same time keep the supporting cost under control, it is important to develop an architecture solution. A service-oriented architecture (SOA) would allow an enterprise to easily provision functions to be accessible by certain types of pervasive channels. A service-oriented architecture would also make it possible to quickly integrate data generated by pervasive devices and make them available in the form of an information service.

In this chapter, we will first look at the communication networks and mobile devices that create the various information-access and information-generation touch points in a pervasive computing environment. Then we will discuss the applications and interaction models for pervasive computing. Finally, we will describe a service-oriented architecture that an enterprise can adopt in order to effectively and efficiently support pervasive computing.

Mobile Communication Networks

Mobile communication technologies range from personal area networks (PANs; a range of about 10 meters) and local area networks (a range of about 100 meters) to wide area networks (WANs; a few kilometers). From a network-topology perspective, most networks are based on a client-server model. A few are based on the peer-to-peer model.

Wireless PANs

A wireless personal area network allows the different devices that a person uses around a cubicle, room, or house to be connected wirelessly. Such devices may

Table 1. Summary of the wireless PANs

Technology	Radio Frequency	Maximum Distance	Data Capacity
Bluetooth	2.4 GHz	10 meters	721 Kbps
HomeRF	2.4 GHz	50 meters	0.4-10 Mbps, depending on distance
ZigBee	2.4 GHz	75 meters	220 Kbps

include the computer, personal digital assistants (PDAs), cell phone, printer, and so forth.

Bluetooth is a global de facto standard for wireless connectivity (Bluetooth SIG, 2005). The technology is named after the 10th-century Danish King Harald, who united Denmark and Norway and traveled extensively.

HomeRF is an early technology for wireless home networking, first marketed in 2000.

The Institute of Electrical Engineers (IEEE) 802.15 wireless-PAN effort (IEEE, 2005a) focuses on the development of common standards for personal area networks or short-distance wireless networks. One technology out of this effort is ZigBee, which is based on the IEEE 802.15.4 standard.

ZigBee is a low-cost, low-power-consumption, wireless communication-standard proposal (ZigBee Alliance, 2005). Formerly known as FireFly, ZigBee is being developed as the streamlined version of HomeRF. A streamlined version would allow most of the functionality with less integration and compatibility issues.

ZigBee's topology allows as many as 250 nodes per network, making the standard ideal for industrial applications. Radio-frequency-based ZigBee is positioned to eventually replace infrared links. To achieve low power consumption, ZigBee designates one of its devices to take on the coordinator role. The coordinator is charged with waking up other devices on the network that are in a sleep mode, moments before packets are sent to them. ZigBee also allows coordinators to talk to one another wirelessly. This will allow for opportunities for wireless sensors to continuously communicate with other sensors and to a centralized system.

For enterprise computing, the wireless PANs are within the corporate firewall. They do not create new requirements for the enterprise architecture to extend access to applications. However, they do require security measures to make sure the device that is receiving information is a recognized device. It also creates an opportunity for the computing infrastructure to potentially know where a particular device, and most likely the associated user, is located. How these are handled will be discussed later in the description of the proposed service-oriented architecture.

Wireless LANs

The set of technical specifications for wireless local area networks (WLANs), labeled 802.11 by IEEE, has led to systems that have exploded in popularity, usability, and affordability. Now wireless LAN can be found in many organizations and public places.

With a wireless LAN, a user's device is connected to the network through wireless access points (APs). APs are inexpensive—many are available for less than $100—and will usually work perfectly with little or no manual configuration.

Wireless LANs use a standard, called IEEE 802.11, that provides a framework for manufactures to develop new wireless devices. The first two standards released for wireless LANs were 802.11b and 802.11a. The 802.11b standard was used in most wireless devices in the early adoption of wireless LAN. A new standard, called 802.11g, combines data-transfer rates equal to 802.11a with the range of an 802.11b network (Geier, 2002). It uses access points that are backward compatible with 802.11b devices.

Wireless technology has become so popular that many new devices, especially laptop computers, have built-in wireless LAN capabilities. Windows XP, Mac OS, and Linux operating systems automatically configure wireless settings, and software such as NetStumbler and Boingo provides automatic connections to whatever WLANs they encounter. What is more, community-based groups have furthered neighborhood area networks (NANs) to share wireless Internet access from one building to the next.

Besides 802.11a/b/g technologies that have shipped products, new technologies are emerging, including 802.11h, 802.11i, and 802.1x. The most important developments for wireless security will be contained in the 802.11i and 802.1x specifications. The 802.11i specification addresses encryption (securing the communication channel), whereas 802.1x will address authentication (verifying individual users, devices, and their access levels).

IEEE 802.1x is another authentication protocol, not an encryption protocol. 802.1x by itself does not fix the existing problems with WLAN security that relate to encryption. Therefore, attackers can still easily read network traffic on 802.1x networks. The 802.11i standard will address communication-channel encryption.

In order to increase the throughput of wireless LANs, a technology called Mimo (multiple input-multiple output) has been developed. Mimo allows for transmission rates of more than 100 Mbps, which is much greater than existing wireless LANs. Presently, wireless LANs use a single antenna operating at only one of a limited number of frequencies (channel) that are shared by all users. Mimo technology allows the use of two or more antennas operating on that channel. Normally, this would cause interference degradation of the signal because the radio waves would

Table 2. Summary of wireless LAN technologies

Technology	Radio Frequency	Maximum Distance	Data Capacity
802.11a	5 GHz	20 meters	54 Mbps
802.11b	2.4 GHz	100 meters	11 Mbps
802.11g	2.4 GHz	100 meters	54 Mbps
802.11i	A security standard for encryption on wireless LANs		
802.11n	Varies	Varies	> 100 Mbps
802.1x	A standard security protocol for user authentication on wireless LANs		

take different paths—called multipath distortion. However, Mimo uses each of these different paths to convey more information. The Mimo technology corrects for the multipath effects. IEEE is standardizing the technology as IEEE 802.11n.

For an enterprise, wireless LAN technologies allow pervasive information access throughout the campus. Employees with authorized mobile devices such as wireless laptops and PDAs will be able to get online wherever they are on the campus.

Table 2 summarizes the wireless LAN technologies.

Wireless MANs

A wireless metropolitan area network (MAN; also referred to as broadband wireless access, or WiMAX) can wirelessly connect business to business within the boundary of a city. It is becoming a cost-effective way to meet escalating business demands for rapid Internet connection and integrated data, voice, and video services.

Wireless MANs can extend existing fixed networks and provide more capacity than cable networks or digital subscriber lines (DSLs). One of the most compelling aspects of the wireless MAN technology is that networks can be created quickly by deploying a small number of fixed-base stations on buildings or poles to create high-capacity wireless access systems.

In the wireless MAN area, IEEE has developed the 802.16 standard (IEEE, 2005b), which was published in April 2002, and has the following features.

- It addresses the "first mile-last mile" connection in wireless metropolitan area networks. It focuses on the efficient use of bandwidth between 10 and 66 GHz.
- It enables interoperability among devices so carriers can use products from multiple vendors. This warrants the availability of lower cost equipment.

- It defines mechanisms that provide for differentiated quality of service (QoS) to support the different needs of different applications. The standard accommodates voice, video, and other data transmissions by using appropriate features.
- It supports adaptive modulation, which effectively balances different data rates and link quality. The modulation method may be adjusted almost instantaneously for optimal data transfer. Adaptive modulation allows efficient use of bandwidth and fits a broader customer base.

The WiMAX technical working group has developed a set of system profiles, standards for protocol-implementation conformance, and test suites (http://www.wimaxforum.org).

One particular technology for WiMAX is non line of sight (NLOS) networking (Shrick, 2002). NLOS networks provide high-speed wireless Internet access to residential and office facilities. NLOS uses self-configuring end points that connect to a PC (personal computer). The end point has small attached antennas and can be mounted anywhere without the need to be oriented like satellite antennas. Two major vendors are Navini Networks and Nokia.

With the wireless MAN technology, enterprises can quickly set up a network to provide wireless access to people in a certain area. It is very useful in situations such as an off-site working session or meeting.

Wireless NANs

Wireless neighborhood area networks are community-owned networks that provide wireless broadband Internet access to users in public areas (Schwartz, 2001). To set up a wireless NAN, community group members lend out access to the Internet by linking wireless LAN connections to high-speed digital subscriber lines or cable modems. These wireless LAN connections create network access points that transmit data for up to a 1-kilometer radius. Anyone possessing a laptop or PDA device equipped with a wireless network card can connect to the Internet via one of these community-established access points.

Wireless NANs have been established in more than 25 cities across the United States. Community-based networks differ from mobile ISPs (Internet service providers) such as MobileStar and Wayport that offer subscribers wireless access to the Internet from hotels, airports, and coffee shops. Wireless NANs extend access to consumers in indoor as well as outdoor areas, and the access is typically offered at no charge. For instance, NYC Wireless (http://www.nycwireless.net) provides Internet access to outdoor public areas in New York City. In addition, this organization is negotiating with Amtrak to bring wireless Internet access to Penn Station.

Enterprises could leverage the existing wireless NANs and equip employees with the right devices and security mechanisms in order to use these wireless networks to securely connect to the corporate network.

Wireless WANs

Wireless wide area networks are commonly known as cellular networks. They refer to the wireless networks used by cell phones.

People characterize the evolution of wireless WAN technology by generation. First generation (1G) started in the late 1970s and was characterized by analog systems. The second generation of wireless technology (2G) started in the 1990s. It is characterized by digital systems with multiple standards and is what most people use today. 2.5G and 3G are expected to be widely available 1 to 3 years from now. 4G is being developed in research labs and is expected to launch as early as 2006.

Wireless WAN originally only offered voice channels. Starting from 2G, people have used modems to transmit data information over the voice network. More recent generations offer both voice and data channels on the same cellular network.

One of the major differentiating factors among the wireless generations is the data transmission speed in which the wireless device can communicate with the Internet. The table below is a comparison of the data transmission rates of the 2G, 2.5G, 3G, and 4G technologies (3Gtoday, 2005). Both 2G and 2.5G include different technologies with different data transmission rates. Global Systems for Mobile Communications (GSM) and Code Division Multiple Access (CDMA) are 2G technologies. General Packet Radio Service (GPRS), CDMA 1x, and Enhanced Data for GSM Environment (EDGE) are 2.5G technologies.

In the United States, cellular carriers Verizon and Sprint use CDMA technology. Cingular uses GSM, GPRS, and EDGE technologies. Both Verizon and Sprint have

Table 3. Data transmission speed of wireless wide area networks

Technology	Maximum	Initial	Typical
2G: GSM	9.6 Kbps	—	—
2G: CDMA	14.4 Kbps	—	—
2.5G: GPRS	115 Kbps	< 28 Kbps	28-56 Kbps
2.5G: CDMA 1x	144 Kbps	32 Kbps	32-64 Kbps
2.5G: EDGE	384 Kbps	64 Kbps	64-128 Kbps
3G	2 Mbps	< 128 Kbps	128-384 Kbps
4G	20 Mbps	TBD	TBD

rolled out their CDMA 1x services, which is 2.5G. Cingular has rolled out GRPS service and is starting to roll out EDGE service in selected markets.

Wireless WANs are available wherever cell phones can be used. For now, they are the most pervasive wireless networks. By subscribing to a service plan, an enterprise user's laptop computer or other mobile device can connect to the Internet through the service provider's cellular towers.

Ultrawideband (UWB)

Traditional radio-frequency technologies send and receive information on particular frequencies, usually licensed from the government. Ultrawideband technology sends signals across the entire radio spectrum in a series of rapid bursts.

Ultrawideband wireless technology can transmit data at over 50 Mbps. A handheld device using this technology consumes 0.05 milliwatts of power as compared to hundreds of milliwatts for today's cell phones. Ultrawideband signals appear to be background noise for receivers of other radio signals. Therefore it does not interfere with other radio signals. Ultrawideband is ideal for delivering very high-speed wireless-network data exchange rates (up to 800 Mbps) across relatively short distances (less than 10 meters) with a low-power source.

Another feature of ultrawideband signals is that they can penetrate walls. Therefore, this technology would allow a wireless device to communicate with a receiver in a different room. This feature can also be used to detect buried bodies, people in a building, or metal objects in concrete.

Mesh Radio and Mess Networks

Mesh radio is a wireless network technology that operates in the 28-GHz range of the radio spectrum and provides high-speed, high-bandwidth connectivity to the Internet (Fox, 2001). A mesh radio network consists of antennas connected in a web-like pattern to a fiber-optic backbone. A single antenna attached to the roof of a building could provide Internet access to all of the subscribers residing in the building. Each node on the network has a small, low-power, directional antenna that is capable of routing traffic for other nodes within a 2.8-kilometer radius. In contrast to other wireless networks, mesh radio avoids many of the line-of-sight issues between the base station and each node on the network. Consequently, the configuration of mesh radio reduces the chance of encountering physical obstructions that could impede access to the network.

Mesh radio networks are being developed in two different ways. CALY Networks has developed a system that utilizes the Internet protocol (IP) as its communication mechanism, while Radiant Networks has created a system that communicates using the asynchronous transfer mode (ATM). Providers of mesh radio services include British Telecommunications, TradeWinds Communications (http://www.tnsconnects.com), and Nsight Teleservices (http://www.nsighttel.com).

Features of mesh radio include the following:

- Provides upload and download data rates of up to 25 Mbps
- Supports up to 600 subscribers per square kilometer without degradation of service
- Provides cost-effective access to broadband services in rural communities or urban areas
- Increases network capacity and resilience as the customer base grows

Different from the mesh radio technology, mesh networks enable wireless devices to work as a peer-to-peer network, using the handsets themselves instead of the radio towers to transmit data (Blackwell, 2002). Each handset would be capable of transmitting data at rates from 6 Mbps to 18 Mbps. This technology can be used for a group of users or devices communicating with each other in a peer-to-peer mode without needing an established wireless network. The technology was developed by Mesh Networks Inc., which has been acquired by Motorola.

Sensor Networks

Motes (also called sensor networks or Smart Dusts; Culler & Hong, 2004) are small sensing and communication devices. They can be used as wireless sensors replacing smoke detectors, thermostats, lighting-level controls, personal-entry switches, and so forth. Motes are built using currently available technology and are inexpensive enough to be deployed in mass quantities. Depending on the sensors and the capacity of the power supply, a mote can be as big as 8 cubic centimeters (the size of a matchbox) or as small as one cubic millimeter.

Motes are the result of a joint effort between Defense Advanced Research Projects Agency (DARPA) and the University of California, Berkeley, research labs. Most initial applications are positioned to helping the military for tasks such as surveillance of war zones, the monitoring of transportation, and the detection of missiles and/or biological weapons. Commercial mote sensors are available from Crossbow Technology.

A mote is typically made up of the following:

- A scanner that can scan and measure information on temperature, light intensity, vibrations, velocity, or pressure changes
- A microcontroller that determines tasks performed by the mote and controls power across the mote to conserve energy
- A power supply that can be small solar cells or large off-the-shelf batteries
- TinyOS, an open-source software platform for the motes. TinyOS enables motes to self-organize themselves into wireless network sensors.
- TinyDB, a small database that stores the information on a mote. With the help of TinyOS, the mote can process the data and send filtered information to a receiver.

These motes enable enterprises to constantly collect important information and send the information to the appropriate server for processing so that the appropriate response can be initiated when necessary. The motes become the generator of pervasive information that reflects the status of business processes or environmental conditions.

Pervasive Devices

Pervasive devices come in different forms and shapes. Compared to a networked computer, some pervasive devices, such as landline or cell phones, are more widely available. Other devices are simply more portable and thus can be easily carried around. Yet other devices are embedded in the environment and are able to deliver specialized information. In terms of their functions, some are for accessing the Internet, some are just for entering information while the user is on the go, and others are for storing large amounts of information and can be easily carried around.

Traditional Telephones, Pagers, and Cell Phones

Traditional landline telephone has been the most pervasive communication device around the world. Voice markup languages such as VoiceXML (voice extensible markup language; Rubio, 2004), together with supporting technologies such as the voice browser and voice gateway, has made the traditional telephone yet another device for connecting the user to the Internet. With speech recognition, users can choose to use touch tone or simply say what they need. Figure 1 shows how a telephone can be used to connect to the Internet.

Figure 1. Voice gateway connects the phone network with the data network

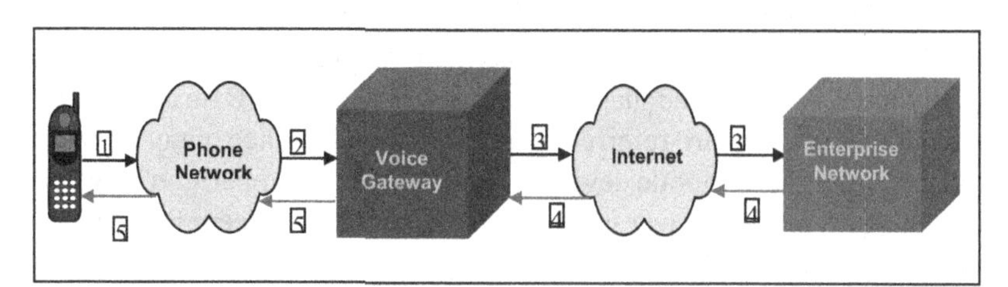

1. The user dials the number from any phone (landline or mobile).
2. The call is routed to the corresponding voice gateway, which maps the phone number to a particular application hosted at the enterprise network.
3. The voice gateway knows the URL (uniform resource locator) of the application. It uses an HTTP (hypertext transfer protocol) request to fetch the first dialog of the application.
4. The enterprise Web server and application server return the dialog to the gateway in the form of a VoiceXML document.
5. The gateway interprets the VoiceXML document, plays the greeting, and asks the user for input. Now the user can use touch tone or speech to provide input. Based on the user input and the application logic as described in the VoiceXML file, the voice gateway decides what dialog to fetch next from the enterprise network.

Pagers allow users to receive alerts with a limited amount of text. With two-way pagers, users can also reply with a text message.

With cell phones (not smart phones), besides the same communication capabilities of a landline telephone, most users can use short-message service (SMS) to send and receive text messages. This is good for near-real-time conversational communications.

Smart Phones, Wireless PDAs, and Blackberry Devices

Smart phones are cells phones that have both voice and data capabilities. Such a cell phone comes with a mini Web browser and thus can be used to access Internet content. However, since the smart phones typically have rather small screens, they can only access pages specifically designed for small screens and coded in a special markup language such as Wireless Markup Language (WML). Some smart phones are equipped with a computing platform such as the Java Virtual Machine that can run applications written in J2ME (Java 2 Micro Edition).

Wireless PDAs typically have larger screens than cell phones and can directly access HTML (hypertext markup language) pages. Some wireless PDAs can also be used to make phone calls, and are referred to as PDA phones. Since many people prefer to carry only one of such mobile device around, there is a competition between PDA phones and smart phones, a war in which the smart phones seem to be winning.

ViewSonic (http://www.viewsonic.com) made a super-sized PDA, called the ViewPad, that offers a regular 800x600-pixel screen. The ViewPad can be a very useful mobile device when regular screen size is a necessity while light weight and zero-boot-up time are also desired.

Blackberry devices made by Research in Motion (http://www.rim.com) has been a big success for enterprise users as they provide a very convenient way for reading and typing e-mails while being away from the office.

Laptop or Tablet PCs with Wireless Access

When size and weight are not inhibitive, mobile users may choose to carry a laptop or tablet PC while on the go. These mobile PCs use wireless cards to connect to either a wireless LAN or wireless WAN. Many such laptops now have built-in wireless LAN cards, and have slots for users to insert a wireless WAN card such as the AirCard made by Sierra Wireless (http://www.sierrawireless.com). An enterprise also needs to be prepared to provide support to mobile users in order to help them connect to the Internet through Wi-Fi hot spots (Hamblen, 2005).

Wireless LAN is often available at a corporation campus, or at public hot spots such as many airports and Starbucks Cafés. Wireless WAN is available wherever cellular service is available for the specific provider that the wireless card is registered with.

IP Phones

IP phones are telephones that use a TCP/IP (transmission-control protocol/Internet protocol) network for transmitting voice information. Since IP phones are attached to the data network, makers of such devices often make the screens larger so that the phones can also be used to access data. What makes IP phones pervasive devices is that a user who is away from his or her own desk can come to any IP phone on the same corporate network, log in to the phone, and make the phone work as his or her own phone. The reason is for this is that an IP phone is identified on the network by an IP address. The mapping between a telephone number and an IP address can be easily changed to make the phone "belong" to a different user.

In terms of the information-access capability, Cisco (http://www.cisco.com) makes IP phones that can access information encoded in a special XML format. Example applications on the phone include retrieving stock quotes, flight departure and arrival information, news, and so forth.

Pingtel (http://www.pingtel.com) developed a phone that runs a Java Virtual Machine. This makes the phone almost as powerful as a computer.

Mitel (http://www.mitel.com) made an IP phone that allows a user to dock a PDA. With this capability, users can go to any such IP phone, dock their PDA into the phone, and immediately have their address books on the PDA available to the telephone. Users can also have their personal preferences transferred from the PDA to the phone and start to use the phone the way they prefer. In addition, users can benefit from new applications on the PDA, such as portable voice mail and dialing by the address book.

The Mitel IP phone seamlessly blends the wired and wireless world for the user so that they are no longer dealing with two separate communication tools. It also provides users with location transparency within the network.

Orbs (Ambient Devices)

Orbs are simple devices that convey information at a glance in a manner that is easy to observe and comprehend (Feder, 2003). Orbs only present a visual indication of the data, not detailed information or actual numbers. Orbs come in different forms. One common orb is a simple globe that changes color and intensity. Other forms include the following:

- Wall panels that adjust color or blink
- Pens, watch bezels, and fobs that change color
- Water tubes that vary the bubble rate
- Pinwheels that change speed

Orbs operate via wireless pager networks under the command of a server. This server gathers pertinent information from sources, including the Web, condenses it to a simple value, and periodically sends the information to the orbs.

Orbs are currently available from several retailers. The wireless service costs about $5 per month per device. Ambient Devices (http://www.ambientdevices.com) sells orbs and provides the communications service.

The information displayed by orbs is configurable. There are currently available data feeds for stock-market movement and weather forecasts.

Input Technologies: Dictation, Anoto Pen, and Projection Keyboard

Two natural ways for mobile users to input information are speech and handwriting.

Speech input can be at two levels: question-and-answer vs. dictation. Question-and-answer speech input is useful for entering structured information where the answers can be predefined using a grammar. Dictation technology allows users to speak freely and tries to recognize what the user has said. Diction technology typically requires a training phase to tune the speech recognizer to each particular speaker in order to achieve high recognition accuracy. Leading dictation products are Dragon NaturallySpeaking from ScanSoft (http://www.scansoft.com) and ViaVoice from IBM (http://www.ibm.com).

The Swedish company Anoto (http://www.anoto.com) invented a technology for pen-based input (McCarthy, 2000). It consists of a digital pen that feels like a regular ballpoint pen, a special paper with patterns of dots printed on it, and a wireless technology such as Bluetooth that sends handwritten information stored in the pen to a computer. As the user writes, the pen not only records what has been written, but also the order in which the user writes it. Anoto has partnered with companies such as Logitech (http://www.logitech.com) and Nokia (http://www.nokia.com) to bring this technology to end users.

For users who want to use a keyboard without carrying one, Canesta (http://www.canesta.com) developed the projection keyboard, in which the image of a keyboard is projected on a surface. By typing on the projection keyboard, information is entered into the associated PDA device.

Application Scenarios

From an enterprise's perspective, the following applications areas are where pervasive computing brings business value.

- Allow employees to stay in touch with phone calls, voice mail, e-mail, and so forth while being away from the office.
- Give employees access to information or transactions via mobile devices while on the road.

- Provide employees with access to the corporate network from anywhere on the Internet (i.e., remote access).
- Send location-based information to employees and customers.
- Monitor device status, perimeter security, and so forth using a wireless sensor network.

Communication: Unified Communication and Instant Communication

With cell phones and pagers, it is not very hard to keep mobile users in touch. But some pervasive communication technologies have reached a higher level. Let us look at two such technologies: unified communication and instant communication.

Unified communications refers to technologies that allow users access to all their phone calls, voice mails, e-mails, faxes, and instant messages as long as they have access to either a phone or a computer. With a computer, a software phone allows the user to make or receive phone calls. Voice-mail messages can be forwarded to the e-mail box as audio files and played on the computer. Fax can be delivered to the e-mail box as images. With a phone, a user can listen to e-mail messages that the system would read using the text-to-speech technology. A user can request a fax to be forwarded to a nearby fax machine.

Unified communications services are offered by most traditional telecommunications technology providers such as Cisco, Avaya, and Nortel.

Instant communication refers to the ability of reaching someone instantly via a wearable communication device. Vocera (http://www.vocera.com) offers a system that uses 802.11b wireless local area networks to allow mobile users to instantly communicate with one another. Each user only needs to have a small wearable device to stay connected. To reach someone, the user would only need to speak a name, a job role, a group, or a location to the system, and the system will take care of the rest. By combining a small wearable device and the speech-recognition capability, Vocera offers a highly usable solution for mobile communication within an organization.

The functions and features of Vocera include the following:

- Instant communication via a small wearable device and speech commands
- Hands-free communication. Except for pressing the button to start and stop a conversation, a user's hands are free during the communication.

- Flexibility in how to specify the recipients. A user can use a name, role, group, or location to tell the system whom to contact.
- The option of having a conversation or leaving a message, for both one-to-one and group communications
- Call controls such as call transfer, blocking, or screening
- Outside calling through the private branch exchange (PBX). The Vocera server can be connected to the PBX to allow users of Vocera to contact people outside the organization.

The Vocera technology has been well received in organizations such as hospitals where users' hands are often busy when they need to communicate with others.

Mobile Access to Information and Applications

Organizations can benefit significantly by allowing mobile access to information and applications. Here are a few examples.

Sales-Force Automation

Salespeople are often on the road. It is important for them to have access to critical business information anywhere at anytime. Pervasive access to information increases their productivity by using their downtime during travel to review information about clients and prospects, about the new products and services they are going to sell, or to recap what has just happened during a sales event when everything is still fresh in their memory. Being able to use smart phones or wireless PDAs to conduct these activities is much more convenient for salespeople as opposed to having to carry a laptop PC.

Dashboard or Project-Portfolio Management

For busy executives, it is very valuable for them to be able to keep up to date on the dashboard while they are away from the office and to take actions when necessary. It is also very helpful for them to be able to look at the portfolio of projects they are watching, update information they have just received during a meeting or conversation, and take notes or actions about a specific project.

Facility Management and Other On-Site Service Applications

Mobile access to information can significantly boost the productivity of on-site service people such as facility- or PC-support staff. With mobile access, they can retrieve ticket information on the spot, update the ticket as soon as they are done with the work, and get the next work order without having to come back to the office. Mobile access also reduces the amount of bookkeeping, which requires a lot of manual intervention, and thus reduces the chance of human errors.

Remote Access to Corporate Network

Allowing employees access to the corporate network from anywhere on the Internet could certainly bring convenience to employees and boost productivity. There are two primary fashions of allowing remote access.

One approach is through a technology called virtual private network, or VPN. This typically requires the user to carry a laptop offered by the employer. Once the user is connected to the Internet, a secure connection (called a VPN tunnel) is established between the laptop and the corporate network after both user and device are authenticated. Then the user will have access to all the information and applications just as if the user were in the office.

The other approach does not require the user to carry a corporate laptop. It simply requires that the user has access to a Web browser. In this case, for security reasons, two-factor authentication is often employed, in which the user not only needs to provide a user ID and password, but also something else, such as the security code generated by a hard token. With this approach, an enterprise can choose which applications to make available for remote access. Terminal service technology offered by Citrix (http://www.citrix.com) can be used to offer browser-based remote access to applications, both Web based and desktop based.

Location-Based Services

A special type of pervasive application is location-based service. With wireless LANs, when a mobile user is in the vicinity of an access point, location-specific information can be delivered to the user's mobile device. With wireless WANs, a user's location can be determined by the cellular tower(s) that the user's handset is communicating with, or by the GPS (Global Positioning System) receiver the user is using. Location-based services include pushing information about local businesses, sending promotions to the user's device based on the user's profile and preferences, and showing meeting agendas and meeting material if the user is on the meeting attendee list for the room at the time.

If location needs to be accurately determined, an ultrasonic location system called the bat system can be used. This 3-D location system uses low power and wireless technology that is relatively inexpensive. An ultrasonic location system is based on the principle of *trilateration*: position finding by the measurement of distances. A short pulse of ultrasound is emitted from a transmitter or bat that is attached to a person or object to be located. On the ceiling are receivers mounted at known points. These receivers can measure the pulse and length of travel.

An ultrasonic location system is composed of three main components.

- **Bats:** Small ultrasonic transmitters worn by an individual or on an object to be located
- **Receivers:** Ultrasonic signal detectors mounted in the ceiling
- **Central controller:** Coordinator of the bats and receivers

To locate a bat, the central controller will send the bat's ID via a 433-MHz bidirectional radio signal. The bat will detect its ID through the embedded receiver and transmit an ultrasonic signal containing a 48-bit code to the receiver in the ceiling. The central controller will measure the elapsed time that it took for the pulse to reach the receiver. The system developed at the AT&T Cambridge facility can provide an accuracy of 3 centimeters.

Overall, location-based services are still in research mode. Once the technology becomes mature and "killer apps" are identified, there could be an explosive adoption.

User-Interaction Models

In the context of pervasive computing, it is usually inconvenient, if not impossible, for the user to enter text using a regular keyboard. Sometimes, it is also inconvenient for the user to read text. Therefore, other input and output mechanisms have to be employed.

Nontraditional input mechanisms include speech recognition, gesture, touch screen, eye gazing, software keyboard, and projection keyboard. Among these, a combination of speech-recognition and pen-based touch-screen input is most natural for most situations. This is also what PDAs and tablet PCs typically offer.

Nontraditional output mechanisms include converting text to speech and using sound, blinking, and vibration to convey information (as in ambient computing described earlier in this chapter).

Multimodal interaction allows a user to choose among different modes of input and output. For mobile users, speech is typically the most convenient way for input, while visual means may still be the most powerful way of seeing the output (especially when the output includes pictures or diagrams).

Kirusa (http://www.kirusa.com) has developed technologies to support multiple levels of multimodal interaction. SMS multimodality allows users to ask a question in voice and have the answers delivered to their mobile devices in the form of an SMS message. Sequential multimodality allows users to use the interaction mode deemed most appropriate for each step of the process. Simultaneous multimodality lets users combine different input and output modes at the same time. For example, for driving directions, a user can say "I need directions from here to there," while pointing to the start and end points.

Both IBM and Microsoft have developed technologies that will support multimodal interaction. IBM's solution is based on the XHTML+VoiceXML (or simply X+V) specification. Microsoft's solution is based on the speech application language tags (SALT) specification, which defines speech input and output tags that can be inserted into traditional Web pages using XHTML.

Besides deciding on what interaction mode to support, much effort is needed to apply user-centered design in order to deliver a good use experience for mobile users (Holtzblatt, 2005).

A Service-Oriented Architecture to Support Pervasive Computing

For an enterprise to leverage pervasive computing, instead of deploying various point solutions, the better way is to build an architecture that is well positioned to support pervasive devices and usage. In order to provide mobile users with maximum access to enterprise information and applications with customized interaction methods and work flow, and at the same time minimize the extra cost in supporting pervasive access, a service-oriented architecture should be established.

The following picture shows a service-oriented architecture that supports pervasive computing. Let us look at this architecture from the top to the bottom.

- Users access applications from different devices. Some devices, such as the regular telephone, have only the voice channel. Some, such as the Blackberry devices, only have the visual display. Others may have both voice and visual channels. The size of the visual display ranges from 1 inch for cell phones, several inches for the PDA and Blackberry, and 15 or more inches for laptops.

- The user devices may be on different network connections, ranging from wireless LAN and wireless WAN to telephone networks.
- Users access applications through the applications' end points. This could be a URL, a phone number, or a start screen stored on the end user's device.
- The pervasive layer sits between the application end points and the SOA layer to provide services to specifically support the mobile users.
    - The device identification engine uses a unique ID to identify the device the user is using. This requires the augmentation of some of the communication protocols to include a universally unique ID (such as the radio frequency identifier, or RFID) of the device that is initiating the request. With this ID, the system can uniquely identify the device and thus have knowledge of its capabilities, the associated user, and so forth. The ID information is also passed to the security service in the SOA layer to help decide whether the user is authorized to access the application.
    - The access-control engine uses information about the device and the communication channel it is coming from to determine the best way to communicate with the device: voice only, visual only, SMS, or some type of multimodal interaction.
    - Based on the desired interaction mode with the user, the content-transformation engine either calls the appropriate version of the application or dynamically transforms the information into the appropriate markup language: HTML, WML, VoiceXML, X+V, and so forth, using the eXtensible Stylesheet Language transformation (XSLT) technology.
    - The location-determination service uses mechanisms built into the networks to determine the geographic location of the user, and then decides whether the information should be tailored based on the location and whether additional location-based information should be pushed to the user.
    - The session-persistence engine uses the device ID and user-identity information to keep the user in the same session while the user is roaming from one network to another, or from disconnected mode to connected mode again during a short period of time. For smart-client applications, where data may be temporarily stored on the device when connection is lost, the session-persistence layer would also take care of synchronizing the data on the device with data on the server.
- The business-application composition layer uses information received from the pervasive layer to determine how to integrate the business services together to best fit the need of this mobile user.
- The SOA layer provides the business services and technical services that are integrated together through the enterprise service bus. The business services

Figure 2. A service-oriented architecture that supports pervasive computing

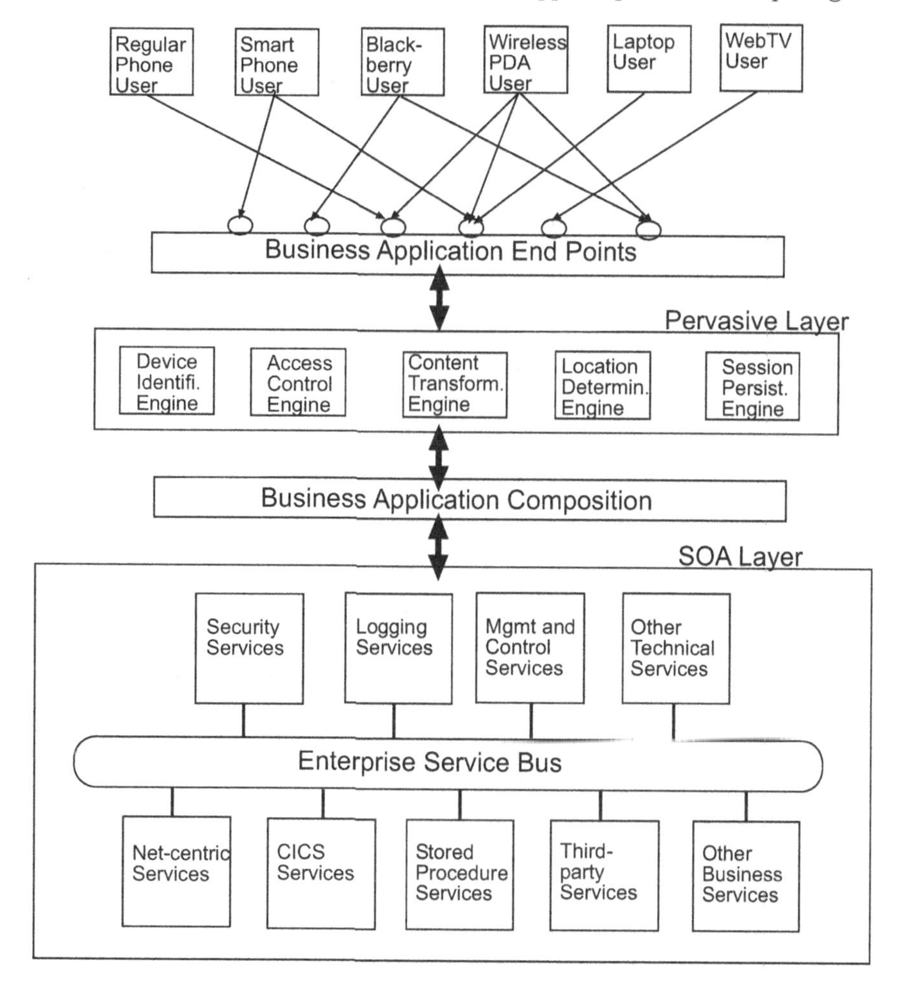

can be built using Net-centric technologies such as Java or Microsoft .NET, or they can be built based on existing legacy business functions such as customer information control system (CICS) transaction and stored procedures. They can also be based on business functions built using third-party tools or existing in-house business modules developed in C or C++, COBOL, and so forth.

After the establishment of such an architecture (which can be gradually built across multiple projects), when building an application that supports pervasive access, business services are either reused or built. When an existing business service is reused, the project team needs to go through the service's specification to verify that the service will work well with the pervasive layer to support all the pervasive channels that the application is going to support. If not, the first thing the project team should try to modify is the pervasive layer. If the issue really lays in the fact that the business service is not a "pure" service, that is, the service is tied to access methods, then the service needs to be either modified or wrapped in order to support the new requirements. When such a modification occurs, the service needs to be made backward compatible, and existing applications that use the service need to be regression tested. Eventually, the service definition needs to be modified to reflect the changes.

With this architecture, when a new pervasive access channel appears, or there is a change in an existing channel, then the only thing that needs to be modified is the pervasive channel. All business services can remain the same.

Future Directions

Moving forward, there needs to be much research and development work on building a system infrastructure that can use different sources of information to judge where the user is, and what devices and interaction modes are available to the user during a pervasive session. This will enable smarter location-based information push to better serve the user.

A related research topic is how to smoothly transition an interaction to a new device and interaction mode as the user changes locations and devices. Some initial work on this subject, referred to as seamless mobility, is being conducted at IBM and other organizations.

Another area that deserves much attention is the proactive delivery of information that users will need based on their profiles and information such as activities on their calendars or to-do lists. This relates to previous research efforts on intelligent personal assistants with integration into the pervasive computing environment.

References

3Gtoday. (2005). Retrieved November 5, 2005, from http://www.3gtoday.com

Blackwell, G. (2002, January 25). Mesh networks: Disruptive technology? *Wi-Fi Planet.* Retrieved October 25, 2005, from http://www.wi-fiplanet.com/columns/article.php/961951

Bluetooth SIG. (2005). *The official Bluetooth wireless info site.* Retrieved November 3, 2005, from http://www.bluetooth.com

Culler, D. E., & Hong, W. (Eds.). (2004). Wireless sensor networks. *Communications of the ACM, 47*(6), 30-57.

Feder, B. (2003, June 10). Glass that glows and gives stock information. *New York Times*, P.C1.

Fox, B. (2001, November 21). "Mesh radio" can deliver super-fast Internet for all. *New Scientist.* Retrieved November 15, 2005, from http://www.newscientist.com/news/news.jsp?id=ns99991593

Geier, J. (2002, April 15). Making the choice: 802.11a or 802.11g. *Wi-Fi Planet.* Retrieved October 16, 2005, from http://www.wi-fiplanet.com/tutorials/article.php/1009431

Hamblen, M. (2005). Wi-Fi fails to connect with mobile users. ComputerWorld, *39*(37), 1, 69.

Henderson, T. (2003, February 3). Vocera communication system: Boldly talking over the wireless LAN. *NetworkWorld.* Retrieved November 12, 2005, from http://www.networkworld.com/reviews/2003/0203rev.html

Holtzblatt, K. (Ed.). (2005). Designing for the mobile device: Experiences, challenges, and methods. *Communications of the ACM, 48*(7), 32-66.

IEEE. (2005a). *IEEE 802.15 Working Groups for WPAN.* Retrieved November 7, 2005, from http://www.ieee802.org/15/

IEEE. (2005b). *The IEEE 802.16 Working Group on Broadband Wireless Access Standards.* Retrieved November 22, 2005, from http://www.wirelessman.org

McCarthy, K. (2000, April 7). Anoto pen will change the world. *The Register.* Retrieved September 14, 2005, from http://www.theregister.co.uk/2000/04/07/anoto_pen_will_change/

Rubio, D. (2004, October 20). VoiceXML promised voice-to-Web convergence. *NewsForge.* Retrieved November 23, 2005, from http://www.newsforge.com/article.pl?sid=04/10/15/1738253

Schrick, B. (2002). Wireless broadband in a box. *IEEE Spectrum.* Retrieved November 19, 2005, from http://www.spectrum.ieee.org/WEBONLY/publicfeature/jun02/wire.html

Schwartz, E. (2001, September 26). Free wireless networking movement gathers speed. *InfoWorld.* Retrieved December 5, 2005, from http://www.infoworld.com/articles/hn/xml/01/09/26/010926hnfreewireless.xml

ZigBee Alliance. (2005). Retrieved November 11, 2005, from http://www.zigbee.org/en/index.asp

Section V

Enterprise Service Computing: Formal Modeling

Chapter XII

A Petri Net-Based Specification Model Towards Verifiable Service Computing

Jia Zhang, Northern Illinois University, USA

Carl K. Chang, Iowa State University, USA

Seong W. Kim, Samsung Advanced Institute of Technology, Korea

Abstract

The emerging paradigm of Web services opens a new way of engineering enterprise Web applications via rapidly developing and deploying Web applications by composing independently published Web-service components to conduct new business transactions. However, how to formally validate and reason about the properties of an enterprise system composed of Web-service components remains a challenge. This chapter introduces an advanced topic of enterprise service computing: the formal verification and validation of enterprise Web services. The authors intro-

duce a Web-services net (WS-Net), which is an executable architectural description language incorporating the semantics of colored petri nets with the style and understandability of the object-oriented concept and Web-services concept. As an architectural model that formalizes the architectural topology and behaviors of each Web-service component as well as the entire system, WS-Net facilitates the simulation, verification, and automated composition of Web services.

Introduction

The emerging paradigm of Web services opens a new way of engineering enterprise Web applications. The key concept is to rapidly develop and deploy Web applications by composing independently published Web-service components to conduct new business transactions. Accordingly, a Web-services-oriented system refers to a Web-application system that contains one ore more Web-service components. In theory, a system containing Web-service components may also contain some parts that are not Web services; however, in reality, every component in a Web-services-oriented system is wrapped as a Web service. Thus, for the rest of this chapter, we will use the terms service components, services, and components interchangeably.

The existing Web-services model favors the creation, registration, discovery, and composition of distributed Web services. In this model, Web services basically adopt a triangular provider-broker-requester operational model, or the so-called services-oriented architecture (SOA), as shown in Figure 1. A service provider publishes services at a public service registry using universal description, discovery, and integration (UDDI; *Universal Description, Discovery, and Integration*, 2004). The public interfaces and binding information of the registered services are clearly defined in the standard Web service description language (WSDL; *Web Services Description Language*, 2004). Such a public service registry generally provides two interfaces: a registry interface serving service providers, and a query interface serving service requesters. As illustrated in Figure 1, published Web services are hosted by the service providers. A service requester queries the service registry for certain services registered, and obtains the binding information of the corresponding service provider. Then the service requester binds to the service provider and remotely invokes the services from the service provider through a lightweight messaging protocol: the simple object access protocol (SOAP; *Simple Object Access Protocol [SOAP] 1.2*, 2003).

Although the paradigm of Web services has been extensively considered as the model of the next generation of distributed computing and Internet computing, how to formally validate and reason about the properties of an enterprise system composed of Web-service components remains a challenge. As a matter of fact,

Figure 1. Service-oriented architecture

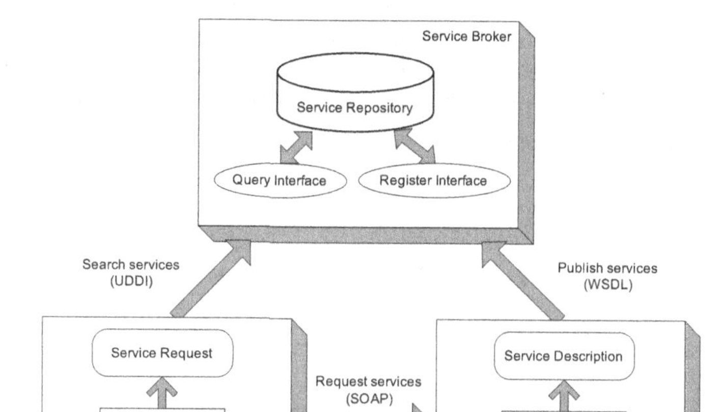

the actual adoption of Web services in industry is quite slow, mainly because there lacks an established way of formally testing Web-services-oriented systems (Zhang, 2005). As a research aiming at facilitating Web-services composition and verification, Web-services net (WS-Net) is an executable architectural description language that incorporates the semantics of colored petri nets (CPNs) with the style and understandability of the object-oriented concept and Web-services concept. WS-Net describes each system component in three hierarchical layers: (a) An interface net declares the set of services that the component provides, (b) an interconnection net specifies the set of services that the component requires to accomplish its mission, and (c) an interoperation net describes the internal operational behaviors of the component. Each component may be either an independent Web service or a composition of multiple Web services, and the whole system can be considered as the highest level component. Thus, WS-Net can be used to validate the intracomponent and hierarchical behaviors of the whole system. As an architectural model that formalizes the architectural topology and behaviors of each Web-service component as well as the entire system, WS-Net facilitates the simulation, verification, and automated composition of Web services. To our best knowledge, our WS-Net is the first attempt to comprehensively map Web-services elements to colored petri nets so that the latter can be used to facilitate the simulation and formal verification and validation of Web-services composition.

The remainder of the chapter is organized as follows. First, we will introduce the state of the art of Web services composition toward services-oriented engineering. Second, we will discuss the challenges of formal Web-services-oriented verification. Third, we will compare options and make selections. Fourth, we will discuss related work. Fifth, we will introduce our WS-Net approach. Finally, we will make conclusions and discuss future work.

State of the Art Web-Services Composition and Challenges

State of the Art of Web-Services Composition

In this section, we will briefly introduce the state of the art of Web-services composition.

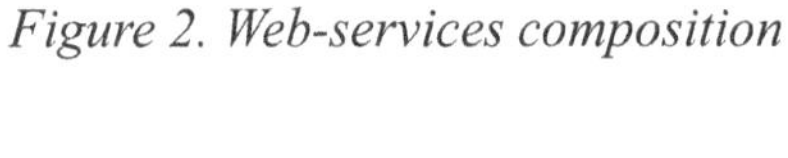

Figure 2. Web-services composition

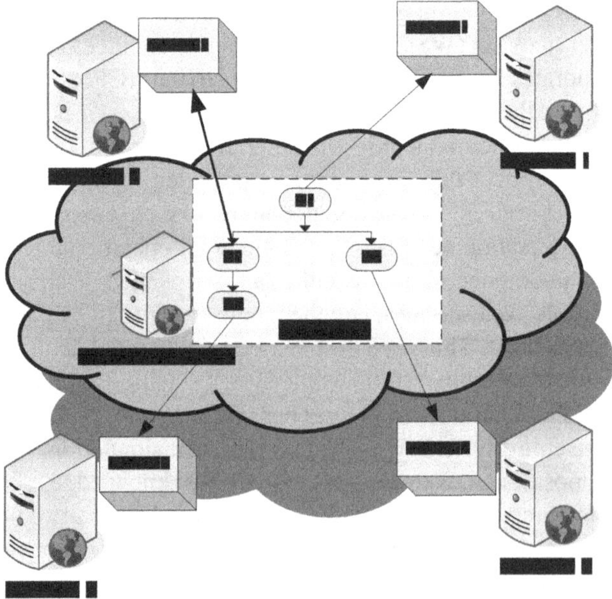

Concept of Web-Services Composition

The SOA model discussed in the introduction describes how to obtain a single Web service. In reality, however, a service requester typically needs to synergistically coordinate and organize multiple Web services into business processes. Web-services composition thus refers to the construction process of composite services from Web services, as shown in Figure 2. It generally involves two procedures: selecting and constructing. The selecting procedure focuses on selecting qualified services, while the constructing procedure focuses on dynamically building flow structures over selected services.

Figure 2 illustrates a simplified Web-services composition. The composition is conducted by a composer residing at the final environment. The composer decides to integrate four existing Web services, namely, S_1, S_2, S_3, and S_4. As shown in Figure 2, there are temporal relationships between the four services. S_1 will run first. Then, depending on the output of S_1, either S_2 or S_3 will be executed. If S_2 is chosen, then S_4 will be executed afterward. As shown in Figure 2, the four service components will be implemented by four Web services that belong to different service providers and reside on different sites. When the composer runs the composite service in the final environment, it will remotely invoke the corresponding Web services according to the predefined scenario.

The construction of a composite Web service can be modeled by specifying a structure of service components using a service flow language, such as the BPEL4WS (business process execution language for Web services; *Specification: Business Process Execution Language for Web Services Version 1.1*, 2003) and Microsoft BizTalk Server (Microsoft, 2003). The structure defines an e-business process model; an invocation of the composite service is treated as an instance of the process model. Examples of the composition model are eFlow (Casati, Ilnicki, Jin, Krishnamoorthy, & Shan, 2000), the scenario-based service composition by Kiwata, Nakano, and Yura (2001), the quality-driven model by Zeng, Benatallah, Dumas, Kalagnanam, and Sheng (2003) and Zeng, Benatallah, Ngu, Dumas, Kalagnanam, and Chang (2004), the constraint-driven composition by Aggarwal, Verma, Miller, and Milnor (2004), and so forth.

With the rapid increase of the number of Web services published on the Internet on a daily basis, the demand for integrating heterogeneous services in an automatic or semiautomatic way becomes urgent. Efforts have been made to automate the service-composition process by employing discovery agents. According to the information provided in a service request, these discovery agents generate a structure of service operations based on some registered services. Commonly used approaches for agents to make decisions are rule-based systems (Ponnekanti & Fox, 2002) and ontology-based approaches (Arpinar, Aleman-Meza, Zhang, & Maduko, 2004). However, to date, fully automated approaches sometimes involve unavoidable, unrealistic assumptions; for example, rule-system-based approaches assume that the service requester knows the exact input and output interfaces of a desired composite service. Liang, Chakarapani, Stanley, Su, Chikkamagalur, and Lam (2004) thus utilized a semiautomatic approach to assist service composition.

As shown in Figure 2, Web-services composition differs from traditional application-component composition in several significant ways. First, unlike in a traditional component composition where component parts are deployed in the same final environment, Web-services composition is a virtual composition in the sense that all participating services will never be physically deployed into the final environment. This is because Web services are hosted by corresponding service providers and can only be used through remote invocations. Second, Web-services composition implies uncertainties. Since Web services are hosted and managed by their own service providers, their availabilities may autonomously change (Zhang, 2005). An available Web service on a specific day may not be available on the next day. Furthermore, the quality of service (QoS) and even functionalities of Web services can also change over time due to changes in adopted technologies or business models from the corresponding service providers. Therefore, Web-services composition may have to be conducted upon a per-usage basis. Third, Web-services composition needs to serve more dynamic changes of user requirements. This is not a new issue; however, users adopt Web services for higher flexibility and adaptability, thus they hold higher expectations than before.

Architecture of Web-Services-Oriented Systems

Just as an architectural diagram is an essential guideline for architectural constructions, software architecture is critical for the success of a Web-services-oriented system

Figure 3. Two dimensions of Web-services composition

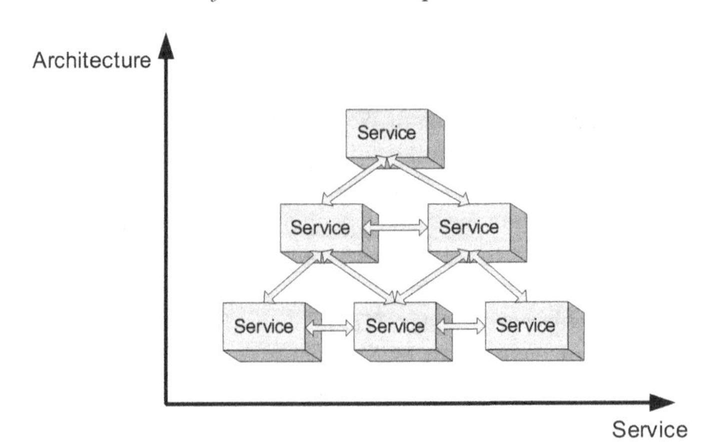

and Web-services composition. According to The American Heritage Dictionary of the English Language (2000), architecture is "the art and science of designing and erecting buildings, a structure of structures collectively, a style and method of design and construction." In other words, software architecture plays an essential role in providing the right insights, triggering the right questions, and offering general tools for thoughts. Thus, architecture description languages (ADLs; Medvidovic & Taylor, 2000) are commonly created to formally specify and define the architectural model of a software system as a guideline before construction.

As shown in Figure 3, Web-services composition is a process of composing multiple Web services as components into a composite Web-services-oriented system. In order to facilitate services composition, the architecture specification (e.g., an ADL) that represents a complex Web-services-oriented system should capture two dimensions of design information: the service dimension and architecture dimension. The service dimension focuses on specifying low-level interconnections between Web services, for example, how to glue different service components together, how to communicate between services, how to orchestrate services, and how to choreograph services. The architecture dimension, on the other hand, focuses on specifying high-level compositional views or architectural views of a final composite service, for example, hierarchical relationships among service components.

In more detail, a Web-services-oriented ADL needs to catch the following three categories of design information: (a) the structural properties and Web-services-component interactions, (b) the behavioral functionality of each major high-level Web-services component, and (c) the behavioral functionality of the entire system. In recent years, researchers from both industry and academia have been developing a number of Web-services-oriented ADLs, typically, WSDL (*Web Services Description Language*, 2004), Web services flow language (WSFL; Leymann, 2001), BPEL4WS (*Specification*, 2003), Web service choreography interface (WSCI; *Web Service Choreography Interface [WSC]), Version 1.0*, 2002), XLANG (Thatte, 2001), and so forth.

Challenges of Formal Verification and Validation of Web-Services and Composition

Although Web-services composition exhibits a boom to enterprise software engineering, it is not clear that this new model guarantees various Web-service components can be composed and integrated seamlessly and properly (Zhang, 2005). In detail, the flexibility of Web-services-centered computing is not without penalty since the value added by this new paradigm can be largely defeated if, for example, selected Web-service components do not thoroughly fulfill the requirements (i.e., functional or nonfunctional), selected Web-service components act errantly in the composed environment, or if selected service components cannot collaborate harmoniously. Therefore, it is necessary to fully test a Web-services-oriented system under various input conditions, and to logically verify certain maintenance and QoS conditions associated with Web services. Without ensuring the trustworthiness of Web-services composition, it is difficult for Web services to be adopted in mission-critical applications. In short, formally verifying and validating the correctness of logic inside a composition of Web-services-oriented systems become critical.

As such, the formal verification of Web-services composition is not a trivial task. The last 50 years of software-development history has witnessed the establishment of an independent research branch called software testing. Software testing involves a wealth of theories, technologies, methodologies, and tools to guide the verification process of a component-based software product. However, formal validation and verification over Web-services composition poses new challenges due to the unique features of Web-services composition.

We cannot simply apply the traditional software-testing technologies to formally measure and test a Web-services-oriented system. First, such a testing needs to be highly efficient. There are times when service components cannot be decided until run time. Testing and making decisions at run time require efficient strategies and techniques. Second, how to test a Web-services-oriented system with limited information is challenging. Current Web-services interfaces expose limited information. Web services are Web components only accessible via interfaces published in standard Web-services-specific interface-definition languages (e.g., WSDL) and accessible via standard network protocols (e.g., SOAP). Third, such a testing may have to be performed on a per-usage basis. Service components are hosted by service providers and invoked remotely; thus, it may be questionable to assume that the services stay at the same quality over time. Fourth, Web-services-oriented testing has to be highly effective. The fundamental hypothesis of the existing testing methods is that exhaustive test cases can be conducted upon the software product if necessary. It is neither feasible nor practical to apply this assumption to Web-services testing. Unlike traditional software products that are deployed into the target environment, Web services require remote accessing. Thus, conducting a significant amount of tests on a Web service implies expensive maintenance of network connections and network transmissions, let alone unpredictable Web conditions such as traffic and safety. Finally, since it is difficult to obtain precise function descriptions of a Web service, most of the time the only feasible and practical way of testing a Web-services-oriented system is through simulation.

In summary, due to the specific distributed nature of Web services, these existing software-testing models and methodologies deserve reinspection in the domain of Web-services composition.

Related Work

Researchers have conducted significant work in the field of Web-services composition description and verification.

Web-Services-Oriented ADLs

In recent years, researchers from both industry and academia have been developing a number of Web-services-oriented ADLs, typically WSDL (*Web Services Description Language*, 2004), WSFL (Leymann, 2001), BPEL4WS (*Specification*, 2003), WSCI (*Web Service Choreography Interface [WSCI], Version 1.0*, 2002), XLANG (Thatte, 2001), and so forth.

The common similarity is that all of these ADLs are built upon the extensible markup language (XML; *XML*, n.d.) technology, which is extensively considered as a universal format for structured documents and data definitions on the Web. Among the ADLs, WSDL is the basis of other work. Intending to formally and precisely define a Web service, WSDL from W3C (World Wide Web Consortium, http://www.w3c.org) is becoming the ad hoc standard for Web-services publication and specification. However, it has been widely admitted that WSDL can only specify limited information of a Web service, such as function names and limited input and output information. In recognition of this problem, researchers have been developing other description languages to extend the power of WSDL to depict system architecture. The following are some outstanding examples.

- WSFL is a WSDL-based language focusing on describing the interactions between Web-service components. WSFL defines the interaction patterns of a collection of Web services, as well as the usage patterns of them in order to achieve a specific business goal.
- BPEL4WS specifies an interoperable integration model aiming to facilitate the automatic integration of Web-service components. BPEL4WS formally defines a business process and process integration.
- WSCI utilizes flow messages to define the relationships and interactions between Web-service components. According to WSCI, a Web-service component exposes both a static interface and a dynamic interface when it participates in a message exchange with other Web-service components.
- XLANG considers the behaviors of a system as an aggregation of the behaviors of each Web-service component. Therefore, XLANG specifies the behaviors of each Web-service component independently. The interactions between Web

services are conducted via message passing, which is expressed as the operations in WSDL.

However, these ADLs either merely focus on static functional descriptions of Web-service components as a whole (e.g., WSDL), or concentrate only on the behavioral integrations between Web-service components (e.g., BPEL4WS). In addition, these ADLs focus on topological descriptions and concentrate on describing interactions between Web-service components. They lack the capability to describe the hierarchical functionality of the components. Moreover, there is little concern about expressing dynamic behaviors of the defined system. Furthermore, none of these current ADLs support dynamic verification and monitoring of the system integrated. In contrast to previous work, our research focuses on supporting both static and dynamic Web-services-based composition.

Chang and Kim (1999) proposed I^3, which is a layered executable architectural model defining component-based systems. However, I^3 is based upon the structural analysis and design technology (SADT) (Ross, 1984), which is a traditional functional decomposition- and data-flow-centered methodology. In contrast to I^3, our WS-Net aims at integrating colored petri nets with the style and understandability of the object-oriented paradigm. In addition, I^3 intends to present a generic specification model oriented to generic component-based software systems. WS-Net, on the other hand, focuses on Web-services-oriented system architecture and seamlessly integrates with WSDL and XML technology. In other words, although WS-Net was strongly influenced by EDDA (Trattnig & Kerner, 1980) and I^3, we have enhanced the state of the art by supporting modern software-engineering philosophy equipped with the object-oriented and component-based notations. Furthermore, WS-Net is applied to Web-services-oriented systems as well as integrated with the ad hoc Web-services standards, such as WSDL and XML.

Formal Verification Work on Services-Oriented Systems

Narayanan and McIlraith (2002) proposed to use petri nets as tools to simulate, verify, and automate Web-services composition. Their Web-services composition refers to the composition of programs into a Web service and does not consider the composition of Web services into new Web services. In contrast to their work, our research covers both inter-Web-services and intra-Web-services composition. In detail, our interconnection net simulates and validates the interactions and composition among Web services. The interoperation net simulates and validates the composition inside a Web service, which may or may not contain other Web services. Moreover, in contrast to their work, which does not provide detailed mapping between Web-service elements and Petri-net elements, our approach provides a direct mapping between the two methodologies. Thus, our approach can be used as a guidance to construct petri nets for Web-services composition. Furthermore, unlike related work, WS-Net itself is an architectural description language that can facilitate hierarchical Web-services composition description and definition, in addition to simulating Web-services composition.

Pi-Calculus, one form of process algebra, was the theoretical basis of the precursors of the ad hoc services-composition definition language BPEL4WS (*Pi-Calculus*, n.d.). The fundamental entity in Pi-Calculus is a process, which can be an empty process, an I/O process, a parallel composition, a recursive definition, or a recursive invocation. Describing services in such an abstract way facilitates reasoning about the composition's correctness through reduction. Pi-Calculus enables the verification of liveness and behavioral properties. Salaun, Bordeaux, and Schaerf (2004) adopted process algebraic notations to describe Web services and the interservice interactions at an abstract level. Then they performed reasoning about the correctness over the services composition through simulation and property verification. They also explored the links between abstract descriptions and concrete descriptions. From the experience out of a sanitary-agency case study, they found that process algebras are adequate to describe and reason about services composition, especially to ensure composition correctness. Furthermore, Ferrara (2004) defined a two-way mapping between abstract specifications written in process algebraic notations and executable Web services written in BPEL4WS. In addition to temporal logic model checking, Bordeaux, Salaun, Berardi, and Mecella (2004) adopted bisimulation analysis to verify the equivalent behaviors between two Web services. In contrast to their work based upon process algebra, our WS-Net has roots in petri nets, which are more suitable to simulate and reason about large-scale Web-services composition and verification. In addition, our WS-Net is an architectural description language that supports hierarchical description and definition.

Some researchers focus on adopting finite-state automata to verify Web-services composition. Bultan, Fu, Hull, and Su (2003) uses finite-state automata to model the conversation specification, thus to verify Web-services composition. Their approach models each service component as a mealy machine, which is a finite-state machine (FSM) with input and output. Service components communicate with each other through asynchronous messages. A conversation between service components is modeled as a sequence of messages. Each service component has a queue to hold messages, while a centralized global "watcher" keeps track of all messages in the whole composite system. Berardi, Calvanese, Giacomo, Lenzerini, and Mecella (2003) proposed to describe a Web service's behavior as an execution tree and then translate it into an FSM. An algorithm was also presented to check whether a possible composition exists; that is, a composition will finish in a finite number of steps. In contrast to their work, based upon finite-state automata, our WS-Net is based upon petri nets, which are more suitable to verify large-scale Web-services composition. In addition, our WS-Net is an architectural description language that supports the hierarchical definition of large-scale Web-services composition and early-stage validation.

Alternative Tools And Methodologies

In this section, we will first discuss briefly alternative tools and methodologies, together with how they can be used to model Web-services composition. By providing their comparisons, we will discuss our selection on petri-net technology.

Currently, there are generally three alternative techniques to formally verify Web-services composition: (a) petri nets, (b) process algebras, and (c) finite-state automata.

Petri Nets

Petri nets are a well-established abstract language to formally model and study system composition (Jensen, 1990). In general, a petri net is a directed, connected, and bipartite graph with two kinds of node types: places and transitions. The nodes are connected via directed arcs, and tokens occupy places. When all the places pointing into a transition contain an adequate number of tokens, the transition is enabled and may fire; thus, the transition removes all of its input tokens and deposits a new set of tokens in its output places.

Web services can be modeled as petri nets by assigning transitions to services, and places to inputs and outputs. Each Web service has its own petri net, which describes service behaviors. Each service may own either one or two places based upon the nature of the service: (a) one input place only, (b) one output place only, or (c) one input place and one output place. At any given time, a service can be in one of the following states: not initiated, ready, running, suspended, or completed. After defining a net for each service, services can be composed by applying various types of composition operators between nets: sequence, choice, iteration, parallel, and so forth. These composition operators will orchestrate nets in different execution patterns.

After modeling Web-services composition with petri nets, one can investigate the generated petri nets to verify system properties, such as deadlocks or nondeterministic status.

Process Algebras

Process algebras are formal description techniques that use the temporal logic model to specify and verify component-based software systems, especially their concurrency and communication. Process algebra is known for its ability to describe dynamic behaviors, compositional modeling, and operational semantics, as well as its ability to reasoning about behavior via model checking and process equivalences. The central unit of process algebras is the process. A process is an encapsulated entity that contains a state. Different processes communicate with each other via interactions. An action (e.g., shipping an order) initiates an interaction. A set of processes can form a larger system.

Process algebras can be adopted to model Web-services composition. Each Web service can be modeled as a process. By studying algebraic services composition, one can verify composition properties such as safety, liveness, and resource management. As a matter of fact, Pi-Calculus, one form of process algebra, was the theoretical basis of the business process modeling language (BPML) and XLANG, two precursors of the ad hoc services-composition definition language BPEL4WS (*Pi-Calculus*, n.d.).

In addition to temporal logic model checking, process algebras contains bisimulation analysis, which can be used to verify whether two processes have equivalent behaviors, that is, whether one service includes behaviors of another (Bordeaux et al., 2004). Thus, bisimulation analysis can be used to decide whether two Web services are replaceable, as well as to decide the redundancy of services.

Finite-State Automata

An FSM or so-called finite automaton is a well-established model of behaviors composed of states, transitions, and actions. A state stores information, a transition indicates a state change and is guarded by a condition that is required to be fulfilled, and an action is a description of an activity to be performed at a given time point. Generally, four types of actions can be identified: entry action, exit action, input action, and transition action.

Finite-state machines can be adopted to represent the aspects of a global composition process. In detail, Web-services composition specification can be described using temporal logic; then the FSM model can be traversed and checked to verify whether the work-flow specification holds. Specially, this approach can be used to verify data consistency, unsafe state avoidance (deadlock), and business-constraint satisfaction. Current research efforts along this direction can be further categorized into two

Figure 4. Comparisons among petri nets, process algebras, and finite-state machines

	PN	PA	FSM
Formal	H	H	H
Theoretical foundation	H	H	H
Analysis ability	H	H	H
Handle complexity	H	-	-
Model static properties	H	H	H
Model dynamic properties	H	-	-
Graphical ability	H	-	-
Simulation	H	-	-
Tools	H	-	-
User effort	L	H	H
Concurrency	H	-	-
Deadlock detection	H	-	-
(a)synchronization	H	-	-
Process concept	I	E	E

groups: conversation specification and automatic services composition (Milanovic & Malek 2004). Research on the former one focuses on using mealy machines to model asynchronous messages between service components, thus, verifying the realizability of a services-composition specification. Research on the latter one focuses on modeling service behaviors as an execution tree, and then translating it into an FSM. The generated FSM then can be checked to verify whether a possible composition exists, that is, whether it can be finished in a finite number of steps.

Comparisons

Each of the three alternative techniques exhibits advantages and disadvantages over simulating and verifying Web-services composition, as shown in Figure 4.

Petri nets are a modeling approach that utilizes graphical representation to specify and simulate events and state changes. It has been widely used as a graphical formal modeling tool for systems involving communication, concurrency, synchronization, and resource sharing. However, petri nets also have disadvantages. First, petri nets are more expressive than finite-state machines (Peterson, 1981). Second, such a kind of graph is easy to understand when the graph contains a small amount of elements. However, if the modeled system is large scale and complex, the net will quickly grow large, and the graph will become too complicated to understand and extract useful information from it. High-level hierarchical petri nets provide scalable support for modeling large-scale systems, such as hierarchical CPNs (HCPNs). Third, there is no direct and explicit mapping between petri-net constructs and Web-services composition. For example, there is no concept of delimited processes in petri nets. Thus, one needs to find an appropriate implicit mapping between petri-net constructs and Web-services concepts; for example, each Web service can be mapped to a transition in petri nets, and its input and output can be mapped to places in Petri nets.

Process algebras can model services composition in a simple and straightforward description. Each Web-service component can be naturally modeled as a process, which is the fundamental entity of process algebras. Process algebra facilitates the formal specification of message exchanges between Web-service components. By studying the interactions between modeled processes, one can verify the services composition in a natural way. However, there exists no tool to automatically investigate and simulate generated algebraic services composition. Manual analysis through algebraic reduction can be both time consuming and error prone.

The same issue exists for finite state automata, although FSMs are natural to model services composition through state modeling and transitions. Enduring the same situation as for process algebras, there is no tool to automatically investigate and simulate generated FSMs for services composition. This approach requires users to capture composition properties with a temporal logic formula. Manual analysis is impractical for dynamic and efficient verification, which is a compulsory requirement for services-composition verification.

In summary, as shown in Figure 4, all petri nets, process algebras, and finite-state machines can perform formal verification and possess analysis ability, and they all have roots in theoretical foundations. With the HCPNs, high-level petri nets can support more complexity. All petri nets, process algebras, and finite-state machines provide powerful mechanisms to model static system properties. However, only petri nets support dynamic properties modeling. In addition, petri nets provide available graphical tools for simulation, while the other two approaches do not. These tools provide powerful facility to analyze concurrency, deadlock, and asynchronous behaviors. Thus, user efforts involved can be relatively low compared to those involved in process algebras and finite-state machines. However, petri nets do not specify an explicit concept of process; thus, modeling process-centric Web-services composition requires special mapping mechanisms between petri-net concepts and Web-services concepts.

Our Selection

In our research, we chose to adopt petri nets due to their combination of (a) rich computational semantics, (b) ability to formally model systems that are characterized as being concurrent, asynchronous, distributed, parallel, nondeterministic, and stochastic, (c) ability to conduct quantitative performance analysis, and (d) availability of graphical simulation tools. Petri nets also have natural representations of changes and concurrency, which can be used to establish a distributed and executable operational semantics of Web services. In addition, petri nets can address off-line analysis tasks such as Web-services static composition, as well as online execution tasks such as deadlock determination and resource satisfaction. Furthermore, petri nets possess a natural way of addressing resource sharing and transportation, which is imperative for the Web-services paradigm. Moreover, petri nets have a set of available graphical simulation tools, represented by Design/CPN (Meta Software Corporation, 1993).

Introduction To WS-Net Approach

In this section, we will introduce our WS-Net. First, we will provide a brief overview of petri nets, focusing on their notions for high-level architectural design. We will also briefly describe a couple of high-level petri-net techniques that will be used in our approach, such as CPNs and HCPNs.

Overview of Petri Nets

The concept of petri nets was coined by Carl Adam Petri in his PhD thesis in 1962 (Petri, 1962). Rooted in a strong mathematical foundation, petri nets is a well-known abstract language to formally model and study systems that are characterized as being concurrent, asynchronous, distributed, parallel, nondeterministic, and/or stochastic (Jensen, 1990). In general, a petri net is a bipartite graph with two kinds of node types: places and transitions. The nodes are connected via directed arcs, and tokens occupy places. Graphically, circles represent places; rectangles represent transitions. All places holding zero or more tokens together exhibit a specific state of the system at a moment. Transitions can change the state of the system: When all the places pointing into a transition contain an adequate number of tokens, the transition is enabled and may fire; thus, the transition removes all of its input tokens and deposits a new set of tokens in its output places.

In other words, transitions are active components. They model activities that can occur (e.g., the transition *fires*), thus changing the state of the system (e.g., the marking of the petri net). Transitions are only allowed to fire if they are *enabled,* meaning that all the preconditions of the activity must be fulfilled (e.g., there are enough tokens available in the input places). When the transition fires, it removes tokens from its input places and adds some at all of its output places. The number of tokens removed or added depends on the cardinality of each arc. The interactive firing of transitions in subsequent markings is called a token game.

Formally, a petri net is a triple PN = (P, T, F) for which the following are true.

1. P is a finite set of places, $P = \{p_1, p_2, \ldots, p_m\}$.
2. T is a finite set of transitions, T = a.
3. F is a set of arcs that represent the flow relation between places and transitions. F contains both input relations that map each transition from a set of places, and output relations that map each transition to a set of places,

   $F = (PXT) \cup (TXP)$.

If we use petri nets to model a system, the transitions model the active part of the system, the places model the passive parts, and the markings describe the system states. The arcs of a graph are classified into three categories: input arcs, output arcs, and inhibitor arcs. Input arcs are arrow-headed arcs from places to transitions, output arcs are arrow-headed arcs from transitions to places, and inhibitor arcs are circle-headed arcs from places to transitions.

CPNs (Jensen, 1990) extend the petri nets to model both the static and dynamic properties of a system. The notation of CPN introduces the notion of token types; namely, tokens are differentiated by colors that may be arbitrary data values. Each place has an associated type, which determines the kind of data that the place may contain. The graphical part of a CPN depicts the static architectural structure of a system. Combined with other powerful elements such as colored tokens and simulation rules, CPN is highly powerful in modeling dynamic behaviors of a system.

Formally, a CPN is a tuple CPN = (∑, P, T, A, N, C, G, E, I) for which the following are true.

1. ∑ is a finite set of nonempty types, called color sets.
2. P is a finite set of places.
3. T is a finite set of transitions.
4. A is a finite set of arcs such that

$P \cap T = P \cap A = T \cap A = \emptyset$.

5. N is a node function. It is defined from A into $(PXT) \cup (TXP)$.
6. C is a color function, which is defined from P into Σ.
7. G is a guard function, which is defined from T into expressions such that

$$\forall t \in T, [Type(G(t)) = Bool \wedge Type(Var(G(t))) \subseteq \Sigma \cdot$$

8. E is an arc expression function, which is defined from A into expressions such that

$$\forall a \in A, [Type(E(a)) = C(p(a))_{MS} \wedge Type(Var(E(a))) \subseteq \Sigma,$$

where p(a) is the place of N(a).

9. I is an initialization function, which is defined from P into closed expressions such that

$$\forall p \in P, [Type(I(p)) = C(p)_{MS}]$$

In order to handle large-scale and complex systems, several variants of petri nets emerged. The HCPNs were created to manage the complexity of large-scale systems by providing facility to specify the hierarchical relationships between nets. They introduce a facility for building petri nets from subnets or modules. The HCPN theory intends to allow the construction of a large-scale model by using a number of small petri nets that are related to each other in a well-defined and well-organized manner.

Earlier researchers have conducted a large amount of work to utilize CPN to model system architecture. EDDA (Trattnig & Kerner, 1980) combines petri nets and SADT technology for high-level system specifications. Although EDDA successfully combines the semantics of petri nets with the syntax of SADT, it lacks the ability to specify modern software systems as it does not embody the object-oriented paradigm and the component-engineering concept (Chang & Kim, 1999). Pinci and Shapiro (1990) presented an automatic mechanism to translate SADT diagrams into HCPNs, and in turn to convert HCPNs into standard ML-executable code. This SADT-like petri-nets-based system specification suffers the same problems faced by EDDA due to the rigid structural nature of SADT and its lack of object-oriented concepts.

WS-Net

As we discussed in the previous sections, the formal verification of Web-services composition is highly in demand, and we decided to adopt the petri-nets technology for the specification and reasoning. Meanwhile, architectural description languages are essential to facilitate services-composition design. Therefore, we combine these two demands: We introduce an ADL that enables the hierarchical formal verification of Web-services-oriented system composition.

In detail, our goal is to enable formal defining and automated reasoning technology to describe, simulate, test, and verify Web-services composition. In other words, we intend to establish a formal modeling and specification framework for Web-service composition. Our approach is to introduce WS-Net, which is a petri-nets-based executable architectural specification language.

WS-Net specifies a Web-services-oriented software system as a set of connected architectural components described as nets. The architectural components correspond to the functional units in the system, and one architectural component may in turn be composed of multiple smaller architectural components. The entire system can be viewed as the highest level architectural component. Each architectural component is either statically or dynamically realized by a Web-service component. Architectural components are connected to each other via XML-message passing through SOAP, the ad hoc transportation standard in the realm of Web services. The message-passing mechanism mediates the interactions between architectural components via the rules, which regulate the component interactions. In our model, we will use the concept of the connector (Shaw, DeLine, Klein, Ross, Young, & Zelesnik, 1995) in CPN to model message passing.

As shown in Figure 5, WS-Net defines each architectural component in a three-aspect specification: (a) the interface net, (b) the interconnection net, and (c) the interoperation net. The interface net declares the services to be provided by each Web-service component, the interconnection net specifies the Web services to be acquired from other Web-service components in order to accomplish its own mission, and the interoperation net describes the functionality of each Web-service component and the entire system in terms of control flow and data flow.

Each component must have one interface-net definition, and it can be accessed only via the defined interface. The definition of the interface net follows that of WSDL. The interconnection net specifies the operations to be acquired from other Web services to perform its execution. It is possible for a service component to not have an interconnection net, in which case the service component is self-sufficient and does not need support from other services to conduct its mission. In the interconnection net, each operation required is further specified by a set of foreign transitions, which represents the operations of other components. In other words, the interface net identifies each component in the system as a unique functional object, and the

Figure 5. WS-Net

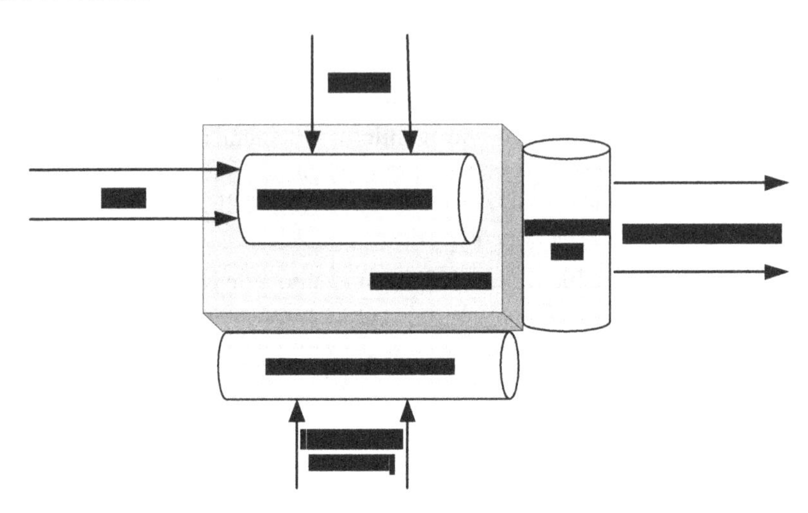

interconnection net specifies the relationships between components. As a result, we can visualize the entire topological view of a system by interconnecting each of the interconnection nets according to our unique component-interconnection technique, which will be discussed in later sections. The interoperation net describes the dynamic behaviors of a service component by focusing on its internal operational nature. The goal of the interoperation net is to dissect each operation into fundamental process units, which, taken together, define the required functional content of the operation. Each transition representing an operation of the component is decomposed into subtransitions representing fundamental process units. Control flow and data flow

Figure 6. Simplified message-queue example

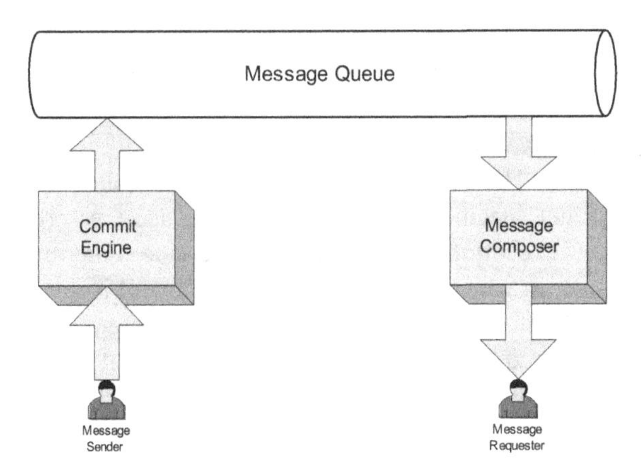

are used for describing intercommunications between process units.

As illustrated in Figure 5, WS-Net covers various aspects of a Web service. The interface net publishes a Web service to the outer world. The interconnection net specifies necessary support. The interoperation net not only handles the internal work flow of the Web service, but also handles incoming requests. As shown in Figure 5, a Web service typically adopts a passive invocation mode, meaning that the Web service will stay idle until triggered (i.e., requested). When a Web service is triggered (i.e., invoked), it will also accept incoming input arguments to start the interoperation net. In summary, the three nets of WS-Net together can represent a Web service in a large-scale system and system composition.

To facilitate our discussions, throughout the rest of the chapter, we will use a simplified, typical Web-services-messaging processing model as an example to elucidate the fundamental idea of WS-Net. As shown in Figure 6, this example illuminates a messaging system that provides a messaging service for users. The system is centered on a central storage place that acts as a message queue. All users submit messages to the message queue and retrieve messages from the message queue. Three distributed Web-service components are identified in the system: the message queue, a commit engine, and a message composer. As shown in Figure 6, if a message sender wants to publish a message, he or she will send the message to the commit engine that stores the message to the message queue. When a user wants to retrieve messages, the message composer will search the message queue on behalf of the user according to his or her profile, fetch messages, and compose them into a package before sending it back to the user. In this example, different components interact with each other via SOAP.

Interface Net

Constructing a WS-Net specification starts from identifying the architectural components from the system specification using a top-down approach. We use the term architectural component here instead of a service component, meaning that an architectural component may refer to the whole system, a subsystem, or a single service component. This generic feature enables the interface net to be applied to various levels of system composition. The interface net of a Web service defines expected responsibilities, or features, of each architectural component by specifying its interface as a set of semantically related services (i.e., operations) provided. The interface here denotes the names of the services provided and their signature information for invoking the services. A Web service can be accessed only through its interface. In an interface net, each service is modeled as a transition in petri-nets notation. Therefore, a transition is called a service transition in WS-Net. The name of a service transition refers to the service to be provided by the corresponding service component. Each service transition has an input place called the input port where the service receives invocations, and an output place called the output port where the result of a service is placed before being returned to the service requester. In other words, a Web service adopts an asynchronous communication mode. The interface net complies with WSDL.

A WS-Net specification of an architectural component j can be denoted as C_i. $C_i \in C$, where C is the set of all Web-service components identified in the software architecture. The interface net of the component C_i can be represented as

$$C_i = \bigcup_j S_{ij}, S_{ij} \in S_i'$$

where S_i is a set of services provided by C_i. Each service $S_{ij} \in S_i$ is represented by a tuple

$$S_{ij} = (PI_{ij}, PO_{ij}, T_{ij}, A_{ij}, c),$$

where PI_{ij} and PO_{ij} are the input port and the output port of S_{ij}, respectively.

Here we use port instead of place for compatibility with WSDL specifications so as to facilitate automatic translation from WSDL specification to WS-Net. T_{ij} represents the service S_{ij} as a transition, A_{ij} represents the input and output arcs of the transi-

Figure 7. Interface net for service component Message Queue

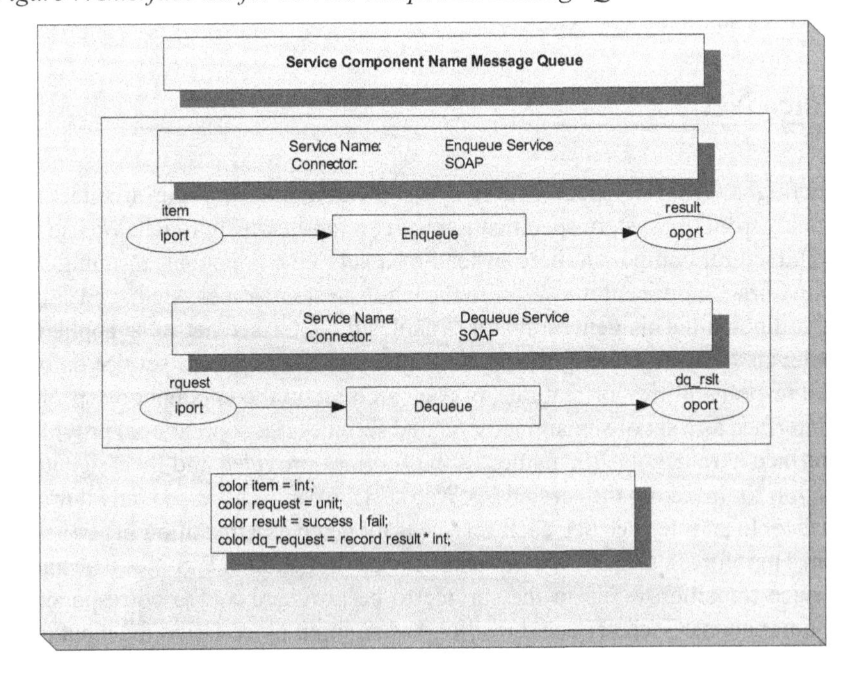

Figure 8. Interface net for Commit Engine

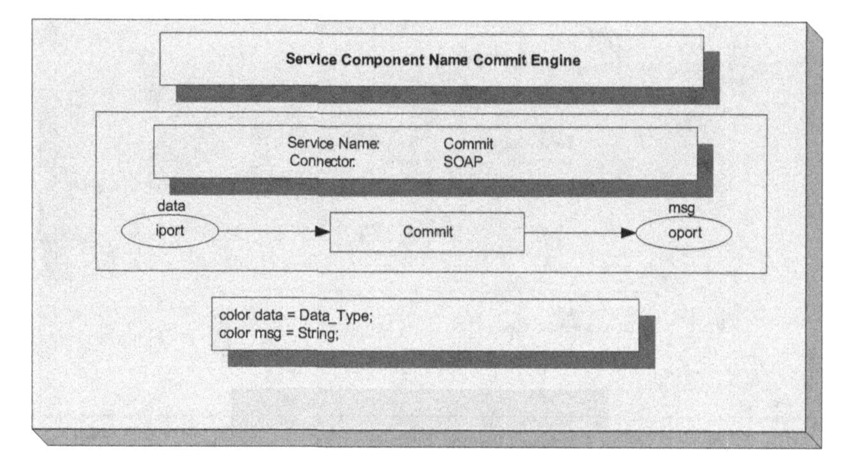

tion T_{ij}, and c is a color function for the place. The color inscriptions of the place represent the signature information of the operations as used in CPN.

The goal of the interface net is to define the provided services that are provided by the service component. The names attached to the service-transition inscription represent the names of the services. Note that the names of the places and transitions are the labels to identify the places and transitions, and they are not considered as semantic inscriptions of petri nets (Meta Software Corporation, 1993). Instead, as defined in CPN (Jensen, 1990), they are used to help designers understand the specifications and to support the hierarchical composition of pages, such as transition substitution and place fusion. The signature information of each service can be described by color inscription on input and output ports (i.e., CPN places). However, it is not important to specify the detailed data structure at this stage of the design. The main purpose of coloring places is to help people understand the usage of a service component at the architectural level in the whole system. Therefore, the only imperative information to be specified is what kind of information needs to be provided to invoke the service of the service component.

Using our message-queue example, the interface-net specification of the message-queue component is illustrated in Figure 7. The message queue provides two services: enqueue service and dequeue service, which accept user-submitted messages and retrieve user message requests, respectively. In order for the services to be invoked, the corresponding input ports of the services must receive proper tokens. The enqueue service receives an item token and returns a Boolean result as either a success or failure. The dequeue service receives a request as a unit token and returns *dq_rslt* in case of success, or *err* in case of failure if there is an empty queue. The unit color set has no predefined service, but is very useful as a placeholder (Jensen, 1990).

Figure 9. Interface net for Composer

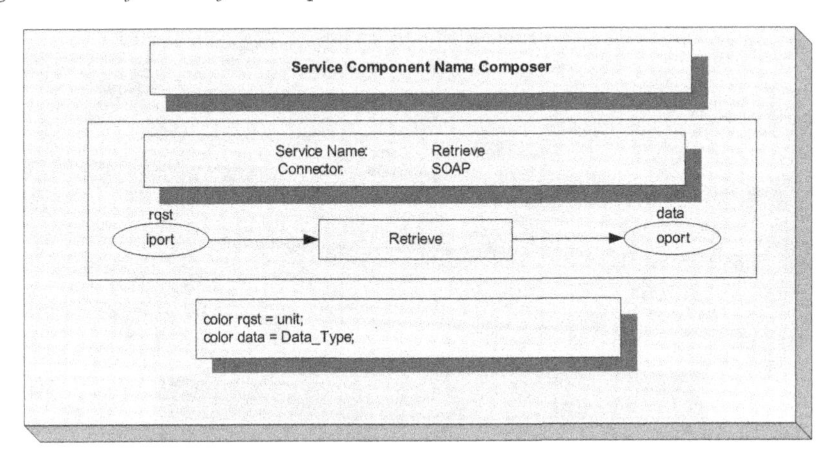

In addition to the inscription for places, transitions, and arcs as used in CPN, WS-Net provides additional inscriptions for both service components and each service provided. In detail, for the service-component inscription, a service-component name is provided to uniquely identify the service component. For a service inscription, service-name and connector-type information are provided. Connector type implies the possible protocols to be used when the service is invoked. In our example, both enqueue and dequeue services are invoked via the SOAP protocol. If multiple protocols can be used, the connector type can have multiple entries. Similar to the name inscription for places and transitions, these inscriptions for service components and services are not considered petri-net inscriptions. Instead, these inscriptions are used to interconnect architectural service components, which will be discussed with the interconnection net in the next section.

In our example, the two service components commit engine and composer (as shown in Figure 6) use the enqueue and dequeue services provided by the message-queue service component. In other words, the commit engine and composer are client components for the message-queue service component. Figure 8 shows the interface net of the commit engine, which provides one service called commit. The commit engine sends data as parameters to the message-queue service component and receives an acknowledgement *msg*. Similarly, as shown in Figure 9, the composer provides one service called retrieve, and sends a request *rqst* to the message-queue service component and receives data as a reply. As we explained previously, the color of the token is used here. We assume that the SOAP protocol is used when the commit engine and the composer service components communicate with the message-queue service component. Note that an interface net specifies only the services provided by the corresponding service component; thus, it does not specify connections between components.

Interconnection Net

In order to describe the relationships between architectural components in a services-oriented system, we need to specify both the provided services (i.e., via the interface-net specification) and required services of the service components. In other words, a service component may require supporting services in order to provide promised services. By specifying all required services of a service component, the interconnection net describes all the possible dependencies upon other service components in the system. The interconnection net is not mandatory, however; instead, it is imperative only when a service component requires foreign services to perform its own commission. For instance, in our example, the message-queue service component does not require any other services as support; therefore, there is no interconnection net associated with the message-queue component. The interconnection net depicts a client-server relationship between components. If service component C_i requests a service from service component C_j, C_i is called a client service component, and C_j is called a server service component. A service component can act as a client service component at some time, and as a server service component at other times.

WS-Net chooses to define required services as foreign services since the services need to be performed by other service components. Conforming to the definitions

Figure 10. A service with n foreign transitions

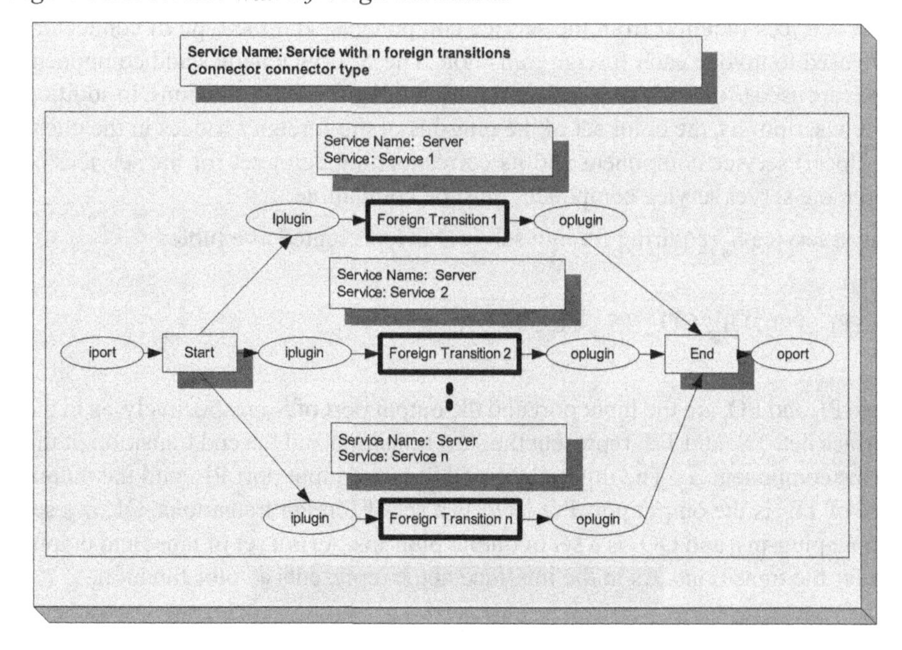

in CPN, in the interconnection net, the required services are specified as a special kind of transition called foreign transition. As in the interface net, the interconnection net specifies an architectural service component as a set of provided services. Each provided service containing foreign transitions is in turn decomposed into a set of required foreign services. A foreign transition is therefore an abstract view of the service provided by another service component. To differentiate it from a local service component, the input and output places of a foreign transition are called input plug-ins and output plug-ins, respectively.

In order to link together a service component and its supporting service components, we introduce two extra transitions for each service component, namely, a start transition and an end transition. As shown in Figure 10, the input port (i.e., place) of the start transition is always the input port of the service component; the output port (i.e., place) of the end transition is always the output port of the service component. A service component may require support from multiple foreign service components before it can perform its execution. Figure 10 illustrates a service that requires *n* foreign transitions. With the introduction of the start transition and end transition, this multiple-support relationship can be denoted using WS-Net specifications. Using the notation we introduced above, if a service requires supporting service from multiple foreign services, it will have a set of input plug-ins and a set of corresponding output plug-ins.

For services that do not require any foreign transitions, the service specification of the interface net will suffice. Foreign transitions also have inscriptions similar to the provided services of a service component. However, as shown in Figure 10, inscriptions of foreign transitions contain names of the remote service components, names of the services required from the service components, and the type of connectors to be used to invoke each foreign transition. These service names and component names are used to identify the services represented by foreign transitions. In addition to the inscriptions, the color set of the plug-ins of the foreign services in the client (i.e., local) service component and its corresponding color set for the services of the remote server service component must be compatible.

Thus, a service S_{ij} requiring foreign services is represented as a tuple:

$$S_{ij} = (PI_{ij}, PO_{ij}, QI_{ij}, QO_{ij}, TS_{ij}, TE_{ij}, TF_{ij}, A_{ij}, c),$$

where PI_{ij} and PO_{ij} are the input port and the output port of S_{ij}, respectively, as in the interface net. TS_{ij} and TE_{ij} represent the start transition and the end transition of the service component S_{ij}. The input place of TS_{ij} is the input port PI_{ij}, and the output place of TE_{ij} is the output port PO_{ij}. TF_{ij} is a set of foreign transitions. QI_{ij} is a set of input plug-ins, and QO_{ij} is a set of output plug-ins. A_{ij} is a set of input and output arcs for the transitions. As in the interface net, c represents a color function.

Figure 11. Interconnection net for Commit Engine

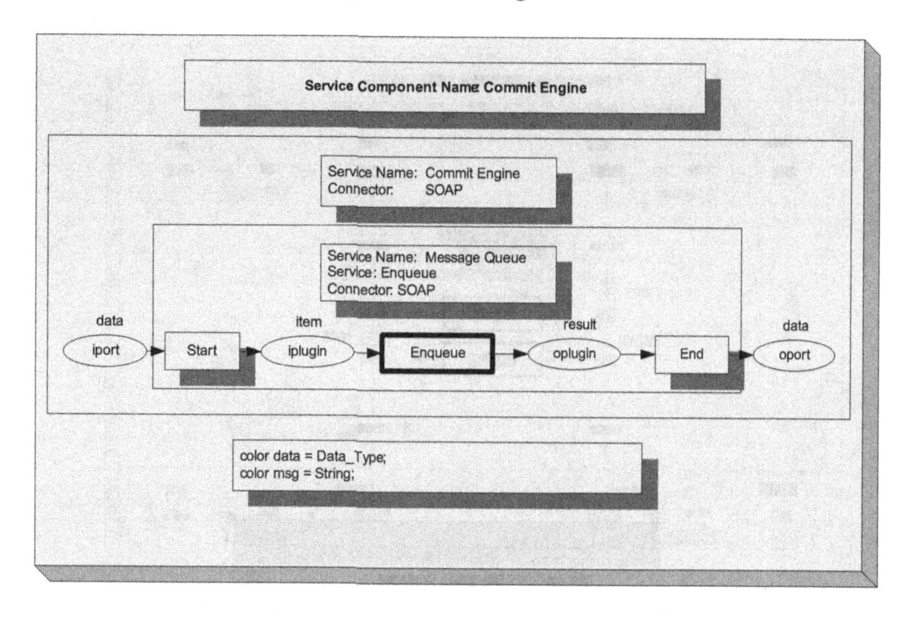

Figure 12. Interconnection net for Integrator

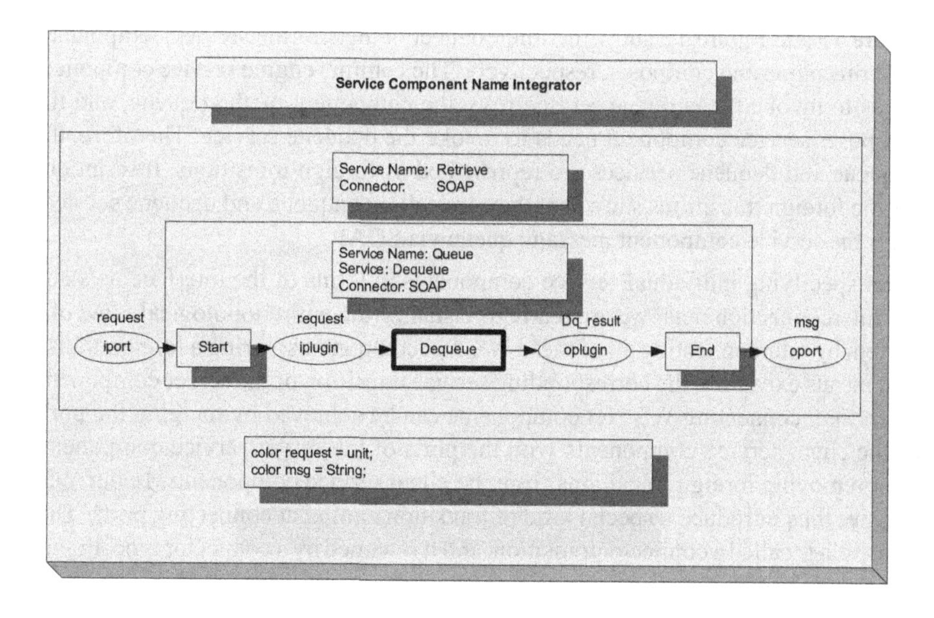

Figure 13. Unfolding interconnection net

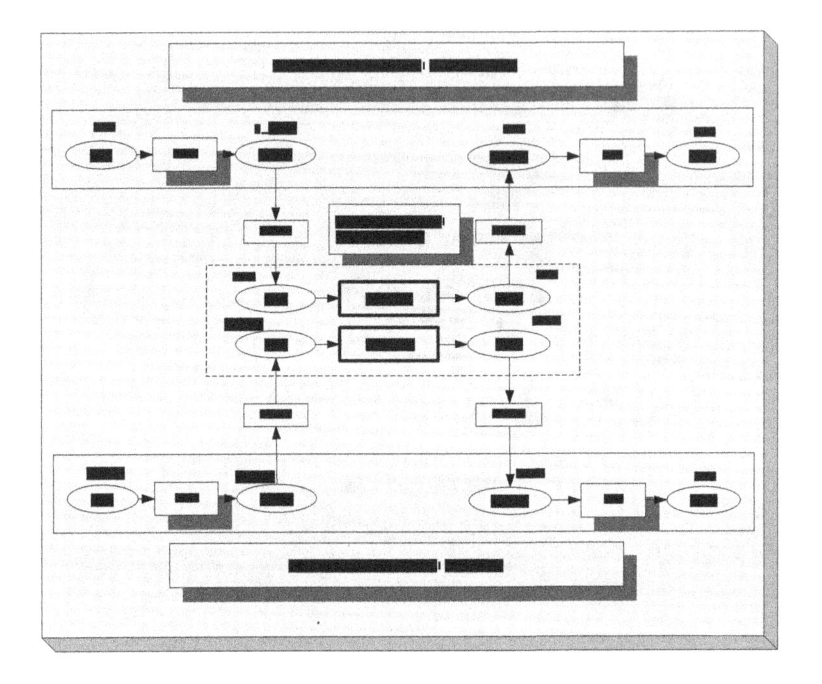

Figure 11 and Figure 12 show the interconnection nets of the service components commit engine and composer, respectively. The commit-engine service component needs to invoke the enqueue service from the component method queue, and the composer service component needs to invoke the dequeue service. Therefore, the enqueue and dequeue services are represented as foreign transitions. Inscriptions for the foreign transitions show that they are calling enqueue and dequeue services from the service component message queue via SOAP.

After specifying individual service components in terms of the interface nets and the interconnection nets, we are ready to visualize the entire topological view of a system by interconnecting all of these WS-Net components. Firing a foreign transition means executing the corresponding service transition of the server component. Therefore, connecting WS-Net components can be achieved by merging the ports of the client service components with the ports of the server service components after removing foreign transitions from the client service components. In our WS-Net, we thus introduce a special kind of transition aiming at connecting ports. This transition is called a connector transition, and it is named by a connector type. Figure

13 shows the connected interconnection net that describes the entire information-communication model by interconnecting the commit engine and the composer with the message queue using SOAP connectors.

In summary, such an interconnection mechanism can be applied across different levels of service-component diagrams. In detail, interconnections can be visualized in two levels: (a) interoperation nets of client-server service components, and (b) the folding and unfolding of interface nets of service components. This is an important feature to visualize very large systems. By applying such visual abstractions, such as replacing large interoperation nets with simpler interconnection nets or even with interface nets, complicated nets can be effectively visualized at various levels of abstraction.

Interoperation Net

The interoperation net describes the dynamic behaviors of a service component by focusing on its operational nature. The goal of the interoperation net is to dissect each service into fundamental process units, which, taken together, define the required functional contents of the service. This is similar to the SADT functional decomposition, where each transition representing the operations of a component is decomposed into subtransitions to represent fundamental operational states. One of the most important differences between the decomposition in our interoperation net and SADT is that the interoperation net uses decomposition as a means of expressing the behaviors of the services provided by an architectural service component rather than functional decomposition for modularization as used in SADT. As in SADT, the control flow and data flow are used to describe the interactions between process units. Note that it is important to distinguish foreign transitions from detailed processes. The foreign transitions along with plug-in places are used to interconnect the interoperation nets to form the entire system view. Like other petri-nets-based high-level design representations, places are used to represent the control or data, and transitions are used to represent processes.

Chang and Kim (1999) found that the straightforward techniques converting functional data flow to petri nets have a potential problem in repeated (persistent) simulations of the nets. To solve this problem, in WS-Net, we distinguish persistent data from transient data. Persistent places are represented as boldface circles. Persistent data items are similar to the data attributes of a class in the object-oriented paradigm. These persistent data items represent the state of a service component, and they exist throughout the lifetime of the service component. On the other hand, transient data items are produced by one process and are immediately consumed by another process. Therefore, transient data items are created only when they are needed and destroyed upon the completion of the service.

Figure 14. First phase of Interoperation net for Queue

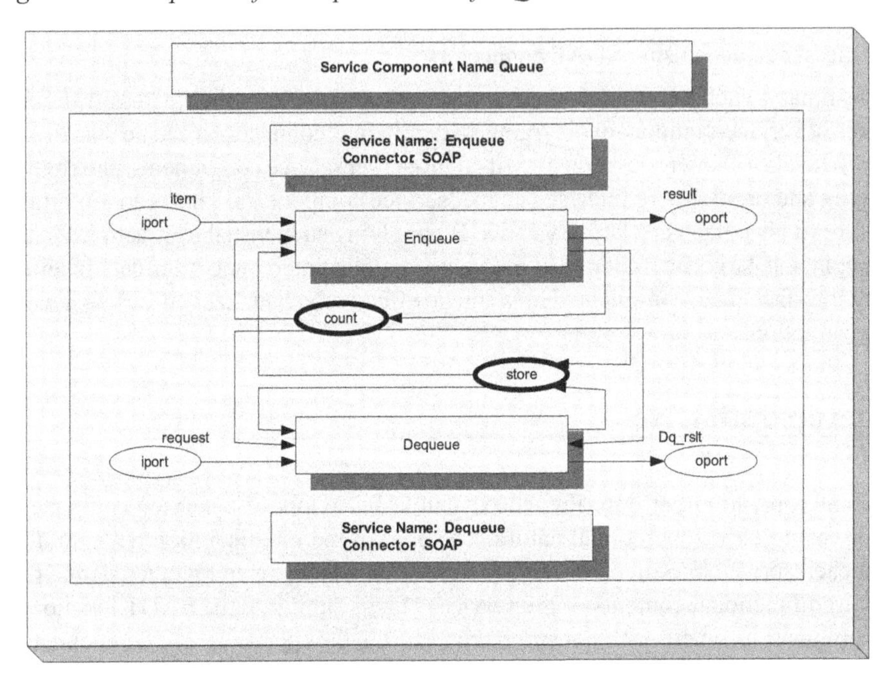

A service $S_{ij} \in S_i$ of service component C_i can be denoted as follows:

$$S_{ij} = (PI_{ij}, PO_{ij}, PT_{ij}, PP_{ij}, QI_{ij}, QO_{ij}, TL_{ij}, TF_{ij}, A_{ij}, c, G, E, IN),$$

where PI_{ij} and PO_{ij} are the input and output ports, and PT_{ij} and PP_{ij} are a set of transient data places and a set of persistent data places, respectively. TL_{ij} is a set of local transitions, and TF_{ij} is a set of foreign transitions. QI_{ij} is a set of input plug-in places serving as input places for the foreign transitions, and QO_{ij} is a set of output plug-in places serving as output places for the foreign transitions. A_{ij} is a set of input and output arcs of the transitions. To describe the functional behaviors of a component, we can use all the inscriptions used in CPN (Jensen, 1990). As before, c is a color function to represent the color sets for the places. G is a guard function for the transitions. E is an arc expression function, and IN is an initialization function for the tokens.

In our example, the message-queue service component has enqueue and dequeue services. The control and data are represented by places, and processes are represented by transitions. Figure 14 shows the first phase of the interoperation description of the message-queue component. The count and storage places are defined as persis-

Figure 15. Interoperation net for Enqueue service

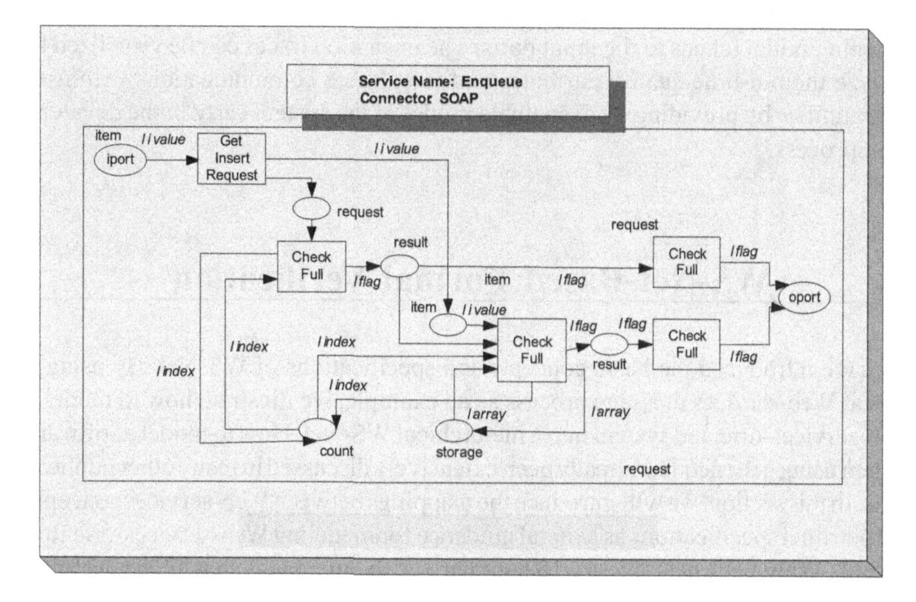

tent data and represented with boldface circles. Since the persistent data may exist throughout the lifetime of a service component, we need to initialize the persistent places with proper tokens for later simulations. Tokens in the transient places are produced as a result of firing transitions. It is common for persistent data items to be shared by other services in the same component. If different services use the same persistent data, they need to be merged using the place-fusion technique defined in high-level petri nets. As shown in Figure 14, count and storage are persistent places of both the enqueue and dequeue services. By merging the persistent places of the two services, the interoperation net for the message-queue component can be completed.

As we further decompose the functional behaviors of each service, we can get a more complex interoperation net. Figure 15 shows a more detailed interoperation net of the service enqueue of the message-queue service component. First, the service receives a request to insert a message into the message queue. Both the content places and store place are checked for whether they are full before an item can be inserted. If the store place is not full, the message can be inserted into the queue. Then the service updates the full flag after the insertion of the message. A petri net can be constructed by mapping each functional process into a transition, and input and output into places. After all the interoperation nets of the architectural service components are specified, we can again visualize the entire system topology by connecting the plug-ins of the client service components with the ports of the server components using the connector transitions.

Connected interoperation nets can be executed under different input scenarios to simulate the behaviors of a services-oriented system. The execution proceeds by assigning initial tokens to the input ports. The execution traces can be visualized to analyze the run-time quality attributes and to enhance communications with user communities by providing an executable model of the system early in the development process.

WS-Net-Based Formal Verification

We have introduced the basic concepts and specifications of WS-Net. By using a typical Web-services message process as an example, we illustrate how to model a Web-services-oriented system into a hierarchical WS-Net. How to model a software system using petri nets has already been extensively discussed in many other publications. In this section, we will introduce the mappings between Web-services concepts and petri-net specifications as general guidance for modeling Web-services-oriented systems using petri nets. Figure 16 summarizes the mappings that are critical due to the fact that petri nets do not explicitly support the concept of process.

As shown in Figure 16, Web services (i.e., service components and services provided) are modeled using transitions. The input and output of a service are modeled by

Figure 16. Mapping between Web-services concepts and petri-net specifications

WS concepts	PN concepts
Messages	Connector
Service	Transition
Input	Input place
Output	Output place
Name	Label
Required service	Foreign transition
Interaction	Input/output places via connector
Data	Token
Signature	Color
Data sharing	Place fusion
Hierarchy	Transition substitute
Persistent data	Persistent place
Transient data	Transient place

two kinds of places, namely, input places and output places, respectively. Messages exchanging between Web services are modeled as connectors in petri nets. In order to enhance readability, labels are used to identify Web services by names. If a Web service requires other services as support, foreign transitions are used to model supporting Web services. Thus, the message interactions between Web services are modeled as data flow between input places and output places via connectors. In order to handle complexity, transition substitution is used to fold and unfold hierarchical services composition into modularized nets according to the architectural structure of a system. In WS-Net, data are modeled by tokens. The concept of color is used to specify the signature and types of data. Persistent data and transient data are differentiated using persistent places and transient places. Data sharing between Web services is modeled by the place-fusion concept in petri nets.

With the mapping mechanisms established, we can turn a Web-services-oriented system into a petri-nets-based WS-Net able to be simulated. With this simulation, we can detect and identify services-composition errors using the analysis mechanisms provided by petri nets. The simulation of the executable system thus can be used to verify the correctness of the system. The interconnection mechanism of WS-Net enables analyzing complicated system composition at different granularities. Associated with the interoperation net, WS-Net provides a structured way to zoom in and out to analyze architectural system composition at various levels.

Using WS-Net, formal verification can be conducted at the design time of system composition in order to detect potential composition errors at early stages, thus allowing the correction of errors as early as possible. Particularly, WS-Net focuses on analyzing important composition criteria, such as reachability, boundedness, and liveness. By examining dead markings, we can verify the reachability of a certain WS-Net, thus verifying whether certain composition protocols (i.e., rules) are enforced and conformed. The state-space analysis can be carried out to detect whether a deadlock possibly exists in the design of the services composition. The visualization of the composition and interactions between Web-service components helps engineers verify compositional message exchanges and synchronizations. WS-Net analysis and simulation can start with an initial marking inputted into a WS-Net model. Running the simulations, we can check whether the service composition will execute as expected, and whether the service composition conforms to the conversational protocols between service components. Furthermore, different markings can be used to feed the constructed WS-Net to verify system behaviors under various situations.

Conclusion

In this chapter, we introduced an advanced topic of service computing: how to formally verify Web-services-oriented system composition. We first introduced the basic concepts of services composition in the context of Web-services technologies. Then we surveyed possible solutions and existing efforts for formally verifying Web-services-oriented system composition. After comparing various options, we introduced WS-Net, an executable petri-nets-based architectural description language to formally describe and verify the architecture of a Web-services-oriented system. The behaviors of such a model can be simulated to detect errors and allow corrections and further refinements. As a result, WS-Net helps enhance the reliability of Web-services-oriented applications. Furthermore, it is compatible with the object-oriented paradigm and component-based concepts. Supporting modern software-engineering philosophies oriented to services computing, WS-Net provides an approach to verify and monitor the dynamic integration of a Web-services-oriented software system. Specification formalism in WS-Net is object oriented, executable, expressive, comprehensive, and yet easy to use. A wide body of theories available for petri nets is thus available for analyzing a system design. To our best knowledge, our WS-Net is the first to comprehensively map Web-services elements to CPNs so that the latter can be used to facilitate the simulation and formal verification and validation of Web-services composition.

However, manually transferring WSDL specifications into the WS-Net specifications is not a trivial job. That is why currently we have only built some simple experiments, for example, the example described in the chapter. In order for WS-Net to monitor and verify real-life applications, the translation from WSDL into WS-Net must be automated.

Meanwhile, since all transient tokens are created by local transitions and all persistent tokens are restored before the completion of the service, repeated simulation of the net is possible. Furthermore, in converting functional data-flow models to petri nets, we also face the concurrency and choice problems (Trattnig & Kerner 1980). These problems need to be addressed properly by system engineers who build the system architecture by using WS-Net.

Despite the challenges WS-Net is facing, our preliminary experiences prove its effectiveness and efficiency in formally verifying Web-services-oriented system composition. WS-Net uses an iterative and top-down process of investigating and examining services composition using the petri-nets vehicle of technology. Future work will focus on building automatic translation tools from Web-services system specification into petri nets' tool-based specifications to automate the simulation of Web-services composition.

References

Aggarwal, R., Verma, K., Miller, J., & Milnor, W. (2004). Constraint driven Web service composition in METEOR-S. *Proceedings of IEEE International Conference on Services Computing (SCC)*, (pp. 23-30).

The American heritage dictionary of the English language (4th ed.). (2000). Houghton Mifflin Company.

Arpinar, B., Aleman-Meza, B., Zhang, R., & Maduko, A. (2004). Ontology-driven Web services composition platform. *Proceedings of IEEE International Conference on E-Commerce Technology (CEC)*, (pp. 146-152).

Berardi, D., Calvanese, D., Giacomo, G. D., Lenzerini, M., & Mecella, M. (2003). Automatic composition of e-services that export their behavior. In *Lecture notes in computer science: Vol. 2910. Proceedings of 1st International Conference on Service-Oriented Computing (ICSOC)* (pp. 43-58). Springer-Verlag.

Bordeaux, L., Salaün, G., Berardi, D., & Mecella, M. (2004). When are 2 Web services compatible? In *Lecture notes in computer science: Vol. 3324. Proceedings of VLDB Workshop on Technologies of E-Services (VLDB-TES)* (pp. 15-28). Springer-Verlag.

Bultan, T., Fu, X., Hull, R., & Su, J. (2003). Conversation specification: A new approach to design and analysis of e-service composition. *Proceedings of the 12th ACM International World Wide Web Conference (WWW)*, (pp. 403-410).

Casati, F., Ilnicki, S., Jin, L., Krishnamoorthy, V., & Shan, M. (2000). *Adaptive and dynamic service composition in eFlow.* Retrieved December 19, 2005, from http://www.hpl.hp.com/techreports/2000/HPL-2000-39.pdf

Chang, C. K., & Kim, S. (1999). I3: A petri-net based specification method for architectural components. *Proceedings of IEEE 23rd Annual International Computer Software and Applications Conference (COMPSAC)* (pp. 396-402).

Ferrara, A. (2004). Web services: A process algebra approach. *Proceedings of the 2nd ACM International Conference on Service Oriented Computing (ICSOC)* (pp. 242-251).

Fu, X., Bultan, T., & Su, J. (2005). Realizability of conversation protocols with message contents. *International Journal of Web Services Research* (JWSR), *2*(4), 72-97.

Jensen, K. (1990). Coloured petri nets: A high level language for system design and analysis. In *Lecture notes in computer science: Advances in petri nets.*

Kiwata, K., Nakano, A., & Yura, S. (2001). Scenario-based service composition method in the open service environment. *Proceedings of 5th International Symposium on Autonomous Decentralized Systems* (pp. 135-140).

Leymann, F. (2001). *Web services flow language (WSFL 1.0)*. IBM Corporation.

Liang, Q., Chakarapani, L. N., Stanley, Y. W., Su, Chikkamagalur, R. N., & Lam, H. (2004). A semi-automatic approach to composite Web services discovery, description and invocation. *International Journal of Web Services Research (JWSR), 1*(4).

Medvidovic, N., & Taylor, R. N. (2000). A classification and comparison framework for software architecture description languages. *IEEE Transactions on Software Engineering, 26*(1), 70-93.

Meta Software Corporation. (1993). *Design/CPN reference manual for X-Windows version 2.0.*

Microsoft. (2003). *BizTalk server 2004: Architecture.* Retrieved December 19, 2005, from http://download.microsoft.com/download/e/6/f/e6fcf394-e03e-4e15-bd80-8c1c127e88e7/BTSArch.doc

Milanovic, N., & Malek, M. (2004). Current solutions for Web service composition. *IEEE Internet Computing, 8*(6), 51-59.

Narayanan, S., & McIlraith, S. A. (2002). Simulation, verification and automated composition of Web services. *Proceedings of the 11th ACM International Conference on World Wide Web (WWW)* (pp. 77-88).

Peterson, J. L. (1981). *Petri net theory and the modeling of systems.* Prentice-Hall International.

Petri, C. A. (1962). *Kommunikation mit automaten.* Unpublished doctoral dissertation, Institut für Instrumentelle Mathematik, Schriften des IIM.

Pi-Calculus. (n.d.). *Automata, state, actions, and interactions.* Retrieved December 19, 2005, from http://www.ebpml.org/pi-calculus.htm

Pinci, V., & Shapiro, R. (1990). An integrated software development methodology based on hierarchical colored petri nets. In *Lecture notes in computer science: Advances in petri nets* (pp. 227-252).

Ponnekanti, S., & Fox, A. (2002). *SWORD: A developer toolkit for building composite Web services.* Proceedings of Alternate Tracks of the ACM 11th International World Wide Web Conference (WWW), Honolulu, HI.

Ross, D. (1984). Application and extensions of SADT. *IEEE Computer, 18*(4), 25-35.

Salaun, G., Bordeaux, L., & Schaerf, M. (2004). Describing and reasoning on Web services using process algebra. *Proceedings of IEEE International Conference on Web Services (ICWS)* (pp. 43-50).

Shaw, M., DeLine, R., Klein, D. V., Ross, T. L., Young, D. M., & Zelesnik, G. (1995). Abstractions for software architecture and tools to support them. *IEEE Transactions on Software Engineering, 21*(4), 314-335.

Simple object access protocol (SOAP) 1.2. (2003). World Wide Web Consortium (W3C). Retrieved December 19, 2005, from http://www.w3.org/TR/soap12-part1/

Specification: Business process execution language for Web services version 1.1. (2003). Retrieved December 19, 2005, from http://www-106.ibm.com/developerworks/webservices/library/ws-bpel/

Thatte, S. (2001). *XLANG: Web services for business process design.* Microsoft Corporation.

Trattnig, W., & Kerner, H. (1980). EDDA: A very high-level programming and specification language in the style of SADT. *Proceedings of IEEE Annual International Computer Software and Applications Conference (COMPSAC)* (pp. 436-443).

Universal description, discovery, and integration. (2004). Retrieved December 19, 2005, from http://www.uddi.org/

Web service choreography interface (WSCI), version 1.0. (2002). Sun Microsystems.

Web services description language. (2004). Retrieved December 19, 2005, from http://www.w3.org/TR/wsdl

XML. (n.d.). Retrieved December 19, 2005, from *http://www.w3.org/XML*

Zeng, L., Benatallah, B., Dumas, M., Kalagnanam, J., & Sheng, Q. (2003). Quality driven Web services composition. *Proceedings of 12th ACM International Conference on World Wide Web (WWW)* (pp. 411-421).

Zeng, L., Benatallah, B., Ngu, A. H. H., Dumas, M., Kalagnanam, J., & Chang, H. (2004). QoS-aware middleware for Web services composition. *IEEE Transactions on Software Engineering, 30*(5), 311-327.

Zhang, J. (2005). Trustworthy Web services: Actions for now. *IEEE IT Professional,* 32-36.

Chapter XIII

Service Computing for Design and Reconfiguration of Integrated E-Supply Chains

Mariagrazia Dotoli, Politecnico di Bari, Italy

Maria Pia Fanti, Politcnico di Bari, Italy

Carlo Meloni, Politecnico di Bari, Italy

Mengchu Zhou, New Jersey Institute of Technology, USA

Abstract

This chapter proposes a three-level decision-support system (DSS) for integrated e-supply-chains (IESCs) network design and reconfiguration based on data and information that can be obtained via Internet- and Web-based computing tools. The IESC is described as a set of consecutive stages connected by communication and transportation links, and the design and reconfiguration aim of the DSS consists of selecting the partners of the stages on the basis of transportation connections and information flows. More precisely, the first DSS level evaluates the performance of all the IESC candidates and singles out the best ones. The second DSS level solves a multicriteria integer linear optimization problem to configure the network. Finally, the third DSS level is devoted to evaluating and validating the solution proposed in the first two modules. The chapter proposes the use of some optimization techniques to synthesize the first two levels and illustrates the decision process by way of a case study.

Introduction

The discipline of service computing covers the science and technology of bridging the gap between business services and information-technology services (Institute of Electrical and Electronics Engineer [IEEE], 2004). Among others, it encompasses a new innovative area: business-process integration and management, also known as enterprise service computing. Enterprise service computing focuses on issues, modeling, methodologies, and the enabling of computing technologies in support of integrated and collaborative enterprise applications. The main task of enterprise service computing is to reengineer operations and integrate international logistics and information technologies in the production process to improve efficiency and minimize risk and cost. This has given rise to the formation of integrated e-supply-chain (IESC) networks, defined as a collection of independent companies possessing complementary skills and integrated with streamlined material, information, and financial flow (Luo, Zhou, & Caudill, 2001; Viswanadham & Gaonkar, 2003). Service computing and Internet- and Web-based electronic marketplaces can provide an inexpensive, secure, and pervasive medium for information transfer between business units in IESC (Gaonkar & Viswanadham, 2001; Luo et al.; Tayur, Ganeshan, & Magazine, 1999). Indeed, by taking advantage of these novel opportunities, companies are able to make smart decisions based on voluminous data flows. Hence, the research community envisages the need of IESC decision support in many areas (Keskinocak, Goodwin, Wu, Akkiraju, & Murthy, 2001). As a result, decision-support systems (DSSs) can be designed to provide effective analysis and comprehension of complex supply chains (SCs). Although several conceptual models for IESC are proposed and discussed in the related literature, research efforts are lagging behind in the development of formal decision models for IESC design (Talluri & Baker, 2002). A systematic way to capture all aspects of SC processes is proposed by Chopra and Meindl (2001) and Biswas and Narahari (2004). This guideline is based on the three levels of the decision hierarchy: the strategic, tactical, and operational ones. Strategic level planning involves SC design, which determines the location, size, and optimal number of suppliers, plants, and distributors to be used in the network. It considers time horizons of a few years and requires approximate and aggregate data and forecast. Tactical level planning basically refers to supply planning, which primarily includes the optimization of the flow of goods and services through a given SC network. Finally, operational-level planning is short-range planning, which involves production scheduling at all plants on an hour-to-hour basis.

Background

The reconfiguration of the supply network over time at the tactical level is essential for a modern business to hold or increase its competitive advantage. Hence, a crucial decision problem of IESC is the network design (Vidal & Goetschalckx, 1997), which determines (a) the number, location, and type of manufacturing plants and warehouses to use, (b) the set of suppliers to select, (c) the transportation channels to employ, and (d) the amount of raw materials and items to produce and ship among suppliers, plants, warehouses, and customers considering bill-of-materials (BOM) relationships. Significant literature deals with the key problem of component selection for network design of the SC, and detailed surveys can be found in Vidal and Goetschalckx, and Shapiro (2001). In the following, we present the research track of decision models for IESC network design, and we show that the current trends go into the direction of using in a more determinant and decisive way modern technologies for enterprises, for example, Internet and service computing.

In particular, Wu, Mao, and Qian (1999) formulate partner selection for distributed agile manufacturing as an integer programming problem to choose one and only one candidate for each task of the production process. They minimize the sum of the costs for performing all the tasks and the transportation costs. However, the method assumes that each IESC task is carried out by one company only: Such a constraint is restrictive and affects the flexibility of the resulting network structure. Moreover, a linear programming model for integrated SC planning is developed by Gaonkar and Viswanadham (2001), who assume that information is freely shared by the SC partners through an Internet-based platform. The linear programming model provides a basis for partner selection such that the cost of operations is minimized. In addition, in a subsequent work, Viswanadham and Gaonkar (2003) study and analyze a multitier SC by using an optimization technique that takes into account capacities and costs. In the work, the authors determined the suboptimal order quantities to be allocated to each of the manufacturers, suppliers, and logistics service providers. A mixed integer programming model is developed for a dynamic manufacturing network, and its objective function maximizes the profit earned by the network subject to various capacity, production, and logistics schedules and flow-balancing constraints. However, the information necessary to describe and characterize the SC can be so complex and large that the definition of constraints appears critical. Hence, Yang, Yu, and Edwin Cheng (2003) propose a strategic production-distribution model for SC design taking into account the BOM constraints represented by logical constraints. In particular, a mixed integer programming model captures the role of BOM in the selection of suppliers to provide the SC structure. Since the work optimizes an objective function considering purchasing, production, and transportation costs, it does not select solutions on the basis of the relative importance of the performance indices.

To face the complexities involved in the SC design process, Talluri and Baker (2002) propose a three-phase mathematical programming approach for SC network design that involves multicriteria efficiency models, and linear and integer programming methods. In particular, Phases I and II design the network, and Phase III addresses operational issues such as the routing of materials. In the same direction, Jang, Jang, Chang, and Park (2002) propose another SC-management system consisting of four modules: an SC network design optimization module, a planning module for production and distribution operations from material suppliers to customers, a model-management module, and a data-management module.

A novel approach to model and optimize IESC networks incorporating e-commerce and electronic linkage is presented by Luo et al. (2001). Interaction and trade-offs between the network components are analyzed and optimized using a fuzzy multi-objective optimization approach. The optimization procedure considers new paradigms for the design of the IESC structure, such as recycling and pollution, which influence the decision process, along with transportation costs and cycle times. The drawback of this approach is that the fuzzy optimization technique provides only one suboptimal network structure and disregards the impact of the solution on the operational-level issues.

Summing up, the proposed models for decisions on SC configuration mainly address the problem of designing a traditional SC, and consider only to some extent the advantages of integrating modern available technologies in the management of these complex systems.

Contribution

Although modern researchers and practitioners agree in asserting the importance of enterprise service computing, few contributions can be found in the recent literature in the field. In order to fill this gap, the chapter proposes a DSS that deploys the opportunities and potential of service computing for effective e-enterprise design and configuration. A DSS is an interactive computer-based system or subsystem that helps decision makers in using information, identifying and modeling problems, and taking decisions (Sprague & Carlson, 1982). The presented DSS addresses IESC network design based on data and information that can be obtained via Internet- and Web-based instruments. Moreover, the DSS selects and updates the IESC network for all stages, each collecting a set of competing companies with analogous characteristics with regard to processes and products, for example, producers, manufacturers, consumers, and so forth. Such a design and reconfiguration process is carried out by the decision structure using the information available on BOM relationships, inventory, transportation cost, distances, and environmental impact. In addition, the DSS is designed with the aim of improving the business agility and flexibility of the IESC for adapting to market fluctuations and evolving as technology advances.

In more detail, in order to face the complexity of the decision process, the IESC structure design is divided into a hierarchy consisting of three decision levels. The candidate-selection level and network-design level take decisions referring to the tactical planning, while the solution-evaluation and -validation level considers operational decision problems and validates the solutions obtained on the first two levels. In particular, the candidate-selection level evaluates the performance of the candidate entities to join the network. On the basis of efficiency criteria considering aggregate data stated by the decision team, the output of this module contributes to create a set of candidate entities connected by links representing transportation and communication. Different procedures can be considered for estimating the relative efficiency of a group of actors (i.e., the candidate members of the IESC stages): Electre (Buchanan, Sheppard, & Vanderpooten, 1999; Mousseau, Slowinski, & Zielniewicz, 2000), the analytic hierarchy process (AHP; Saaty, Vargas, & Dellmann, 2003), the data envelopment analysis (DEA; Thanassoulis, 2001), and many more. As an example and for the sake of brevity, here we consider only the results obtained by the Electre method.

Furthermore, the second level receives the stages of the IESC by the first DSS layer and provides as output one or more network structures. More precisely, the IESC network is described by the digraph proposed in Luo et al. (2001), where nodes are partners and edges are links. Different costs are assigned to each link (edge) so that the performance indices can be obtained by the digraph analysis. The data analyzed at this level consider transportation and information connections among stages, cost, and transport environment impact. In order to configure the IESC network and to select appropriate links among the stages, some multicriteria objective functions are defined, and suited constraints are introduced on the basis of the digraph structure. Hence, a multicriteria integer linear optimization problem is solved.

While the first and second layers help in tactical decision making and produce some possible structures of the IESC network, the third level is devoted to evaluating and validating the solution proposed in the first two levels, taking into account operational issues. In particular, this layer determines and studies the evolution of the IESC and evaluates appropriate performance indices using low-level analytical models or simulation models. The validation process performed at this layer allows us to verify the IESC structures obtained at the higher levels and helps in the final selection of the network.

The chapter focuses on the synthesis of the first two higher levels of the DSS (candidate selection and network design) with the perspective of realizing the third level. The DSS realizes the IESC network on the basis of appropriate optimization tools that deploy the available knowledge base and provide a set of solutions on the basis of the determined performance indices. Moreover, modern information technology supports the communication of different partners and enables the information flow within the value-added chain (Chryssolouris, Makris, Xanthakis, & Mourtzis, 2004). This peculiarity is the key of the flexibility and the business agility of the DSS, which can easily modify and improve the structure of the IESC for each market change.

The proposed methodology contributes to the development of the discipline of enterprise service computing with particular regard to Web-based services in two aspects. First, electronic links devoted to information and commercial services play an important role in IESC design decisions. Second, the proposed DSS may be implemented by employing Web-based technologies and remote modeling and calculus services (see, for instance, Gaonkar & Viswanadham, 2001; Mustajoki et al., 1999) in order to facilitate or enable the participation of different decision makers and develop and boost the collaboration and resolution of the various trade-offs characterizing the different phases of the design and reconfiguration of an IESC. Indeed, Web services offer, nowadays, simple ways to carry out different phases of the decision process leading to an IESC design, for example, the constitution of the decision-making team, acquiring and evaluation of historical data, development of performance indices and recognition of best practices, development of models for configuration of networks, tackling of optimization and decision-analysis issues, and implementation and validation of models. Moreover, the proposed method supports an effective team-oriented process for operational decision making, characterized by flexibility in response to market changes or organizational changes, while the use of Web computing and modeling tools facilitates both its adoption and plain use in a decentralized and collaborative way.

This chapter is organized as follows. First it describes the structure of the proposed DSS and presents a methodology to obtain the DSS candidate-selection module. Next we synthesize the DSS network design module and the optimization model employed in the decision strategy. In addition, the technique is applied to a case study. Finally, the chapter discusses the DSS solution and evaluation module, and then summarizes the conclusions.

The Structure of the Decision-Support System for Integrated E-Supply-Chains

Description of the Integrated E-Supply-Chain

Consider a generic distributed manufacturing process that requires a number of component suppliers, subassembly manufacturers, logistics service providers, and users located at different geographical sites. The distributed manufacturing system is arranged as an IESC composed of a sequence of stages connected by material transporters and by information exchange. More precisely, partners belonging to different stages can exchange information on their product schedules and cost. The considered IESC contains different stages: raw-material supply, intermediate supply, manufacturing, distribution, retail, customers, and demanufacturing or recycling. After the demanufacturing stage, recovered material, components, or energy feedback to suitable supply-chain stages are considered. We denote the IESC stages by the set $ST = \{P_1, \ldots, P_k, \ldots, P_{N_S}\}$, where N_S is the number of stages. Each stage P_k is described as a set of partners representing different candidates (or actors) of the SC.

The Hierarchical Structure of the Decision-Support System

Analyzing both the characteristics of the candidate entities able to join the IESC and the performance of members of the network is a crucial step of the decision process related to the configuration of the overall system. As a decision maker is often a heterogeneous team of experts, the proposed analysis tools have to be quite simple to use and understand (Shapiro, 2001), as well as applicable to different SC stages. Moreover, Web services provide an efficient and simple means to support decision makers in taking decentralized and multidisciplinary decisions.

In this section, we describe a three-level approach to design the DSS for configuring (or reconfiguring) the IESC. More precisely, the DSS decomposes the design of the IESC into three hierarchical levels; then it proposes different solutions by analyzing various sets of data and considering different scenarios. Figure 1 shows the hierarchical structure of the DSS for IESC configuration and depicts the three levels of the design procedure. A specific module is devoted to each decision stage, and the interactions between modules allow us to obtain and refine proposed configurations for the IESC (see Figure 1). The resulting DSS may be implemented as a shared and remote platform dedicated to enterprise service computing. The levels of the hierarchical DSS are described as follows.

Figure 1. The DSS structure for the IESC configuration

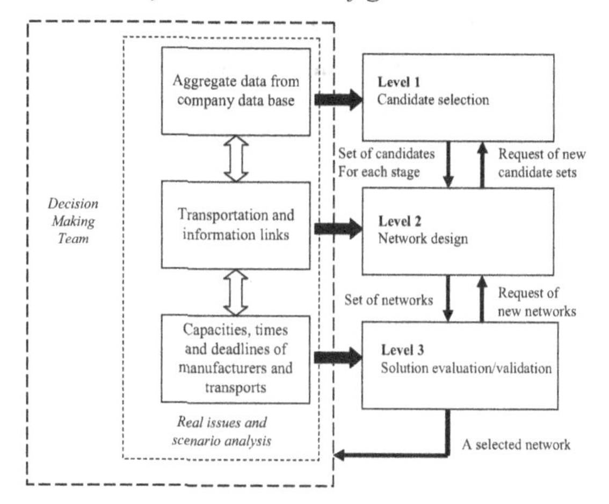

- **First level: Candidate-selection module:** Some multicriteria data-analysis techniques are applied to the company's database in order to create a pool of ranked candidates to join a supply-chain project. At this DSS level, the criteria to select partners are based on aggregate performance indices that are used to analyze the efficiency of the possible members. Hence, we consider different procedures for estimating the relative efficiency of a group of SC partners. The output of this module produces a ranking and classification of the entities considered as candidates for each IESC (see Figure 1).
- **Second level: Network-design module:** This DSS module receives the set of IESC candidates that compose each stage from the first level. Moreover, the transportation and information links that connect the candidates have to be specified. More precisely, each link is characterized by a structural description (the partners that the link connects) and by some relevant performance indices (distances, transportation cost, etc.). This decision level has to select the actors of the stages and the transportation and information links that have to connect the actors. To this aim, an integer multicriteria linear optimization problem is stated, and a set of nondominated solutions are generated. Each solution proposes an IESC network structure that minimizes a set of performance indices, for example, transportation cost, CO_2 emission, cycle time, and energy. Note that while some of the selected indices may be typical indicators of an IESC operation, others are important for measuring the chain's environmental impact and sustainability. However, the optimization procedure may provide an unsatisfactory set of networks. This situation can happen because the former DSS level supplies this module with an insufficient number of stage candidates or with candidates not properly connected. Thus, with reference to Figure 1,

in such cases the network-design module (Level 2) requires a new application of the candidate-selection module (Level 1).

- **Third level: Solution-evaluation and -validation module:** This DSS module receives as input the set of IESC networks determined by the second level with the associated performance indices (see Figure 1). Moreover, this level gets and takes into account the capacities of each partner and transportation facility, the times employed to process and transport the products, and the production deadlines. Hence, at this level, the IESC is modeled addressing tactical and operational issues. Accordingly, analytical or simulation models provide operational performance indices to evaluate and validate the solutions generated by the previous levels. Therefore, this module allows us to validate the IESC network structures obtained at the previous levels in order to evaluate some lower level performance indices such as lead times, utilization, inventory levels, and so on (Viswanadham & Raghavan, 2000). As a result, this DSS stage verifies whether the received system structures are suited to perform the IESC tasks. In case the verification is not satisfactory, it is possible to feed information back to the higher level to change the IESC network appropriately (see Figure 1). For example, if the production capacities provided by the proposed networks are not sufficient, then Level 3 asks Level 2 for IESC networks having a larger number of manufacturers. More precisely, when the verification fails, Level 3 has to single out the stages of the IESC networks that must be modified by Level 2.

At this point, the reasons leading to the choice of the three hierarchical levels can be clear: The DSS structure follows an intuitive procedure to find the optimal network. The first level selects the available actors, the second one builds the possible networks, and the last one chooses the suitable network on the basis of operational performance indices. The proposed decision structure is able to operate both statically, that is, to design a new network, and dynamically, that is, when unsuccessful validations from Level 3 or changes in the market context call for a reconfiguration of the SC. Moreover, the decision team should analyze different scenarios so that the applied optimization methods are able to yield a family of solutions according to the required objectives of the problem.

Candidate-Selection Module

The DSS candidate-selection module is devoted to analyzing the performance and efficiency of the candidates of each IESC stage. This level of the DSS is modeled as a candidate-ranking problem, that is, a multicriteria optimization problem in which a number of criteria are defined to measure the impact on the IESC. We remark that the selection of the criteria and determination of the corresponding efficiency is a hard task for the decision team (Buchanan et al., 1999). Moreover, it is not possible to provide a specific methodology to accomplish this work, which is based on experience in the particular field in which the IESC operates and on the peculiar stage. However, Web services and service-computing technology can support experts by the use of Internet data and communication. Indeed, the criteria necessary for the decisions have to be determined by the team of decision makers. Such a process may be eventually performed with the support of an Internet-based platform, possibly via a remote and collaborative evaluation. Similarly, the corresponding performance indices may be obtained as 0-to-100 scores by a Web-based service collecting and elaborating data in a remote way.

We can provide just a trace to follow in order to obtain the output of this module: a current ranking (on the basis of the efficiency criteria stated by the decision team) of the considered entities, so that for each stage of the SC an ordered list of candidates is provided. In the following, the structure and the tasks of this module are described by four steps, which are iterated for each IESC stage.

Step 1. Single out the candidate set of the generic stage denoted by $P_k^c=\{n_i^c\}$. The set P_k^c collects all the candidates n_i^c that may compose stage P_k of the SC.

Step 2. Define the most relevant criteria that measure the impact of each alternative. To this aim, the following types of criteria can be used (Beamon, 1999; Buchanan et al., 1999): financial (F), including cost and financial return; risk management (RM), including the risk of plant failure and damage following natural disasters; environmental (E), including the effect on relationships with resource partners and on access to resources; flexibility (FL), representing the capacity of a candidate to adjust market requests; operation times (T), representing the ability to respect the decided deadlines; and quality (Q), evaluating the products and provided service.

Step 3. Determine a normalized score on a 0-to-100 scale for each candidate with reference to each criterion. These data, where each alternative is assessed using each criterion, produce a table of impacts, referred to as performance indices.

Step 4. Perform an optimization process to provide a final ranking of the partners in P_k^c on the basis of criteria and scores obtained at the previous steps. At this point, the designer is ready to select the candidates in a subset $P_k \subset P_k^c$ denoting the actors of the *k*th stage of the IESC.

To implement this decision level, several well-known multiattribute decision-making (MADM) methodologies can be adopted, for example, the Electre method (Buchanan et al., 1999; Mousseau et al., 2000), the analytic hierarchy process (Saaty et al., 2003), the data envelopment analysis (Thanassoulis, 2001), and many others, such as, for instance, fuzzy methods (Farina & Amato, 2004). The customary best practice is to apply different MADM methodologies to the given problem so that different rankings of candidates are obtained: The set of candidates that receives the best scores in each classification represents the "best" candidates with a good safety margin (see, for example, Dotoli, Fanti, Meloni, & Zhou, 2005). Indeed, bearing in mind the idea of a Web-based DSS implementation, the described module may consider multiple optimization methods to remove subjectivity from the decision. Similarly, the different decision parameters of the techniques may get rid of polarizations by interviewing various decision makers. Moreover, the DSS lets us carry out different investigations of the given data so that the decision makers may obtain diverse answers to each decision question while being eventually oblivious of the details of the underlying optimization techniques. For the sake of brevity and to show a simple example of this DSS module, the candidate ranking and classification is here performed using the Electre method.

Candidate Selection with Electre

The first step of the procedure determines the candidate set of the generic stage P_k^c. All candidates to join the IESC *k*th stage are identified and their information is collected.

Having defined the most relevant criteria in the second step (e.g., F, RM, E, FL, T, Q), the third step assigns the scores to each candidate. Hence, a performance table reports the obtained performance indices.

The fourth step determines the final ranking of the candidates to each IESC stage with the selected MADM technique, that is, Electre. In particular, the Electre method is a multiple-criteria sorting method originally developed to integrate imprecision and uncertainty in decision making by using the so-called thresholds of indifference, preference, and veto. A further feature distinguishes Electre from many MADM solution methods: It is fundamentally noncompensatory. In other words, low scores with reference to some criteria cannot be compensated by high scores on other criteria. To apply the Electre method, the decision makers are required to define a table collecting the indifference, preference, and veto thresholds as well as a set of weight-importance coefficients. While thresholds model the noncompensatory nature of the method, the weights deal with preference information, reflecting the relative importance of each criterion according to the decision-making team (Mousseau et al., 2000). The thresholds and weights are subjective: Once the performance is agreed upon by all decision makers, then the subjective inputs of thresholds and weights can be processed. Hence, the indifference, preference, and veto thresholds together with the weights are defined for the considered IESC with respect to each criterion taken into account.

Using the defined thresholds, the Electre method seeks for an outranking relation of the candidates resulting from the intersection of the results of an ascending and descending distillation process (Buchanan et al., 1999; Mousseau et al., 2000). According to the ranking results, the decision makers select the required number of actors for the *k*th stage, for example, via a Web-based platform (Gaonkar & Viswanadham, 2001), to be included in the SC network starting from the top one in the ordered list of candidates. The procedure is then iterated for each stage.

Network-Design Module

This module applies optimization algorithms to the IESC resulting from the previous level of the DSS and determines the optimal subnetworks according to the objectives and constraints indicated by the decision maker. Moreover, the user can select different objectives: operative cost functions, cycle time, energy savings, environmental cost, and so forth. In addition, by imposing the appropriate constraints, the designer can ask for a certain number of manufacturers (or customers), the presence of a specific link in the solution, and the presence of a particular subnet (e.g., in the case of SC expansion) in the solution.

Similar to the previous DSS level, the present module may also benefit from the use of a Web-based platform with the previously discussed advantages, that is, an automatic, decentralized, collaborative, and less subjective implementation of the strategy.

The Network Description

Denoting the IESC as the stage set $\mathbf{ST} = \{\mathbf{P}_1, \ldots, \mathbf{P}_k, \ldots, \mathbf{P}_{N_s}\}$, each stage P_k is described as a set of s_k partners representing different actors of the IESC, i.e., $\mathbf{P_k} = \{n_{i_k}, n_{i_k+1}, n_{i_k+2} \ldots, n_{i_k+s_k-1}\}$, where i_k is the generic index such that:

$$i_k = \sum_{h=1}^{k-1} s_h$$

with k=2, 3, …, N_s and i_1=1. We suppose that there are *N* partners in the system. In addition, the (*k*-i)th stage ((*k*+i)th stage) is called the upstream (downstream) stage

Figure 2. The structure of a generic IESC network

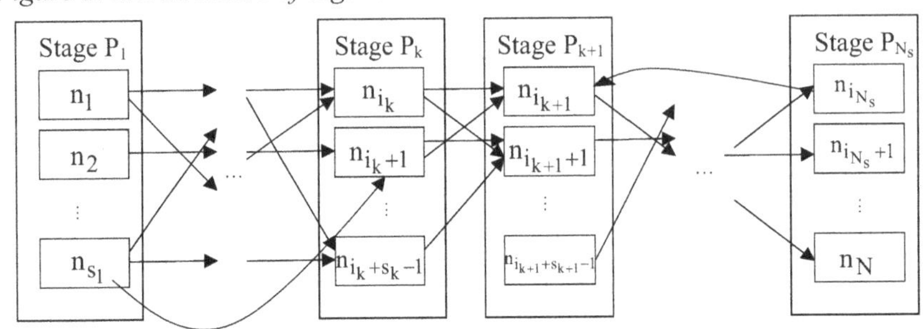

of P_k, with i>0. Moreover, the BOM of stage P_k is a set of components required for processes in the *k*th stage and produced by upstream stages.

Furthermore, the partners of different IESC stages can be connected by two types of links: material-flow links (m-links) and information links (e-links). More precisely, an m-link represents the physical transportation link between two partners, and multiple m-links are allowed between two partners to model different transportation modes or split delivery routes. In addition, an e-link models e-business relationships between business entities for streamlining the material flow efficiently and effectively. We assume that two partners can be connected by m-links and/or by e-links, and that an e-link may connect two partners of the SC without the presence of an m-link. Hence, the proposed structure is able to extend the traditional SC into a more sustainable and integrated production system. Note that IESC partners belonging to the same stage are not connected by links since in the considered model, material and information flow through different stages. Formally, the m-links of stage P_k are denoted by the set $\mathbf{L}_{\mathbf{m}_k} = \{m_{ij}\}$, where m_{ij} is an m-link starting from $n_i \in P_k$ and ending in $n_j \in P_h$, with $P_k, P_h \in ST$ and $k \neq h$. The set $\mathbf{L}_\mathbf{m} = \cup \mathbf{L}_{\mathbf{m}_k}$ denotes the overall material-flow link set. Analogously, we denote the e-links set of stage P_k by $\mathbf{L}_{\mathbf{e}_k} = \{e_{ij}\}$, where e_{ij} is an e-link starting from $n_i \in P_k$ and ending in $n_j \in P_h$, with $P_k, P_h \in ST$ and $k \neq h$. The set $\mathbf{L}_\mathbf{e} = \cup \mathbf{L}_{\mathbf{e}_k}$ denotes the e-links set and $\mathbf{L} = \mathbf{L}_\mathbf{m} \cup \mathbf{L}_\mathbf{e}$ is the complete set of links of the IESC. Figure 2 depicts a generic IESC network exhibiting a succession of stages; each stage is composed of a set of partners (actors). In addition, stages are connected by different links that can represent here m- or e-links. Note that for the sake of readability, only some IESC links are represented in Figure 2 and their nature (m- or e-link) is omitted.

Moreover, we complete the description of the IESC network by introducing the set of performance indices $\mathbf{M} = \{M_1, M_2, \ldots, M_{N_M}\}$, where each element $M_q \in M$ corresponds to a performance measure. Typical indices include cost, transportation and process time, product quality, energy consumption, and environmental impact (Beamon, 1999; Luo et al., 2001). A performance value is assigned to each link considering m- and e-links: $M_q(m_{ij})$ ($M_q(e_{ij})$) with q=1, …, N_M denotes the value of the performance measure M_q associated with the link $m_{ij} \in \mathbf{L}_{\mathbf{m}_k}$ ($e_{ij} \in \mathbf{L}_{\mathbf{e}_k}$). Particularly, an e-link speeds up the communication process and thus reduces the response

Figure 3. The digraph associated to a generic IESC

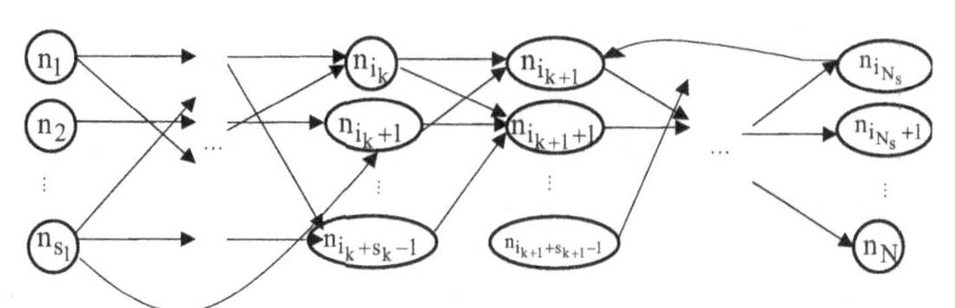

time affecting performance measures such as cost and productivity. Hence, if two partners (e.g., $n_i \in P_k$ and $n_j \in P_h$) are connected both by an e-link and m-link (i.e., m_{ij} and e_{ij}), the performance measures are suitably updated and are associated with the m-link m_{ij} only. On the other hand, if just an e-link connects two IESC actors, then it models the fact that no material flow is possible between the two partners. So, the performance measure assigned to the e-link is the cost of information, such as Internet portals, Web sites, and electronic databases.

The Digraph Definition

To exhibit the interactions among the IESC stages, we define a direct graph D=(N,E). The node set N represents the complete partner set of the network, and each node $n_i \in N$ for i=1, …, N is associated with partner $n_i \in P_k$ for $k \in \{1,\ldots,N_S\}$ of the IESC network. For the sake of simplicity, the same symbols indicate nodes and partners. Moreover, the edge set E is such that an arc y_{ij} directed from n_i to n_j is in E if there exists an m-link $m_{ij} \in L_m$ and/or e-link $e_{ij} \in L_e$. We denote with E the number of edges in D. Figure 3 shows the digraph associated with the generic IESC network in Figure 2. For the sake of simplicity, the digraph edges are not labeled. Note that there is a one-to-one correspondence between the IESC structure (e.g., the example in Figure 2) and its digraph (e.g., the example in Figure 3). Moreover, once the overall IESC structure is known, an additional table can then collect all the information on transportation ad communication links (i.e., m- and e-links).

The Optimization Model for the Network-Design Module

This section proposes two alternative optimization models presented in the related literature in order to design the IESC network. The first optimization model was presented by the authors and leads to an integer linear programming (ILP) problem (Dotoli, Fanti, Meloni, & Zhou, 2006); the second technique was presented by one of the authors and uses a fuzzy optimization model (Luo et al., 2001). The two procedures start from the knowledge of the digraph D=(N,E), which takes into account all the possible actors belonging to the IESC stages and all the possible links that can connect the considered partners. Hence, an optimization technique has to select a subdigraph D*=(N*,E*), with $N^*\subset N$ and $E^*\subset E$, that corresponds to an IESC responding to structural constraints and exhibiting optimal or suboptimal performance indices. To this aim, the optimization problems are established considering a single-criterion or a multicriteria optimization subject to a set of constraints obtained by the structure of the IESC.

In order to state the optimization model and to use a clear mathematic notation, we introduce the decision variable vector $\boldsymbol{x}=[x_1 x_2 \ldots x_E]^T$, where each element $x_h\in\{0,1\}$ with h=1, …, E is associated with an edge $y_{ij}\in E$. More precisely, the value of x_h indicates the presence (x_h=1) or the absence (x_h=0) of the edge $y_{ij}\in E$ in the solution digraph (Hillier & Lieberman, 2005). The objective functions are obtained considering the performance index functions M_q that assign to each m-link m_{ij} and e-link e_{ij} the value $M_q(m_{ij})$ and $M_q(e_{ij})$, respectively, and to each m- and e-link m_{ij} and e_{ij} the comprehensive value $M_q(m_{ij})$. Consequently, value $M_q(m_{ij})$ or $M_q(e_{ij})$ is associated with edge $y_{ij}\in E$ and the corresponding variable x_h. Let us indicate with c^q the column vector of E entries where the hth entry is $c^q_h=M_q(m_{ij})$ or $c^q_h=M_q(e_{ij})$ associated with the edge $y_{ij}\in E$ and the decision variable x_h labeling y_{ij}.

The Optimization Model Solved by ILP

The objective of the model is to minimize a single-criterion or multicriteria cost function subject to the constraints that we characterize as BOM, path, mutual-exclusion, and structural constraints. The optimization problem is as follows:

$$z=\min \phi(\boldsymbol{x}) \tag{1}$$

subject to

$$\boldsymbol{A}\boldsymbol{x}\geq\boldsymbol{B} \tag{2}$$

$$x_h\in\{0,1\} \text{ for } h=1, \ldots, E, \tag{3}$$

where A is the constraint matrix of dimension $v\times$E, $\boldsymbol{B}$ is a v-entry vector of integers, and v represents the number of constraints. Minimizing the objective function $\phi(\boldsymbol{x})$ means to minimize either only one performance index (Problem 1) or a subset of the chosen performance indices (Problem 2); that is, $\phi(\boldsymbol{x})$ represents either a single-criterion or multicriteria objective function.

The Objective Function Definition

Problem 1. The single-criterion problem minimizes only one performance index (for instance, M_q), so that the objective function is defined as follows:

$$\phi(\boldsymbol{x})=f_q(\boldsymbol{x})=(\boldsymbol{c}^q)^T\boldsymbol{x}. \quad (4)$$

Each solution $\boldsymbol{x}^*$ of the ILP problem of Equations 1 to 4 for a particular vector $\mathbf{c}^q$ corresponds to a possible IESC structure. More precisely, the optimal solution vector $\boldsymbol{x}^*$ selects a subdigraph D*=(N*,E*) of D. If the *h*th entry of $\boldsymbol{x}^*$ is $x^*_h=1$ and x_h is associated to the edge y_{ij} connecting from n_i to n_j, then the solution selects the edges $y_{ij}\in E^*$ and the nodes $n_i,n_j\in N^*$. In other words, the optimal IESC with respect to index M_q is described by D*, exhibiting the actors (nodes) and the links (edges) selected in the IESC design.

Problem 2. The multicriteria objective function is defined as follows:

$$\phi(\boldsymbol{x})=f_q(\boldsymbol{x})=\boldsymbol{C}\boldsymbol{x}, \quad (5)$$

where

$$C=\begin{bmatrix}(c^{q1})^T\\ \ldots\\ (c^{qQ})^T\end{bmatrix}$$

is a $q_Q\times E$ criteria matrix and $\boldsymbol{c}^{q1}$ to $\boldsymbol{c}^{qQ}$ are E-entry vectors associated with performance indices M_{q1} to M_{qQ}, respectively.

The set of solutions of the multicriteria ILP problem of Equations 1 to 3, and 5 provides the maximal Pareto face of the solutions set (Ehrgott, 2000). More precisely, we obtain a subset of solutions $\boldsymbol{X}^*=\{\boldsymbol{x}^*_i\}$ where each $\boldsymbol{x}^*_i\in\boldsymbol{X}^*$ is a Pareto optimal solution corresponding to a subdigraph D^*_i of D and to an IESC structure.

The Constraints Definition

BOM constraints. Each candidate supplier can provide a subset of materials (or components) to each producer. Analogously, a retailer actor provides a subset of products to a consumer. Generally, the BOM of each partner of the manufacturer stage or consumer stage is described as a list of specified materials needed by the stage. We can distinguish between two cases: (a) the manufacturer and consumer stages include only one actor each, and (b) the manufacturer and consumer stages present more than one partner each. The BOM constraints for the first case are defined as follows, and the reader can refer to Dotoli et al. (2006) for the constraints of the second case.

The actor of the considered stage P_k (e.g., manufacturer or consumer) has to obtain all the BOM components. For instance, let us suppose that the manufacturer requires all the components necessary to assemble a final product. In addition, suppose that each of these products can be equally shipped by means of three different edges respectively labeled with variables x_1, x_2, and x_3. This condition can be written as an inequality in 0 to 1 variables: $x_1+x_2+x_3\geq 1$. Hence, if we suppose that there are v_1 BOM constraints, then the following inequality constraints are formulated:

$$A_1 x \geq \mathbf{1}, \qquad (6)$$

where $\mathbf{1}$ is a v_1-entry vector with all elements equal to 1 and A_1 is a $v_1\times E$ constraint matrix defined by

$$A_1 = \begin{bmatrix} a_1^1 \\ a_1^2 \\ \dots \\ a_1^{v1} \end{bmatrix}, \qquad (7)$$

with $a_1^i=[a_1(i,1)\ a_1(i,2)\dots a_1(i,E)]$ representing the ith row of A_1 and $a_1(i,j)\in\{0,1\}$ for j=1, …, E and i=1, …, v_1.

Path constraints. It is necessary to select in the digraph at least a path starting from a node of the producers and ending at the nodes of the consumers. To impose this condition, we associate with the digraph the N×E incidence matrix I_M where each element is $I_M(i,j)\in\{-1,0,1\}$. More precisely the following apply.

$I_M(i,j)=0$ if the arc labeled by x_j does not belong to node n_i.

$I_M(i,j)=-1$ if the arc labeled by x_j starts from node n_i.

$I_M(i,j)=1$ if the arc labeled by x_j ends in node n_i.

Moreover, to define a constraint that imposes the presence of a path starting from node n_h and ending at node n_W, we introduce an N vector $\boldsymbol{b}_{h,w}=[b_1\ b_2\ \dots\ b_N]^T$ with $b_h=-1$, $b_W=1$, and $b_p=0$ for p≠h,w and p=1, …, N. The constraint is written as

$$\boldsymbol{I}_M \boldsymbol{x} \geq \boldsymbol{b}_{h,w}. \qquad (8)$$

Hence, the constraint submatrix and the left-side vector are

$$A_2 = \begin{bmatrix} I_M \\ \dots \\ I_M \end{bmatrix} \text{ and } B_2 = \begin{bmatrix} b_{h1,w1} \\ \dots \\ b_{hi,wi} \end{bmatrix}. \qquad (9)$$

Mutual-exclusion constraints. In some cases it is necessary to choose only one actor in a stage. In other words, the condition "at most one edge among those labeled with variables x_1, x_2, and x_3 can be in the solution digraph" is written with the following inequality in 0 to 1 variables: $x_1+x_2+x_3 \leq 1$. On the other hand, the condition "one and only one edge among those labeled with variables x_1, x_2, and x_3 has to be in the solution digraph" is expressed by $x_1+x_2+x_3=1$. Hence, v_3 mutual-exclusion constraints can be expressed by the following equation:

$$\boldsymbol{A}_3 \boldsymbol{x} \leq \boldsymbol{1}, \qquad (10)$$

where **1** is a v_3-entry vector with all elements equal to 1, and matrix A_3 is a constraint matrix defined by:

$$A_3 = \begin{bmatrix} a_3^1 \\ a_3^2 \\ \dots \\ a_3^{v3} \end{bmatrix}, \qquad (11)$$

with $a_3^i=[a^i_3(i,1)\ a^i_3(i,2)\ \dots\ a^i_3(i,E)]$ and $a^i_3(i,j) \in \{0,1\}$ for j=1, …, E and i=1, …, v_3.

Structural constraints. Some particular structural constraints are related to a digraph. For example, the following condition can be imposed: "If the edge corresponding to x_1 belongs to the solution digraph, then the edges corresponding to x_2 or x_3 belong to the solution digraph." This condition is expressed by the constraint $x_2+x_3 \geq x_1$. A second type of structural condition is "the edge corresponding to x_2 belongs to the solution digraph if and only if (iff) the edge labeled with variable x_1 belongs to the solution digraph," that is, $x_2=x_1$. Therefore, the v_4 structural constraints can be expressed as follows:

$$\boldsymbol{A_4 x \geq 0} \tag{12}$$

$$A_4 = \begin{bmatrix} a_4^1 \\ a_4^2 \\ \ldots \\ a_4^{v4} \end{bmatrix}, \tag{13}$$

with $a_4^i=[a_4^i(i,1)\ a_4^i(i,2)\ \ldots\ a_4^i(i,E)]$ and $a_4^i(i,j)\in\{0,1,-1\}$ for j=1, …, E and i=1, …, v_4.

Finally, the constraint matrix and the specification vector are as follows:

$$A = \begin{bmatrix} A_1 \\ A_2 \\ -A_3 \\ A_4 \end{bmatrix}$$

and

$$B = \begin{bmatrix} 1 \\ B_2 \\ -1 \\ 0 \end{bmatrix} \tag{14}$$

The optimization models of Equations 1 and 4 or 5 subject to constraints in Equations 2 and 3 can be solved by applying standard algorithmic approaches for ILP problems (Hillier & Lieberman, 2005). However, we find that the defined constraints give to the model a particular structure that plays an important role in the effective solution of the problem. Indeed, the constraints have a logical structure that falls into the case of the shortest-path problem (Hillier & Lieberman). Consequently, the experimental results show that the linear programming (LP) relaxation of the problem gives integer (0,1) solutions. Hence, the proposed model is suitable to be straightforwardly solved by applying standard algorithmic approaches for LP problems. Indeed, we also find that the integer solutions can be obtained by the standard two-phase simplex method (Mangini, 2003). On the other hand, the case study considered in the following section models a high-level and aggregate real IESC, exhibiting a typical optimization-problem dimension.

The Fuzzy-Optimization Model

This section discusses the fuzzy-optimization model for IESC network design proposed in Luo et al. (2001). Since the IESC data may be uncertain or information may be vague or imprecise, fuzzy multicriteria optimization may be successfully employed to design the IESC structure on the basis of competing objectives. Indeed, fuzzy logic provides a natural framework to incorporate qualitative knowledge with quantitative information such as real data (Bellman & Zadeh, 1970).

Given the IESC network obtained by Level 1 (candidate-selection module) of the proposed DSS (see Figure 1) and the associated digraph D, the fuzzy goals correspond to the defined objective functions $f_q(x)=(c^q)^T x$ with q=1, …, N_M. If the tolerance of fuzzy constraints is assigned, then the membership functions of the fuzzy objectives can be established as $\mu_q(x)$ for q=1, …, N_M. In particular, such membership functions are piecewise linear and can be defined via a payoff table of positive ideal solutions that optimize each objective function independently (Luo et al., 2001).

The feasible solution set is defined by the interaction of the fuzzy objective sets and characterized by the membership function *w*:

$$w = \min\left(\max\left(\mu_1(\mathbf{x}),...,\mu_{N_M}(\mathbf{x})\right)\right) \quad (15)$$

subject to

$$\max\left(\mu_1(\mathbf{x}),...,\mu_{N_M}(\mathbf{x})\right) \le \mu_q(\mathbf{x}) \text{ for each } q=1,\dots,N_M, \quad (16)$$

$$x_h \in \{0,1\} \text{ for } h=1,\dots,E. \quad (17)$$

To solve this problem, the maxi-min operator (Bellman & Zadeh, 1970) and the fuzzy programming method (Luo et al., 2001) are used.

The recalled fuzzy multiobjective programming method may be successfully employed for optimizing an IESC network thanks to its property of providing integer solutions and to its low computational complexity. However, the model in Equations 15 to 17 provides as a solution only one optimized subnet of the original IESC that meets multiple goals.

The Case Study

To illustrate the network design optimization procedure, we consider a case study inspired by an example proposed in Luo et al. (2001). The target product is a typical desktop computer system consisting of a computer, hard-disk driver, monitor, keyboard, and mouse. The SC is composed of N_S=6 stages: suppliers, manufacturers, distributors, retailers, consumers, and recyclers. In the following, we apply and discuss the different steps of the proposed design procedure for this case study.

Candidate Selection

The first module of the DSS structure in Figure 1 has to perform the candidate selection for each stage of the IESC case study. As an example, we focus on the partner selection for the second stage, that is, the manufacturers. Obviously, the methodology proposed for manufacturer selection is applicable to each stage of the IESC.

Table 1. Performance matrix of manufacturers with scores for each criterion

	n_1^c	n_2^c	n_3^c	n_4^c	n_5^c	n_6^c	n_7^c	n_8^c	n_9^c	n_{10}^c	n_{11}^c	n_{12}^c	n_{13}^c	n_{14}^c	n_{15}^c
E	80	70	20	10	90	35	35	60	30	30	30	65	70	40	80
RM	30	15	60	80	70	40	90	30	75	10	60	20	30	85	40
F	10	90	85	55	30	60	40	15	20	10	35	45	25	40	20
FL	40	10	30	55	75	80	15	90	45	40	20	40	20	30	20
T	65	30	80	30	80	10	45	45	20	30	90	35	40	15	35
Q	85	80	60	50	65	50	60	15	10	40	15	60	45	60	30

Table 2. Thresholds and weights for the Electre method

	E	RM	F	FL	T	Q
Indifference threshold	10	25	20	5	10	20
Preference threshold	20	45	30	10	20	30
Veto threshold	35	85	60	30	60	60
Weights	1.0	0.6	0.8	0.6	0.6	0.6

Table 3. Manufacturers' rankings according to the Electre method for the case study

Position	1	2	3	5	6	7	8	9	10	12	13
Manufacturer	n_5^c	n_1^c	n_3^c, n_{12}^c	n_8^c	n_6^c	n_2^c	n_{13}^c	n_{15}^c	n_7^c,n_{11}^c	n_{10}^c	n_4^c, n_9^c, n_{14}^c

Figure 4. The stages of the IESC network for the case study

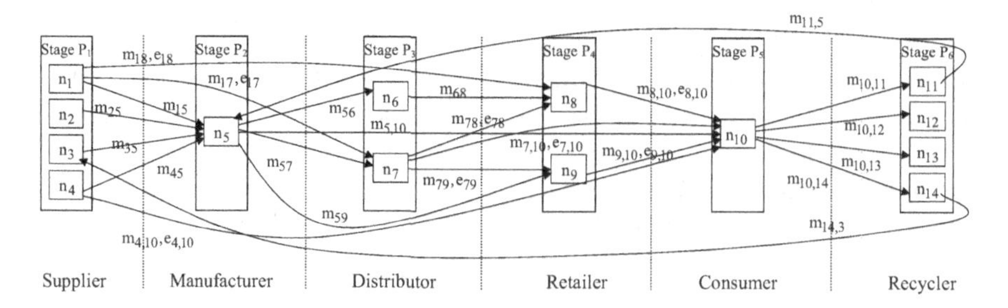

The first step of the procedure determines the candidate set of the second stage, for example, $P_2^c=\{n_1^c,n_2^c,\ldots,n_{15}^c\}$, where we assume that 15 candidates are competing to join the second stage of the IESC.

In the second step of the procedure, the decision makers define the most relevant criteria for the selection: F, RM, E, FL, T, and Q. Obviously, such a choice can only be based on experience and expert knowledge of the IESC processes, products, and actors. Then, in the third step, a data-analysis system assigns the scores to each candidate. Table 1 reports the performance matrix assigned to each alternative manufacturer.

Subsequently, since the Electre method is employed, the decision makers assign the thresholds and weights for the case study as shown in Table 2. Using the thresholds of Table 2, the Electre method seeks for an outranking relation. Table 3 shows the final ranking of the candidates, obtained with a Matlab implementation of the method that employs the intrinsic characteristic of the Matlab programming environment to operate with matrices (MathWorks Inc., 2002). The reader is referred to Mousseau et al. (2000) for a discussion on the definition of the decision parameters required by the Electre method, and to a previous work by the authors (Dotoli et al., 2005) for further insights on the provided example.

According to the results in Table 3, the decision maker selects $P_2=\{n_5^c\}$ if one manufacturer only is to be included in the network. On the contrary, if several manufacturers have to be incorporated in the IESC, a corresponding number of candidates are selected from Table 3 starting from the one with the highest posi-

Figure 5. The digraph associated with the IESC of the case study

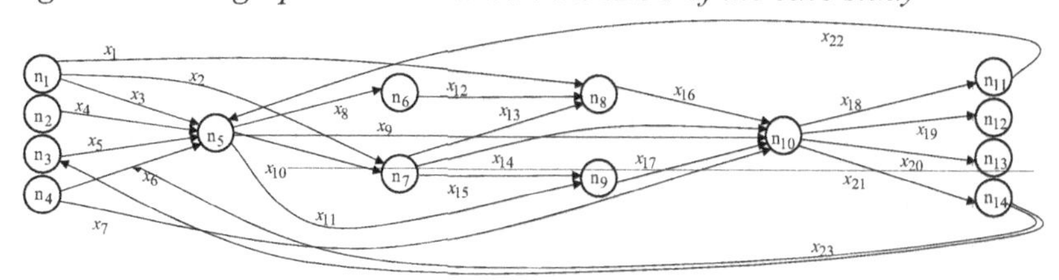

tion. For instance, if two manufacturers are to be included in the IESC, the decision maker selects $P_2=\{n_1^c,n_5^c\}$. Note that the former choice is made in the following so that one manufacturer only is selected.

The IESC network obtained after the iteration of the candidate-selection technique for each stage is depicted in Figure 4, while its digraph is shown in Figure 5, composed as follows: four suppliers, one manufacturer, two distributors, two retailers, one consumer, and four recyclers, for a total of N=14 partners. The data for the IESC are reported in Table 4 (Luo et al., 2001), showing the value of each performance index M_q with q=1, …, 4 associated with the links of the considered IESC.

More precisely, the adopted performance indices are total cost (M_1), energy (M_2), CO_2 emission (M_3), and cycle time (M_4). We indicate generically by cycle time

Table 4. Data sheet for the network links in the case study

Links	Edges	Variables	Cost (M_1) in $	Energy (M_2) in MJ	CO_2 emission (M_3) in KgCE	Cycle time (M_4) in hours
m_{18}, e_{18}	y_{18}	x_1	41.80	359.00	0.87	19.30
m_{17}, $e_{1,7}$	y_{17}	x_2	46.70	332.00	0.74	16.80
m_{15}	y_{15}	x_3	319.00	1479.00	2.21	12.50
m_{25}	y_{25}	x_4	308.00	1776.00	2.19	12.80
m_{35}	y_{35}	x_5	238.00	1540.00	3.10	16.20
m_{45}	y_{45}	x_6	246.00	1409.00	1.47	10.20
$m_{4,10}$, $e_{4,10}$	$y_{4,10}$	x_7	53.90	369.00	30.20	5.30
m_{56}	y_{56}	x_8	448.00	3618.00	8.74	19.20
$m_{5,10}$	$y_{5,10}$	x_9	379.00	3542.00	296.00	4.20
m_{57}	y_{57}	x_{10}	358.00	2885.00	6.26	16.20
m_{59}	y_{59}	x_{11}	358.00	3259.00	223.00	3.90
m_{68}	y_{68}	x_{12}	20.89	13.40	0.87	121.70
m_{78}, e_{78}	y_{78}	x_{13}	25.20	16.40	1.10	123.00
$m_{7,10}$, $e_{7,10}$	$y_{7,10}$	x_{14}	22.90	35.10	2.58	65.80
m_{79}, e_{79}	y_{79}	x_{15}	20.70	9.18	0.59	61.30
$m_{8,10}$, $e_{8,10}$	$y_{8,10}$	x_{16}	64.00	90.40	0.56	120.30
$m_{9,10}$, $e_{9,10}$	$y_{9,10}$	x_{17}	58.10	4.68	0.13	100.00
$m_{10,11}$	$y_{10,11}$	x_{18}	0.42	4.80	0.37	0.80
$m_{10,12}$	$y_{10,12}$	x_{19}	0.42	4.80	0.37	0.80
$m_{10,13}$	$y_{10,13}$	x_{20}	0.42	4.80	0.37	0.80
$m_{10,14}$	$y_{10,14}$	x_{21}	0.42	4.80	0.37	0.80
$m_{11,5}$	$y_{11,5}$	x_{22}	-18.00	-11.00	0.74	4.80
$m_{14,3}$	$y_{14,3}$	x_{23}	-28.00	-6.60	1.10	6.50

associated with an m-link the related time required by the transportation and/or the production process. The considered performance index values are reported in Table 4 and depend on the type of link (m- and e-link, or *m*-link only), the distance between the connected SC partners, the transportation mode (truck, car, airplane, etc.), and the type of material to be transported. In particular, the cost and energy performance indices reported in the last two rows of Table 4, respectively associated with links $m_{11,5}$ and $m_{14,3}$, are negative. In fact, in the recycler stage P_6, partner n_{11} is a demanufacturer with an output link $m_{11,5}$ connecting to manufacturer n_5, and partner n_{14} is a material recoverer with an output link $m_{14,3}$ connecting to supplier n_3 (see Figure 4). Hence, the total cost and energy associated with links $m_{11,5}$ and $m_{14,3}$ are negative; that is, they correspond to recycling materials and parts.

According to the data in Table 4, the IESC in Figure 4 exhibits the m- and e-links m_{18}, e_{18}, m_{17}, e_{17}, $m_{4,10}$, $e_{4,10}$, m_{78}, e_{78}, m_{79}, e_{79}, $m_{7,10}$, $e_{7,10}$, $m_{9,10}$, $e_{9,10}$, $m_{8,10}$, and $e_{8,10}$, while the remaining links are m-links. Moreover, the associated digraph D=(N,E) depicted in Figure 5 has N=14 nodes and E=23 edges. Obviously, edges y_{18}, y_{17}, $y_{4,10}$, y_{78}, y_{79}, $y_{7,10}$, $y_{9,10}$, and $y_{8,10}$ are associated both with m- and e-links, and the remaining edges of the digraphs are associated with m-links only. Moreover, each edge in E is labeled by its corresponding variable x_h with h=1, …, E used in the optimization procedure and defined in the previous section.

Optimization Model

Various computational experiments are performed to minimize cost, energy consumption, CO_2 emission, and total lead time (TLT). In particular, the TLT is defined as the total time elapsed from the instant at which the raw material begins its travel until the instant the finished product is delivered to consumers. Furthermore, a multiobjective function for Problem 2 is chosen. The solutions are obtained via the well-known two-phase simplex method in the Matlab framework (MathWorks Inc., 2002; Venkataraman, 2001). The first step of the optimization is to define the model constraints. Then Problem 1 or Problem 2 is solved.

BOM constraints. The component supplier constraints are obtained assuming that the BOM of the second stage in Figure 4, representing the manufacturer, is the following: computer (C), hard-disk driver (H), monitor (M), and keyboard and mouse (K). We suppose that C is produced by n_1 and n_2; H is produced by n_1, n_2, and n_3; M is produced by n_2, n_3, and n_4; and K is produced by n_3 and n_4 (Luo et al., 2001). Hence, with reference to Figure 5, the constraints imposed on the variables labeling the edges are as follows:

$$\begin{aligned} &x_3 + x_4 \geq 1 \\ &x_3 + x_4 + x_5 \geq 1 \\ &x_4 + x_5 + x_6 \geq 1 \\ &x_5 + x_6 \geq 1 \end{aligned} \quad (18)$$

Path constraints. The case study includes only one manufacturer and only one consumer (node n_5 of stage P_2 and n_{10} of P_5, respectively, in Figure 4). Hence, a path between nodes n_5 and n_{10} is needed. Consequently, we build the N×E incidence matrix I_M associated with digraph D. Moreover, we define the 23-vector $\boldsymbol{b}_{5,10}=[b_1\ b_2\ \dots\ b_{23}]$ with $b_5=-1$ $b_{10}=1$ and $b_p=0$ for p≠5, 10 and p=1, …, 23. The constraint that imposes the presence of a path starting from node n_5 and ending in node n_{10} is written as follows:

$$\boldsymbol{I}_M\,\boldsymbol{x} \geq \boldsymbol{b}_{5,10}. \quad (19)$$

Mutual-exclusion constraints. It is assumed that one and only one partner is to be included in the IESC recycler stage (stage P_6 in Figure 4). Furthermore, only one type of commerce has to be present between the second and third stages, and one and only one m and e-link has to be present among the first stage and the others. Hence, with reference to Figure 5, the mutual-exclusion constraints are the following:

$$\begin{aligned} &x_{18} + x_{19} + x_{20} + x_{21} \leq 1 \\ &x_{13} + x_{14} + x_{15} \leq 1 \\ &x_1 + x_2 + x_7 = 1 \end{aligned} \quad . \quad (20)$$

Table 5. The values of objective functions f_1, f_2, f_3, and f_4 for Problem 1

	f_1 ($)	f_2 (MJ)	f_3 (KgCE)	f_4 (hours)
min f_1	**946.02**	6566.30	16.34	98.20
min f_2	1030.92	**6112.66**	12.51	190.00
min f_3	1037.50	6415.86	**11.38**	190.30
min f_4	997.90	6799.00	329.88	**16.70**

Figure 6. Solution digraph of min(f_1)

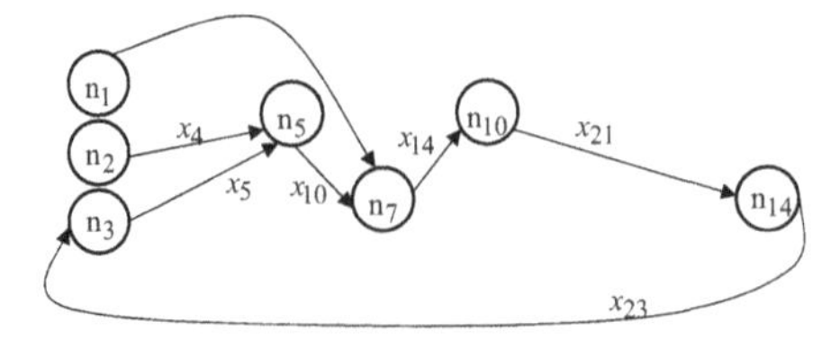

Figure 7. Solution digraph of min(f_2)

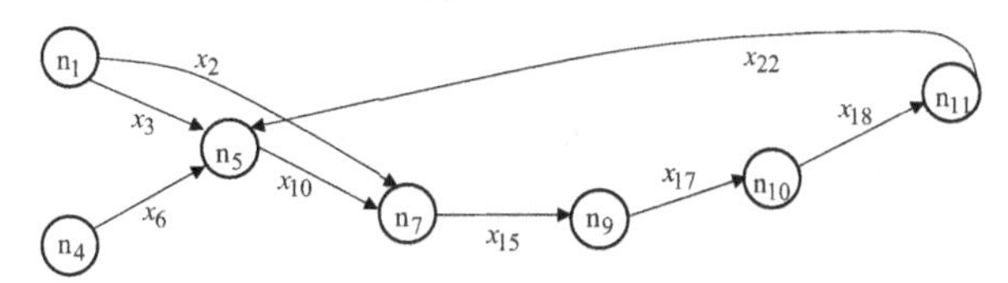

Figure 8. Solution digraph of min(f_3)

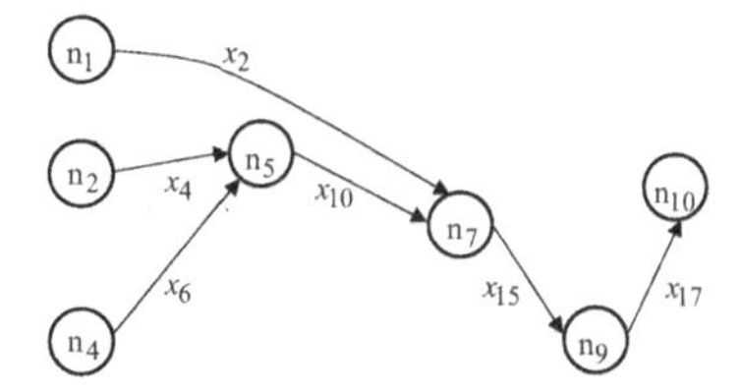

Figure 9. Solution digraph of min(f_4)

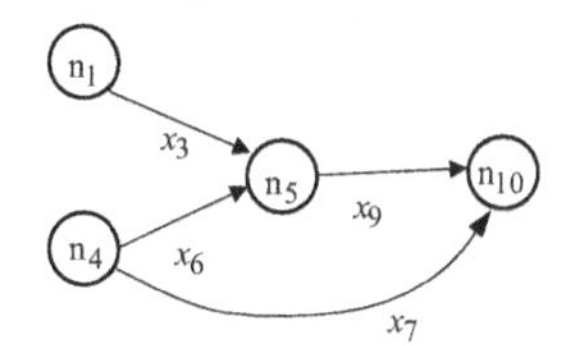

Structural constraints. The constraints derived from the digraph in Figure 5 are as follows:

$$\begin{aligned}
&x_{22} - x_{18} = 0 \\
&x_{23} - x_{21} = 0 \\
&x_5 - x_{23} \geq 0 \\
&x_{16} - x_1 \geq 0 \\
&x_{13} + x_{14} + x_{15} - x_2 \geq 0 \\
&x_{16} - x_{13} \geq 0 \\
&x_{17} - x_{15} \geq 0
\end{aligned} \qquad (21)$$

For example, the first constraint of Equation 21 means that the edge corresponding to x_{22} is selected if and only if the edge labeled by x_{18} is selected. In addition, the third constraint of Equation 21 means that if the edge labeled by x_{23} is selected, then the edge corresponding to x_5 is selected.

Figure 10. The digraph structure of a traditional SC composed of m-links

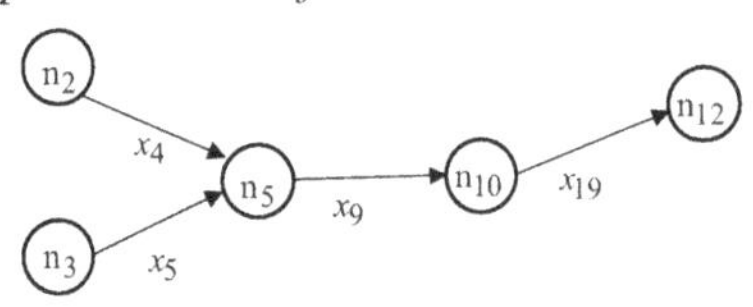

Solution of Problem 1

Problem 1 is solved with respect to four objectives: cost, energy, CO_2 emission, and TLT. The corresponding objective functions are denoted by f_1 to f_4, respectively. The obtained subdigraphs are presented in Figures 6 to 9, and the corresponding objective functions are given in Table 5.

Comparing our results with the solutions obtained by the fuzzy optimization and reported in Luo et al. (2001), we note the following two aspects.

First, while the results of the optimization problems min(f_2) and min(f_4) provide the same results as the fuzzy optimization, the minimization of the objective functions f_1 and f_3 does not provide the same digraphs obtained by the fuzzy optimization. Indeed, the fuzzy optimization can lead to suboptimal solutions: The optimal value of cost and CO_2 emission obtained with ILP is f_1=\$946.02 and f_3=11.38 KgCE respectively, while the fuzzy optimization performed in Luo et al. (2001) determines two solutions with f_1=\$951.00 and f_3=14.10 KgCE. Consequently, the ILP approach with a single-criterion objective function guarantees optimal solutions.

Second, in Luo et al. (2001), the authors use the same structure of BOM constraints for all the considered performance indices, but such a structure is not suited to the TLT performance measure. Indeed, the cycle time associated with BOM constraints is not the sum of the corresponding edge performance indices but the maximum among the performance indices. For example, if we choose edges y_{25} and y_{35} corresponding to variables x_4 and x_5 as the BOM for P_2, the corresponding cycle time cannot be computed as $M_4(m_{25})+M_4(m_{35})$, but as the maximum between $M_4(m_{25})$ and $M_4(m_{35})$: In such a case, the constraint becomes nonlinear. Hence, to obtain a more rigorous model but with linear constraints, we modify the constraints of Equation 18

Figure 11. Solution digraph of min(f_1) imposing a fixed structure of m- and e-links

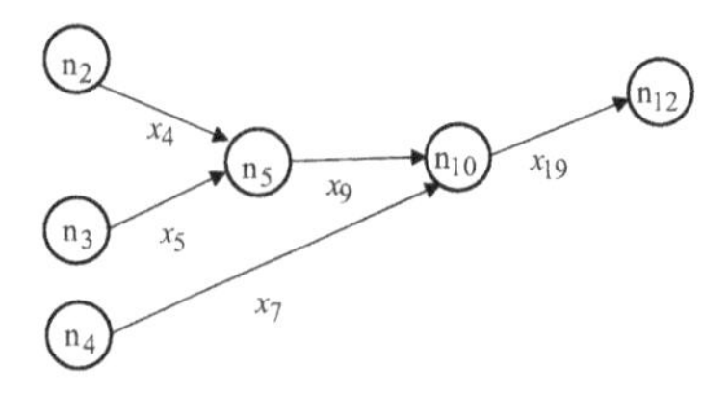

Figure 12. Digraph representing solution x_D of Problem 2

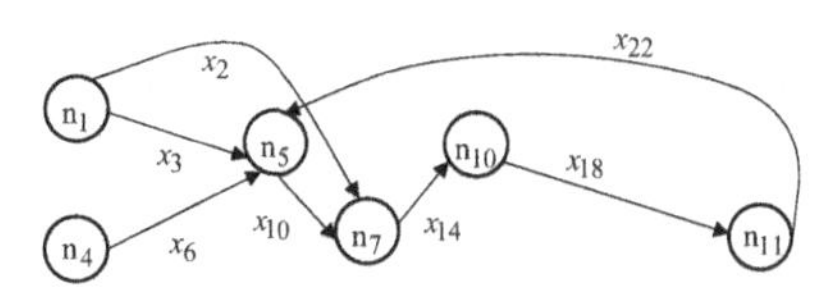

Table 6. The values of the performance indices for Problem 2: Optimal solutions for multiobjective function of cost, energy, and CO_2 emission ($min(f_5)$)

Solutions	Cost ($)	Energy (MJ)	CO_2 (KgCE)	TLT (hours)	Variables indices x_h=1
x_A	**946.02**	6566.30	16.34	98.20	2,4,5,10,14,21,23
x_B	957.02	6269.30	16.36	**98.20**	2,3,5,10,14,21,23
x_C	964.02	6430.90	14.35	94.80	2,4,6,10,14,18,22
x_D	975.02	6133.90	14.37	94.50	2,3,6,10,14,18,22
x_E	981.60	6437.10	13.24	94.80	2,4,6,10,14
x_F	992.60	6140.10	13.26	94.50	2,3,6,10,14
x_G	1030.92	**6112.66**	12.51	190.00	2,3,6,10,15,17,18,22
x_H	1037.50	6415.86	**11.38**	190.30	2,4,6,10,15,17
x_I	1048.50	6118.86	11.40	190.00	2,3,6,10,15,17

to transform the nonlinear BOM constraints for the TLT in suited linear constraints (Mangini, 2003). However, since the cycle times assigned to links $m_{ij} \in L_m$ do not differ much, in this particular case the assumption used in Luo et al. is admissible, and we obtain the same solution digraph of problem $min(f_4)$ (see Figure 9).

Finally, the following example shows that the presented optimization method can improve the reconfigurability of the network. In particular, let us consider a traditional SC that has a network composed of m-links only (see Figure 10). Moreover, the designer has to add the e-links in order to introduce e-commerce and e-business fixtures in the network structure optimizing the cost. Consequently, Problem 1 is solved by selecting cost as a performance index subject to the constraints in Equations 18 to 21, and the following mutual-exclusion constraints that impose the initial structure of the SC:

$$x_4=x_5=x_9=x_{19}=1. \quad (22)$$

The resulting IESC network is depicted in Figure 11 and exhibits the cost of $979.32.

Solution of Problem 2

The multiobjective optimization problem is solved considering the following performance indices: cost, energy, and CO_2 emission (f_5). According to the previous remarks, we do not consider the cycle time in the multiobjective optimization, but we compute the TLT values of the problem solutions. Table 6 reports the ILP solutions and the corresponding performance indices. The results show the efficiency of the proposed method, which is able to provide a set of optimal solutions. For example, solution x_A, obtained by minimizing objective function f_5, is equal to the solution obtained by minimizing f_1 (compare the first row of Table 6 with Figure 6). Table 6 shows that solution x_A exhibits a large value of CO_2 emission. On the other hand, minimizing objective function f_5 provides solution x_D in Table 6, featuring satisfactory values of cost, energy, CO_2 emission, and TLT. In other words, the benefits of using multicriteria optimization are due to the fact that the method enables us to choose among several near-optimal solutions. The digraph corresponding to solution x_D in Table 6 is depicted in Figure 12; the other solution digraphs may easily be obtained from the last column in Table 6.

Comparing the results obtained solving the ILP problem with the fuzzy-optimization results, we note that the presented optimization method provides a set of near-optimal solutions instead of only one suboptimal solution. Hence, the designer can be guided by priorities and preferences to choose a satisfactory IESC network, improving system flexibility and agility. Note that solution x_B of Table 6 is the solution obtained by fuzzy optimization (Luo et al., 2001).

Solution-Evaluation and Validation Module

The purpose of this DSS module is to evaluate alternative IESC network configurations obtained from the higher levels with respect to operational performances representing resources (cost, utilization, and inventory), output (quality and lead time), and flexibility (lead time and its variability; Beamon, 1999; Persson & Olhager, 2002). At this level of the decision process, it is necessary to increase the understanding of the interrelationships among parameters, relevant for describing the IESC at the operational level, such as operation and transportation times, global capacities of manufacturing facilities, pull demand from retailers, and the push of raw material from suppliers. Similar to the previously described DSS modules, in this level the process of data collection and elaboration may be simplified by remote and collaborative evaluation using a Web-based platform.

In order to capture these relationships, analytical models and simulation models can be alternatively used. In particular, analytical models include discrete event models that are particularly suitable for the verification of distributed manufacturing systems. In such a modeling approach, the evolution of the system depends on the complex interaction of the timing of various discrete events such as arrivals of components at the suppliers, departures of trucks from the suppliers, the beginning of assembly operations at the manufacturers, arrivals of finished goods at the customers, payment approvals by the sellers, and so forth (Viswanadham & Raghavan, 2000). Despite the appropriateness of discrete event models to represent IESC, they cannot be detailed enough to handle all the relevant parameters of complex supply-chain systems. Hence, simulation can represent a more general and efficient instrument to evaluate alternative SC designs and to validate an IESC network configuration (Jansen, van Weert, Beulens, & Huime, 2001). Very attractive general-purpose simulation packages are now available to model a manufacturing enterprise, for example, ARENA, SIMPROCESS, and Taylor II (Viswanadham & Raghavan).

Summing up, comparing different IESC network design solutions and analyzing the system behavior in the presence of additional details or uncertainties allow us to determine the performance of a given solution at the operational level. Hence, the DSS to configure the IESC is closed by this module, which is able to evaluate and validate the optimal or near-optimal solution. As specified, if the third level results are not satisfying, it may be necessary to select different solutions among the network structures obtained at the previous levels in order to improve the IESC performance.

The obtained DSS results in a closed-loop procedure. Moreover, if the proposed DSS is equipped with an agile data-acquisition and -elaboration tool, it may be constantly employed to confirm or modify the IESC configuration upon variations of the conjectured scenarios or changes in the context of the real market.

Conclusion

This chapter focuses on the application of enterprise service computing to determine and optimize the configuration of IESCs, that is, business strategies incorporating the power of e-commerce to streamline the manufacturing processes. An IESC system has a more complex structure than a traditional SC system since it embraces the e-business strategy to establish information links and integrates end-of-life processes into the entire SC structure. In particular, a hierarchical DSS is presented to design and reconfigure an IESC based on data and information that can be obtained via Internet and Web-based instruments.

More specifically, the proposed DSS is composed of three hierarchical levels differing in data requirements, performance-index utilization, and output solutions, which are comprehensively reviewed and discussed with regard to the related literature. In particular, the first level (candidate-selection module) uses aggregate performance indices and optimization techniques to obtain a rank of possible candidates for each stage of the IESC. In the second level (network-design module), the structure of the IESC is modeled by a digraph, describing the actors of the stages, and the material and information links connecting the stages. An integer multicriteria optimization model proposes different structures for the IESC on the basis of a set of chosen performance indices and cost. Finally, the third level (solution-evaluation and -validation module) analyzes and evaluates the optimal or suboptimal solution networks obtained at the previous levels by comparing operational performance indices. The evaluation of the performance measures tests the design of the IESC as obtained from the candidate-selection and network-design levels. It can provide some feedback for IESC modifications if the performance indices are not satisfactory at the third level.

A case study based on enterprise service computing via the presented decision structure is reported in detail. In addition, the integer multicriteria optimization methodology is applied to the case study and is compared with a fuzzy-optimization method presented in the related literature.

The third module is not described in detail and will be the subject of future research. Moreover, a further future perspective is a Web-based implementation of the proposed DSS for a shared and remote platform dedicated to enterprise service computing.

Acknowledgments

This work was supported in part by the Italian Ministry for University and Research (MIUR) under Project No. 2003090090, and NSFC under Grant No. 60228004 and 60334020.

References

Beamon, B. M. (1999). Measuring supply chain performance. *International Journal of Operations and Production Management, 19*(3), 257-292.

Bellman, R. E., & Zadeh, L. A. (1970). Decision-making in a fuzzy environment. *Management Science, 17*(4), 141-164.

Biswas, S., & Narahari, Y. (2004). Object oriented modelling and decision support for supply chains. *European Journal of Operational Research, 153*(3), 704-726.

Buchanan, J., Sheppard, P., & Vanderpooten, D. (1999). *Project ranking using ELECTRE III* (Tech. Rep. No. 99-01). Waikato, New Zealand: University of Waikato, Department of Management Systems.

Chryssolouris, G., Makris, S., Xanthakis, V., & Mourtzis, D. (2004). Towards the Internet-based supply chain management for the ship repair industry. *International Journal of Computer Integrated Manufacturing, 17*(1), 45-57.

Dotoli, M., Fanti, M. P., Meloni, C., & Zhou, M. C. (2005). A multi-level approach for network design of integrated supply chains. *International Journal of Production Research, 43*(20), 4267-4287.

Dotoli, M., Fanti, M. P., Meloni, C., & Zhou, M. C. (2006). Design and optimization of integrated e-supply chain for agile and environmentally conscious manufacturing. *IEEE Transactions on Systems Man and Cybernetics, Part A, 36*(1), 62-75.

Ehrgott, M. (2000). *Multicriteria optimization.* Berlin, Germany: Springer-Verlag.

Farina, M., & Amato, P. (2004). A fuzzy definition of optimality for many-criteria optimization problems. *IEEE Transactions on Systems, Man and Cybernetics, Part A, 34*(3), 315-326.

Gaonkar, R., & Viswanadham, N. (2001). Collaboration and information sharing in global contract manufacturing networks. *IEEE/ASME Transactions on Mechatronics, 6*(4), 366-376.

Hillier, F. S., & Lieberman, G. J. (2005). *Introduction to operations research.* Singapore: McGraw Hill.

Instutute of Electrical and Electronics Engineer (IEEE). (2004). *IEEE Technical Community for Services Computing Newsletter, 1*, 1.

Jang, Y.-J., Jang, S.-Y., Chang, B.-M., & Park, J. (2002). A combined model of network design and production/distribution planning for a supply network. *Computers & Industrial Engineering, 43*(1-2), 263-281.

Jansen, D. R., van Weert, A., Beulens, A. J. M., & Huirne, R. B. M. (2001). Simulation model of multi-compartment distribution in the catering supply chain. *European Journal of Operational Research, 133*(1), 210-224.

Keskinocak, P., Goodwin, R., Wu, F., Akkiraju, R., & Murthy, S. (2001). Decision support for managing an electronic supply chain. *Electronic Commerce Research, 1*(1-2), 15-31.

Luo, Y., Zhou, M. C., & Caudill, R. J. (2001). An integrated e-supply chain model for agile and environmentally conscious manufacturing. *IEEE/ASME Transactions on Mechatronics, 6*(4), 377-386.

Mangini, A. (2003). *Modelli e algoritmi per la configurazione di una catena logistico-produttiva.* Unpublished master's thesis, Politecnico di Bari, Bari, Italy.

MathWorks Inc. (2002). *Matlab release notes for release 13.* Natick, MA: Author.

Mousseau, V., Slowinski, R., & Zielniewicz, P. (2000). A user-oriented implementation of the ELECTRE TRI method integrating preference elicitation support. *Computers & Operations Research, 27*(7-8), 757-777.

Mustajoki, J., & Hämäläinen, R. P. (2000). Web-Hipre: Global decision support by value tree and AHP analysis. *Information Systems and Operational Research, 38*(3), 208-220.

Persson, F., & Olhager, J. (2002). Performance simulation of supply chain designs. *International Journal of Production Economics, 77*(3), 231-245.

Saaty, T. L., Vargas, L. G., & Dellmann, K. (2003). The allocation of intangible resource: The analytic hierarchy process and linear programming. *Socio-Economic Planning Sciences, 37*(3), 169-184.

Shapiro, J. F. (2001). *Modeling the supply chain.* Pacific Grove, CA: Duxbury Press.

Shapiro, J. F. (2004). Challenges of strategic supply chain planning and modeling. *Computers and Chemical Engineering, 28*(6-7), 855-861.

Sprague, R. H., Jr., & Carlson, E. D. (1982). *Building effective decision support systems.* Englewood Cliffs, NJ: Prentice-Hall.

Talluri, S., & Baker, R. J. (2002). A multi-phase mathematical programming approach for effective supply chain design. *European Journal of Operational Research, 141*(3), 544-558.

Tayur, S., Ganeshan, R., & Magazine, M. (1999). *Quantitative models for supply chain management.* Norwell, MA: Kluwer Academic Publisher.

Thanassoulis, E. (2001). *Introduction to the theory and application of data envelopment analysis.* Dordrecht, Netherlands: Kluwer Academic Publisher.

Venkataraman, P. (2001). *Applied optimization with MATLAB programming.* New York: Wiley Interscience.

Vidal, C. J., & Goetschalckx, M. (1997). Strategic production-distribution models: A critical review with emphasis on global supply chain models. *European Journal of Operational Research, 98*(1), 1-18.

Viswanadham, N., & Gaonkar, R. S. (2003). Partner selection and synchronized planning in dynamic manufacturing networks. *IEEE Transactions on Robotics and Automation, 19*(1), 117-130.

Viswanadham, N., & Raghavan, N. R. S. (2000). Performance analysis and design of supply chains: A petri net approach. *Journal of the Operations Research Society, 51*(10), 1158-1169.

Wu, N., Mao, N., & Qian, Y. (1999). An approach to partner selection in agile manufacturing. *Journal of Intelligent Manufacturing, 10*(6), 519-529.

Yang, H., Yu, Z., & Edwin Cheng, T. C. (2003). A strategic model for supply chain design with logical constraints: Formulation and solution. *Computers of Operations Research, 30*(14), 2135-2155.

Section VI

Enterprise Service Computing: Best Practices and Deployment

Chapter XIV

Best Practice in Leveraging E-Business Technologies to Achieve Business Agility

Ehap H. Sabri, University of Texas at Dallas, USA

Abstract

This chapter explains the best practice in implementing e-business Technologies to achieve business cost reduction and business agility. Many companies started to realize that gaining competitive advantage is no longer feasible by only managing their own organizations; it also requires getting involved in the management of all upstream supply organizations as well as the downstream network. E-business technologies present huge opportunities that are already being tapped by several companies and supply chains. Although the benefits of implementing e-business technologies are clear, enterprises struggle in integrating e-business technologies into supply-chain operations. The author illustrates the strategic and operational impact of e-business technologies on supply chains and explains the performance benefits and challenges firms should expect in implementing these technologies. Also, the author provides the best-practice framework in leveraging e-business applications to support process improvements in order to eliminate non-value-added activities and provide real-time visibility and velocity for the supply chain. Finally, this chapter presents the future trends of using e-business in transformation programs.

Introduction

Executives realized that producing high-quality products is not enough in today's competitive environment; the new challenge is to get products to customers when and where they need them, exactly the way they want them, with a competitive price and in a cost-effective manner. Many factors are making this challenge more complicated; globalization, increased complexity of supply chains (SCs) with outsourcing and the move to mass customization and build-to-order (BTO) environments, the need for a shorter time to market to gain competitive advantage, and the shift from vertical to horizontal supply chains make an efficient integration with suppliers and customers more critical.

E-business technologies address the above challenges by enabling enterprises to collaborate with their internal and external suppliers and customers, providing visibility, automating the paper-driven business processes, and interconnecting inventory, logistics, and planning systems.

The way to survive the competition in today's business world is to stay ahead of competitors. Leveraging e-business technologies effectively is key to staying competitive and achieving business agility. Although the benefits of implementing e-business technologies are clear, enterprises struggle with integrating e-business technologies into supply-chain operations. Decision makers find themselves asking the most fundamental questions. How can we do it? What is the best practice? Does it apply to us? Does technology add value? If so, what is the best way to quantify it and then maximize it? Since many have failed in achieving value, how can we make sure that we will not be one of them and will be able to minimize the risk? What does senior management need to do to support transformation initiatives? This chapter gives powerful tools for answering these questions.

This chapter addresses the strategic and operational impact of e-business technologies on supply chains and explains the performance benefits and challenges firms should expect in implementing e-business technologies. Also, it provides the best-practice framework in leveraging e-business technologies to support process improvements in order to achieve cost reduction and velocity for the supply chain. This framework includes a practical and effective return-on-investment (ROI) model to calculate the benefits of e-business transformation programs.

The objectives of this chapter can be summarized as follows.

- Provide a good understanding of the challenges in today's business environment
- Identify the impact of e-business technologies on enterprise processes
- Highlight the benefits of implementing e-business technologies

- Provide guidelines and a framework for implementing e-business technologies successfully

The concepts in this chapter are presented in an easy-to-understand manner that is intended for any reader interested in learning about e-business technologies. Because e-business can be leveraged by several functions within the organization, this chapter has been written for the wide audience that is interested in learning how to leverage e-business techniques in improving processes and slashing waste. This chapter provides strategies for senior managers to use in planning for transformation programs, and also provides middle mangers with tools to effectively manage and implement the best practice. Graduate students can use this chapter to gain an excellent understanding of how e-business technologies work, and then use this knowledge to either extend the research in this field or implement the concepts learned from this chapter in the industry.

Background

Valencia and Sabri (2005) stated that the widespread use of the Internet has turned the eyes of many companies to the numerous solutions that the Internet provides. E-business technologies have helped many companies in improving their overall processes and performances.

On the other hand, Handfield and Nichols (2002) mentioned that integrated supply-chain management (SCM) is now recognized as a strategy to achieve competitive advantage. When pressed to identify how to achieve this strategy, however, the path forward for executives is not clear. Numerous solution providers offer the "silver bullets" to supply-chain integration, yet the results are never guaranteed." Devaraj and Kohli (2002) mentioned that executives are concerned that even when there is promise of a payoff, the assumptions may change and payoffs may never be realized.

Rigby, Reichheld, and Schefter (2002) mentioned that, in a survey conducted by a CRM (customer-relationship management) forum, when asked what went wrong with their CRM projects, 4% of the managers cited software problems, 1% said they received bad advice, but 87% pinned the failure of their CRM projects on the lack of adequate change management.

Any transformation program should encompass three specific phases: initial enablement, followed by implementation, and finally ongoing support and maintenance. An effective transformation plan must support these three phases and address all the challenges around selecting the right strategy, change management, and software provider; maintaining upper management buy-in; managing by metrics; and rolling out a maintenance strategy.

Valencia and Sabri (2005) mentioned the several guidelines that are important to consider when e-business technologies are implemented. These guidelines can be grouped as follows.

- **Initial enablement:**
  - Synchronize e-business objectives with corporate initiatives
  - Use e-business as a driver for significant process improvement
  - Identify operational and financial benefit metrics
  - Clearly identify the process and the scope of the transformation program
  - Obtain consensus from all process stakeholders
- **Implementation:**
  - Simplify processes by eliminating non-value-added activities
  - Deploy in small phases with compelling ROI
  - Consider proactive enforcement of to-be processes in the solution
  - Enable exception-based problem resolution
- **Ongoing support and maintenance:**
  - Define the ongoing process for capturing, monitoring, and analyzing metrics data

E-Business Definition

E-business can be defined as the use of the Internet to facilitate business-to-business (B2B) or business-to-consumer (B2C) transactions and includes all operations before and after the sale. E-business can also be defined as the adoption of the Internet to enable real-time supply-chain collaboration and integration of the planning and execution of the front-end and back-end processes and systems. E-business has already impacted the industry significantly by providing important benefits like cost reduction, visibility improvement, lower inventory, streamlined processes, better response time, faster time to market, better asset utilization, higher shareholder value, fulfillment lead-time reduction, flexibility improvement, revenue increases by penetrating new markets, improvement in customer satisfaction, and better standards of living.

E-business, or the Internet computing model, has emerged as perhaps the most compelling enabler for supply-chain integration. Because it is open, standard based, and virtually ubiquitous, businesses can use the Internet to gain global visibility across their extended network of trading partners and respond quickly to changing business conditions such as customer demand and resource variability.

The Internet is considered to be the enabler for e-business technologies, while e-business is the enabler for supply-chain collaboration and integration. E-business technologies can support different environments: business to employee (B2E), B2C, and B2B. The B2E environment links the ERP (enterprise resource planning), SCM, warehousing, shipping, and human-resources systems together into a Web-based system. B2E focuses on the internal transactions of a company and affects the internal supply-chain process. The B2C environment allows customers to place, track, and change orders online, and allows sellers to gather information about the consumer in real time. B2C is sometimes referred to in the literature as e-commerce. Other literature differentiates between the direct business customer and the end customer (consumer). The B2B environment links B2E and B2C to the systems of the suppliers and affects the external supply-chain processes. The best practice is to have all of the three environments (B2E, B2B, and B2C) under one portal for seamless information transfer between them.

E-business technologies can also be divided into two types: (a) process-focused technologies like online auctions and (b) infrastructure-focused technologies like Web services (XML [extensible markup language], SOAP [simple object access protocol], UDDI [universal description, discovery, and integration], and WSDL [Web service description language]), the wireless application protocol (WAP), the global positioning system (GPS), bar coding, and radio-frequency identification (RFID) to transmit the data into computer applications. Web services are built on platform-independent standards that enable e-business applications to share information over the Internet between internal and external systems.

XML provides the best way for companies to pass data and coordinate services over the Internet. In essence, XML is a file format that allows users to include definitions of terms and processing rules within the same file (Harmon, 2003). SOAP and UDDI are additional software protocols, where SOAP is a protocol that enables one computer to locate and send an XML file to another computer, and UDDI is a protocol that allows one company to query another company's computers to determine how certain kinds of data are formatted. WSDL is the language in which the UDDI protocol is implemented.

There are still issues to be resolved, and many groups are working on middleware and security standards to make XML more flexible and secure (Harmon).

Today's Business Challenges

Executives face many challenges today in every aspect of their operations and enterprise integration. The following are considered to be the top 10 challenges.

- The need to be customer oriented and at the same time manage cost more efficiently
- Information delay, which is considered to be a typical concern. Since the processes are the ones that realize the information flow between partners, enterprises need to redesign their business processes to address this challenge.
- Globalization, which intensifies the competition and makes competitive advantage crucial
- The increased complexity of supply chains and the growing need for a tighter control of over it. The supply chain can be defined as a network of facilities (manufacturing plants, assembly plants, distribution centers, warehouses, etc.) that performs the function of the transformation of raw materials into intermediate products and then finished products, and finally the distribution of these products to customers. The increased complexity may result from the involvement of one company in multiple supply chains, the growing complexity of products, the growing complexity of managing information flows, and the increasing trends of third-party logistics and subassembly (contract) manufacturers.
- Long and unpredictable product life cycles. It is important to shorten the product-introduction cycle and be faster to react to the market needs to gain competitive advantage.
- The shift from vertical integration to horizontal supply chains, which calls for more efficient collaboration with suppliers and customers. The new trend is for companies to buy out competitors in the same business or merge with them (consolidation) instead of buying their suppliers (i.e., to expand vertically instead of horizontally); this requires the synchronized addressing of collaboration, planning, and procurement issues.
- Outsourcing and having suppliers across the world. Companies nowadays outsource assembly work, information-systems management, call centers, service-parts repair and management, and product engineering to contractors. The bigger challenge is to decide what to outsource, and how to make sure that customer satisfaction, delivery service, or quality are not impacted.
- Expensive cost structure, especially when companies are facing intensified competition. Related to this is the increase in shipping costs due to outsourcing.

- The disruption to the supply chain from demand and supply mainly caused by the supply-chain dynamics.
- Supporting the redesigned processes with leading-edge technology that is easy to integrate, cheap to maintain, fast to achieve results, and low in risk

In order to address the previous challenges in today's environment, companies should look for certain key enablers to implement in their operations. These key enablers can be grouped as follows.

1. **Cross-organizational collaboration:** This can reduce lead time, improve efficiency, reduce quality risk, and streamline processes.
2. **Flexibility:** Flexibility should be built into product designs and manufacturing processes to become more customer oriented and mitigate the challenge of supply and demand fluctuations by providing the ability to change plans quickly. Flexibility is measured based on the ability to shift production load (flexible capacity), change production volumes and product mix, and modify products to meet new market needs. It is important to mention that cost-reduction initiatives usually inversely impact flexibility.
3. **Visibility:** Real-time visibility reduces the uncertainty and nervousness in the supply chain, reduces safety stock (i.e., reduces cost), and increases customer satisfaction (i.e., increases revenue) by presenting the real picture and providing the ability to solve potential problems ahead of time. It will also reduce the impact of disruption to the supply chain caused by demand and supply.
4. **Process innovation:** This enabler addresses the following types of questions. How can companies reengineer their processes to increase speed when introducing new products? What supply-chain-process improvements and tools should companies invest in to gain competitive advantage? How can companies restructure their supply chains to reduce cost and increase profitability across their total global network? How can companies reduce their supplier base?
5. **Risk management:** This enabler addresses the new risks related to product quality and service delivery that arise from outsourcing, locating supply-chain activities across the world, and the acceleration of new product introductions. It also addresses the security concerns after September 11, 2001.
6. **Technology:** Technology should be adopted to enable and support the new or modified processes, and reduce the supply-chain complexity.

E-business is the major driver to implement the first, third, and last enablers, while best practice is the major driver for the rest of the enablers. These two drivers (e-business and best practice) will be discussed later in this chapter. The following section will provide the benefits of implementing the best practice, which is enabled by the right e-business technology.

The Impact of E-Business Technologies on SCM, CRM, & SRM Superprocesses

The e-business technologies available through the Internet, combined with ongoing advances in advanced planning and scheduling (APS) software development, have given SCM an enormous boost and helped in maximizing business agility. APS is a constraint-based planning logic that emerged in the early 1990s, and it is considered to be the major breakthrough after MRP (material resource planning) logic. SCM is defined as the process of optimizing the flow of goods, services, and information along the supply chain from supplier to customer. It is also the process to strategize, plan, and execute business processes across facilities and business units. It focuses on the internal supply chain, which is under the direct control of the enterprise.

Information at all points along the supply chain is captured and well presented by e-business technologies (in a B2E environment), enabling better decision making in SCM businesses processes, especially regarding maintaining appropriate inventory levels and the efficient movement of products to the next production or distribution process, and supporting sales and operations planning (S&OP). E-business has replaced excess inventory with accurate, inexpensive, and real-time information. Also, real-time information about the supply and the ability to collaborate with customers on the forecast have helped producers to balance and match supply with demand. All the above have enabled many companies to transform their supply operations from build to stock (BTS) and mass production to BTO and mass customization. Anderson (2003) defined build to order as the on-demand production of standard products, while mass customization is the on-demand production of tailored products.

Supplier-relationship management (SRM) is the process of supporting supplier partnerships in the supply chain, and coordinating processes across product development, sourcing, purchasing, and supply coordination within a company and across companies. CRM is the process of covering all customer needs throughout the phases of customer interaction. CRM is also defined as the process of allocating organizational resources to activities that have the greatest return on profitable customer relationships. Since the best practice of SRM and CRM are built around supply-chain integration, e-business is considered as the enabler for these superprocesses. The B2B environment is typically leveraged for SRM, while the B2C environment is typically leveraged for CRM.

Organizations utilize e-business solutions (a solution is a combination of best-practice processes and enabled technology) in the SRM area to eliminate paperwork, streamline shipment and payments, reduce the cycle time of finding and acquiring suppliers, easily monitor contract terms, leverage spend consolidation by supplier and part rationalization, increase supplier awareness during the design and production phases, and automate the procurement process.

On the other hand, organizations utilize e-business solutions in the CRM area to automate order receiving; to capture, analyze, and leverage customer information; and to reduce order-to-delivery cycle time and expediting costs.

We will drill down in the next section to detail the SCM, SRM, and CRM business processes and highlight the impact e-business has on them. We will mention the related pain points and show the potential benefits of addressing them using e-business technologies in addition to several success stories in this area.

SRM, SCM, AND CRM business processes; Pain Points; benefits of adopting e-business technology; and success stories

Each superprocess (like SCM, SRM, and CRM) consists of several business processes. Each key business process is defined (see Table 2), and typical challenges in the current practice (as-is process) are highlighted. The chapter shows how e-business technology and its different application levels can address these challenges, and finally typical benefits resulting from adopting e-business technology are mentioned.

Key Business Processes

Table 2 shows the key business processes in each superprocess of SCM, SRM, and CRM. Although the name and scope of these processes may vary across different industries, this table and the following process definitions give companies a good idea on how to map their business to this table and understand the impact of e-business on their own processes. The criteria used to link certain business processes to the appropriate superprocess are the following.

- If a business process has a direct relationship with suppliers based on the best practice, it would belong to the SRM superprocess.
- If a business process has a direct relationship with customers based on the best practice, it would belong to the CRM superprocess.
- If a business process has a direct relationship with neither suppliers nor customers, and it helps in balancing supply and demand, it would belong to the SCM superprocess.
- If a business process is in the gray area where it spans across two superprocesses, its own subprocesses should be evaluated, and the business process

will be linked to the corresponding superprocess to which the majority of the subprocesses belong to.

In the second dimension of Table 2, the criteria that have been used to decide the level a certain business process belongs to are as follows.

- If a business process is used mainly to generate strategies to run the business, it would belong to the strategic level.
- If a business process is used mainly to generate a plan to operationalize the strategies, it would belong to the tactical level.
- If a business process is used mainly to execute the plans to realize the strategies, it would belong to the operational and execution level.

The strategic level is considered to be the long-term process, the tactical level is the midterm process, and the operational level is the short-term process, which includes execution (for simplicity). Going up the levels from the operational to the tactical to the strategic, the percentage of cost savings goes up and the impact of decisions on the success of the organization is higher.

Table 2. Business Process Distribution

		Superprocess		
Level		SRM	SCM	CRM
	Strategic	Strategic Sourcing	Supply-Chain Design	Marketing Management
	Tactical	Product Design	Sales & Operations Planning	Selling Management
	Operational	Procurement	Order Fulfillment	Customer-Service Management

Strategic Sourcing

Strategic sourcing is defined as the process of identifying the best sourcing strategy to reduce cost and raw-material supply risk, and to achieve a long-term relationship with suppliers. One of the major challenges (pain points) in this process is the inability to identify and manage the total supplier spend and demand because of complex, disconnected purchasing systems. Another disconnection is between the engineering and purchasing systems, which causes the lack of visibility to the design engineers of the approved vender list (AVL). Another related challenge is the inability to consider supplier performance during sourcing decisions by design engineers or purchasing analysts.

Product Design

Product design is the process of collaboration among companies, partners, and suppliers to share product design, schedules, and constraints to arrive at a single bill of material for a finished product effectively. It is a critical process due to its ability to facilitate bringing innovative and profitable products to the market quickly, and to ensure product quality. The product design consists of three phases: concept, development, and pilot. It also includes product-engineering changes (product revisions) as a subprocess. These changes can be due to component-cost change, product improvements, process change, quality corrective actions, material shortages, or product obsolescence. Product revisions involve design engineers, procurement, suppliers, manufacturing and process engineers, contract manufacturers, and service support. Product design is tightly integrated with PDM (product-definition management), which is the database for all designed parts.

Product design has several challenges like intensified competition, which increases the need to introduce new products to the market more quickly; complex products, which make optimizing the design more challenging; frequent design changes, which increase prototype-part cost due to the late involvement of suppliers in the design phase; and the inability to identify the right products to launch or fund, the right suppliers to collaborate with during the design, and the right standard items to reuse. Also, subcontracting and outsourcing extends the need for real-time collaboration with partners and suppliers.

Procurement

Procurement is the process of executing the selected sourcing strategies by performing a request for quote, reverse-auction, bid-analysis, and contract-processing work flows to select the source of supply, and then managing all daily activities of procurement with the selected suppliers. It is also the process of managing two-way real-time communications with suppliers regarding the part or raw-material supply to achieve and execute a synchronized procurement plan.

There are three key pain points for this process. First is the lack of intelligent visibility and consistent supplier performance throughout the life cycle of the purchase order. This results from the inability to conduct reliable bid analysis considering different criteria in addition to the price, the lack of early problem detection for the inbound material and mismatch-resolution framework, a costly and difficult-to-maintain EDI (electronic data interchange) because different versions of the software may result in transmission errors, and the inability to capture supplier performance data.

The second key pain point is the many non-value-added and time-consuming (paper-based) activities in this process, like manual RFQ and contracting processes, the checking of shipment status, and paper invoices. The high cost of both expedition and mismatch resolution between the receipt and the invoice is the third key pain point.

Supply-Chain Design

SC design is a long-term planning process to optimize the supply-chain network or configuration and material flow. The primary objective of this process is to determine the most cost-effective SC configuration, which includes facilities; supplies; customers; products; and methods of controlling inventory, purchasing, and distribution on one hand; and the flow of goods throughout the SC on the other hand. Sabri and Beamon (2000) mentioned that strategic supply-chain design concerns are the location of facilities (plants and distribution centers), the allocation of capacity and technology requirements to facilities, the assignment of products to facilities, and the distribution of products between facilities and to customer regions.

The major pain points or challenges of this process are the inability of the inflexible supply-chain configuration to react efficiently to the variability in demand and supply, or to the introduction of new products into the market. The shift from mass production to customized products forces companies to rethink about their physical SC configurations, and nonrepeatable and manual processes.

Sales and Operations Planning

The APICS (American Production and Inventory Control Society) dictionary defines sales and operations planning as "a process that provides Management the ability to strategically direct its businesses to achieve competitive advantage on a continuous basis by integrating customer-focused marketing plans for new and existing products with the management of the supply chain." The main objective of this process is to balance demand and supply, which is not an easy exercise due to the dynamic nature and continuous fluctuation of demand and supply. Historically, supply and demand balancing has been done reactively by sales, marketing, and operations teams allocating constrained products to customer orders, expediting product shipments, or reconfiguring products to create the required models. This business process consists of several subprocesses like demand forecasting, inventory planning, master planning, and revenue planning.

Some of the major pain points of this process are the inability to generate a unified demand plan and to reach consensus across multiple departments, demand and supply volatility, the lack of historical data related to supply and transportation lead-time variability, and the lack of visibility downstream and upstream the supply chain after unexpected events happen.

Order Fulfillment

Order fulfillment is a process of providing accurate, optimal, and reliable delivery dates for sales orders by matching supply (on-hand, in-transit, or planned inventory) with demand while respecting delivery transportation constraints and sales channels or distribution centers' allocations. It is the process of promising what the company can deliver, and delivering on every promise. It is tightly integrated with the selling-management process under CRM.

The major pain points for order fulfillment are the inability to provide visibility across the supply chain to ensure on-time delivery to the customer, and the inability to provide flexibility in meeting customer's expectations. The Internet market forces companies to deal with small shipment sizes more frequently, and changes the destination of deliveries to residential areas.

Marketing Management

Marketing management is the process of creating effective marketing programs to increase revenue, and to increase demand for the products in profitable market segments. It is also the process of generating new customers in a profit-effective manner and providing the best mix of products and services. The marketing process is an input to the demand forecasting subprocess to communicate the projected demand upstream for planning and forecasting purposes. The traditional marketing process is company centric, where phone or face-to-face meetings are the primary marketing media.

In the past, manufacturing controlled the market by determining the price, quality, specifications, and delivery parameters of their products. Company-centric organizations were organized as isolated departments, each dedicated to specific fulfillment functions along the supply chain (Curran & Ladd, 2000).

The other main pain points for this process are the inability to identify appropriate market segments for incentives and promotions, the inability to create effective marketing programs to achieve revenue and profitability objectives while considering supply-chain constraints, and lost revenue due to ineffective new-product launches.

Selling Management

It is the process of helping and guiding the customer to decide on what to buy, and providing accurate and reliable information on the price, delivery, and configuration options. Selling management is the trigger point for sales order processing, which is a subprocess under order fulfillment.

The main pain points in the current practice are the inability to match offerings to customer needs profitably through intelligent pricing, the configuration and availability checking, and the inability to provide the sales agents with guided selling, pricing, and order-promising information to ensure accurate order generation and provide the best product-service bundle to the customer.

Sabri (2005) summarized the challenges that the retail industry faces in the area of selling management.

1. Managing the complicated pricing process
2. Managing the growing product catalog with the challenge of limited space
3. The need for speed in the complex supply network of short-life products
4. Managing promotions and discounts effectively, and the need for markdown optimization
5. Managing product assortments, phasing out products, and seasonality

Customer-Service Management

Customer-service management usually starts through the signing up of a service agreement with the customer, which includes several contract items like discounts on replacement parts, guaranteed response time, technicians' hourly rates, and support time (Curran & Keller, 2000). This process consists of service planning and scheduling, service contract management, service order processing, replacement-part delivery, damaged-part recycling or returns, call centers, and billing. The call center or help desk is the traditional channel for receiving the service order notification, which represents a request for a customer-service activity that can be used to plan specific tasks related to the usage of spare parts and resources, allocate resources to the service task, monitor the performance of the conducted service tasks, and settle service costs.

Some of the CSM-process pain points are the complexity of today's products, which makes managing this process more difficult; customer dissatisfaction due to service-part shortage; inaccurate forecasting for service parts that are considered of high-dollar value, and slow-moving items; and the fact that customer service is becoming a competitive advantage due to the intensified competition and the need to compensate for revenue losses in a flat economy.

E-Business Application Levels and Benefits

Application Levels of E-Business

Before discussing how e-business can address the aforementioned pain points and provide benefits, let us first discuss the different application levels of e-business technology as shown in Figure 1.

There are four application levels of e-business as shown in Figure 1. The higher one goes in the pyramid, the more impact these levels have on the success of the organization and business agility. It is important to mention that achieving the higher levels depends on mastering the lower levels. In Level 1 (the base of the pyramid), information is extracted from data. Then, knowledge is extracted from

Figure 1. Application levels of e-business

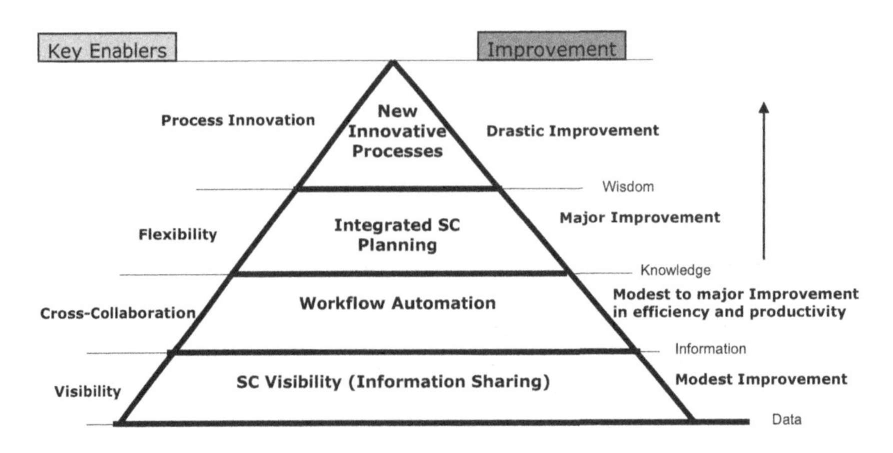

information in Level 2, and wisdom is obtained in Level 3; this enables companies to accelerate to Level 4.

Level 1: SC Visibility

SC visibility refers to sharing data across different participants of the supply chain, and presenting the needed information extracted from data to all participants, depending on their roles, online and on a real-time basis. Some examples of data include forecast data, inventory pictures, capacity plans, promotion plans, shipment schedules, and production schedules. SC visibility addresses the challenge of supply and demand uncertainty by providing visibility to supply events like supplier shortages, and demand events like changes to customer orders. SC visibility provides several benefits like forward (proactive) visibility, early problem detection, and increased productivity and efficiency.

E-business software providers like i2 Technologies support several important capabilities for this level, like a centralized view into critical supply-chain information, and multiple modes of notification (e-mail, e-mail digests, pagers, or cell phones).

Savi Technologies is an example of a company that makes use of RFID technologies to track individual products, containers, and transportation vehicles as they move through the supply chain. The information is put on the Internet so that real-time visibility of movements can be obtained (Lee & Whang, 2001).

Another example for providing visibility to in-transit products and for improving the order-fulfillment process is of a contract trucking company that uses a two-way mobile satellite-communication and position-reporting system to monitor the location of its trucks in order to improve performance under just-in-time programs (Ballou, 2004).

Level 2: Work-Flow Automation

Work-flow automation refers to the automation and streamlining of activities between supply-chain participants. For example, the request for a quote (a subprocess of procurement) and the related activities between buyers and suppliers can be automated and tightly coupled to increase productivity and reduce cycle time. Another example is the automation of the request for information (a subprocess of product design) between engineers in the buying organization and suppliers; here, productivity and quicker-time-to-market gains are achieved. In Level 2, shared information (from Level 1) is taken one step further by collaborating on it or resolving mismatch problems (exceptions). In this level, exceptions are prioritized so that the most important supply-chain disruptions are dealt with in the quickest and most optimal manner; this provides the SC partners the ability to respond to problems in real time to minimize the impact of disruptions on the supply chain, which means cost-effective, speedy, reliable, and almost-error-free SC activities.

The automation of critical (core competency) business processes should be done after improving the as-is process by eliminating nonvalue activities, simplifying and streamlining processes, and removing barriers (disconnects) between processes or functions. This means the automation should be done for the redesigned modified to-be process, and the software needs to be customized to follow and support the process to maintain the competitive advantage. On the other hand, automating noncore competency processes can be done by either using out-of-the-box work flows provided by software companies by which such processes are changed accordingly (no value in tailoring the software to fit the process), or by outsourcing the management of these processes to a third party.

A good example of work-flow automation is the application of electronic exchange portals in the area of procurement, such as Covisint for the automobile industry, e2open for the electronics industry, and Transora for the grocery industry.

Level 3: Integrated SC Planning

The integrated-SC-planning level allows companies to respond quickly and effectively to unplanned supply and demand events that may disrupt information and material flow in the supply chain as one unit. It allows a company to plan based on real-time execution data, and execute based on an up-to-date plan. Integrated SC planning provides a process-centric view coordinating different business subprocesses like product introduction, forecasting, replenishment, manufacturing, fulfillment, and procurement with suppliers and customers, while enabling event management. For example, it supports event-triggered planning and replanning. This level blends information gathered from users using collaboration in Level 2 and multiple transactions and planning systems to allow the exchange of knowledge by the SC partners and create synchronized plans and one global view of the supply chain.

Each supply-chain member (buyer, supplier, carrier, third-party logistics, contract manufacturer, etc.) often operates independently and only responds to immediate requirements. If the Internet is integrated with the SC planning process, SC members can share needed information on a real-time basis, and react quickly and efficiently to changes in demand, material shortages, transportation delays, and supplier inability to replenish. One example is the collaborative planning, forecasting, and replenishment (CPFR) initiative.

McDonald's Japan is a good example of the successful use of CPFR. McDonald's Japan established a process around the Internet whereby stores, marketing, distribution centers, and suppliers would communicate and collaborate via the company's Web site to agree on order sizes and supply-replenishment delivery schedules (Ballou, 2004).

TaylorMade (a large golf supplier) leveraged integrated SC planning to improve the order-fulfillment process. TaylorMade adopted Provia Software as the warehouse-management system and integrated it smoothly with i2's planning and fulfillment systems to prioritize orders based on service level, order volume, promised delivery date, and transport mode (Bowman, 2002).

Level 4: New Innovative Processes

Once companies master e-business application levels, they start to think of adopting new strategies and models for conducting business, seeking not only incremental improvements, but drastic ones. They might seek to reengineer (redesign) their processes to leverage the most out of e-business technologies. Sometimes, companies start to define new processes, seeking new business opportunities or trying to penetrate new markets and customer segments that were neither apparent nor possible prior to the e-business. Companies seek the new-generation business models to achieve competitive advantage and significant benefits. One example is what Dell Computer did when it adopted the build-to-order strategy and provided flexible configuration capability for customers online. The following are examples that show the range of possibilities for companies that pioneered in these areas.

Example 1: Mass Customization

The Internet and e-business technologies facilitate mass customization and allow customers to configure specific order options tailored to their preferences.

Mass customization is the centerpiece of a strategy that woke the big golf supplier TaylorMade and propelled it ahead of the competition in terms of agility and innovation. Today, TaylorMade can customize virtually any aspect of a club. The results to date are impressive (Bowman, 2002).

Example 2: Public Marketplaces

The Internet and e-business technologies helped many companies do business online using a secured specialized Web site. One example is World Chemical Exchange, providing a global market for chemical and plastic manufacturers and buyers. More than 2,500 members can now conduct around-the-clock trading of chemicals and plastics of all types (Lee & Whang, 2001).

Example 3: Supply-Chain Redesign

A good example is what many remote discount computer-hardware and -supply houses did to compete with local retail stores. Many of them used the Internet technologies as a strategy to compress the order cycle time and improve the order-fulfillment process: A customer enters the order through the company's Web site, the inventory and payment are checked, and the order is filled from the warehouse and shipped using UPS, FedEx, or other carriers directly to the end customer.

Example 4: Value-Added Replenishment Programs

Companies as part of lean initiatives are trying to focus on value-added activities to cut waste in the supply chain and reduce overhead cost. Therefore, manufacturers are moving away from making products to stock and sell them later. They are moving away from procuring based only on forecast. Vendor-managed inventory (VMI) is a replenishment program that helps companies achieve their objectives. VMI delays the ownership of goods until the last possible moment and delegates managing the stock to the supplier.

Western Publishing is using a VMI program in its Golden Book lines. It develops a relationship with its retailers in which these retailers give Western point-of-sale data. Ownership of the inventory shifts to the retailer once the product is shipped (Ballou, 2004).

Kanban replenishment is another program in which replenishing parts is based on part consumption. It avoids the inaccuracy in forecasting and eliminates the need for inventory.

Example 5: Online Retailing

Amazon.com understood e-business technologies very well. It has based its business model around it. Amazon.com depends on its efficient supply chain to satisfy customer needs worldwide. It mastered the selling-management process by improving the Web shopping experience through providing quick and reliable promises, and suggesting product bundles, among many other features. This makes Amazon.com one of the biggest and early adopters of e-business technologies.

Benefits of Adopting E-Business Application Levels

Tables 3, 4, and 5 illustrate how the four application levels of e-business can address the challenges of SCM, SRM, and CRM business processes that were mentioned in the beginning of this section. These tables also show the potential benefits of adopting e-business strategies.

Table 3. The impact of e-business application levels on SRM processes

SRM Business Processes				
E-Business Application Levels	**Strategic Sourcing**	**Product Design**	**Procurement**	**Benefits**
SC Visibility	Sharing AVL with design and procurement departments	Real-time visibility on engineering change requests (ECRs)	Sharing supplier and shipment information, real-time exception visibility, audit-trail notification, alerts, and tracking	• Reducing part-inventory obsoleteness • Improving inventory turns • Reducing safety stock • Reducing expedition cost
Work-Flow Automation	A single user interface for design, sourcing, and procurement with flexible and configurable work flows	Shared design workbench	Automated procurement subprocesses, bid analysis, and resolution work flow	• Reducing design rework • Reducing process cycle time • Improving productivity
Integrated SC Planning	Consolidation of enterprise spend/demand across separate systems	Tightly integrated to PDM and AVL	Synchronized replenishment, supporting different replenishment types, and matching execution documents like purchase orders, ASN, and invoices	• Increasing reuse of existing parts in the design • Improving on-time delivery

Table 3. continued

New Innovative Processes	Analyzing supplier and SC performance (slice and dice by site, commodity, time, supplier, and KPI)	Design collaboration	Auctions, marketplace exchanges	• Reducing development cost • Improving time to market • Reducing part/ raw-material cost • Improving quality

Table 4. The impact of e-business application levels on SCM processes

	SCM Business Processes			
E-Business Application Levels	**Supply-Chain Design**	**Sales and Operations Planning**	**Order Fulfillment**	**Benefits**
SC Visibility	Providing an aggregated view on the SC performance and strategic information	Real-time visibility to unexpected events in the SC and audit-trail data	Real-time SC visibility for the order-delivery life cycle including contract manufacturers, distribution centers, and logistic providers	• Reducing uncertainty and safety stock • Early issue detection
Work-Flow Automation	Consistent process with friendly user interface	Unified demand plan across different departments	Exception work-flow resolution for demand changes and fulfillment delays	• Increasing efficiency • Fast response
Integrated SC Planning	Integration with strategic sourcing to reduce supplier base	Synchronized marketing, sales, production, and procurement plans	CPFR	• Speed • Accuracy
New Innovative Processes	SC redesign	Mass customization	Build to order	• Flexibility • Penetrating new markets • Customer satisfaction

Table 5. The impact of e-business application levels on CRM processes

E-Business Application Levels	Marketing Management	Selling Management	Customer-Service Management	Benefits
SC Visibility (Information Sharing)	Capturing feedback from the customers, providing a mix of products and service offerings customized to customer needs	Providing up-sell and cross-sell product recommendations and product bundles, flexible pricing models for markdown and rebates	Providing service order status and highlighting exceptions	• Publicizing product information • Increasing customer satisfaction • Reduce Inv.
Work-Flow Automation	Capturing log records for every visit of a user in the Web servers' log file, including pages visited, duration of the visit, and whether there was a purchase, demand collaboration with customers	Product configuration, quotation processing	Service order logging, billing of services	• Better prediction of customer demand • Improving response time • Improving productivity
Integrated SC Planning	Considering the supply-chain constraints while executing the marketing campaigns, providing customer profiling and segmentation	Supporting different channels for order capturing (Web based, call center, EDI, phone, e-mail, or personnel meeting)	Warranty check, service order processing, integrating the call center	• Increasing revenues and profit • Creating new market/ distribution channels • Accurate promising date

Table 5. continued

New Innovative Processes	Real-time profiling that tracks the user click stream, allows the analysis of customer behavior, and makes instantaneous adjustments to the site's promotional offers and Web pages	Online flexible configuration and real-time promise date	Dealing with products and services as one package during selling	• Long-term relationship and trust with the customer • Gaining competitive advantage

Tables 3, 4, and 5 show the operational and financial benefits of adopting e-business application levels. The operational benefits can be grouped under inventory reductions, cycle-time reductions, productivity increase, supplier performance improvement, and customer-service-level increase. The financial benefits are driven from the operational benefits and can be grouped as follows.

- Cost reduction due to cost savings. The tight integration of supply-chain processes reduces the cost and time needed to exchange transactions and allows efficient procurement, which helps the purchasing staff to focus more on strategic activities like building supplier relationships than managing day-to-day transactions.
- Revenue growth and profit increase due to increased customer satisfaction by delivering on every promise and responding quickly to customer needs, and the ability to penetrate new markets.
- Better asset utilization by replacing inventory with real-time visibility
- Higher shareholder value due to growing profit.

The next section will provide the needed guidelines to implement e-business technologies successfully. Finally, a case study will be presented.

A Framework for Successful Implementation of E-Business Technologies

Many companies are struggling with implementing e-business technologies and achieving the promised value or ROI. In addition, companies are looking for guidelines and strategies for ongoing operational management and support after the go-live, which includes rolling more customers, suppliers, and new business units when implementing e-business solutions to improve SRM, SCM, and CRM superprocesses.

According to a survey of 451 senior executives, one in every five users reported that their CRM initiatives not only had failed to deliver profitable growth, but also had damaged long-standing customer relationships (Rigby et al., 2002).

Currently, there is uncertainty and doubt among organizations regarding the new Internet technologies, and although the appeal for best practice and the benefits of implementing e-business technologies are clear, enterprises struggle in integrating them into supply-chain operations because they are encountered by many challenges like the inability to master change management, the need for new skills to support processes that span across suppliers and partners, the need for e-business strategy and continuous upper management support, the lack of comprehensive metrics and continuous monitoring, and the inability to select the right software-providing partner.

Figure 2. Framework for implementing e-business transformation programs

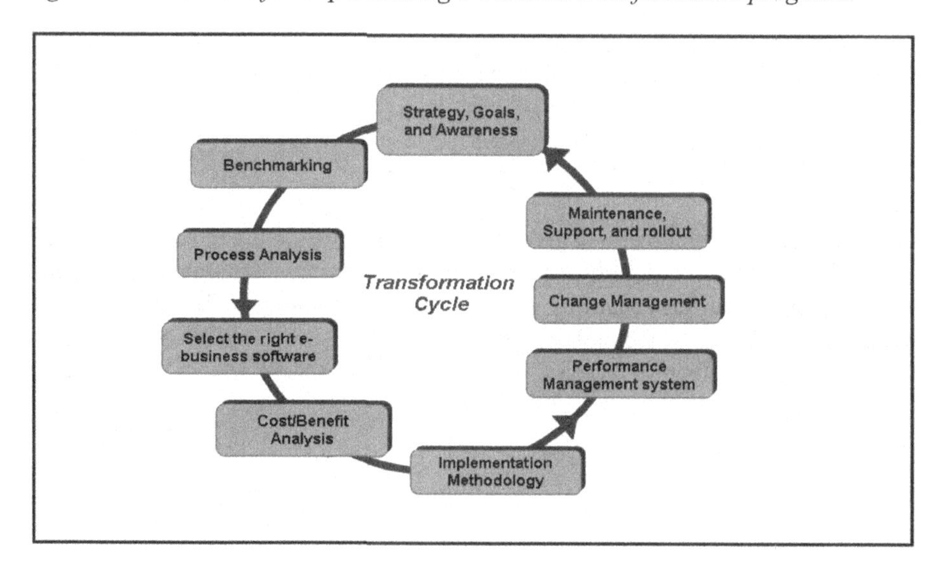

Figure 2 is proposed to address these challenges and provide best-practice guidelines to implement e-business program transformation successfully.

E-Business Strategy and Goals

A clear strategy is the first step for a successful transformation. Executives need to understand the big picture, the interactions between all the processes, and the e-business applications to help them in creating an e-business strategy.

Benchmarking

Benchmarking is the process of comparing and evaluating a firm or a supply chain against others in the industry to help in identifying the gaps and areas of improvement. Benchmarking is used to validate the potential benefit and gain in performance measures from implementing e-business applications.

Process Analysis

The purpose of process analysis (also called design and requirements) is to use modeling (process mapping) methods to analyze "as-is" business processes, capture the existing challenges and pain points in the current process and the supply chain, design and validate the to-be process improvements against best-practice benchmarks, determine the extent of process and technology changes possible in the currently existing systems, and identify the additional software (application) capabilities that are required to support the to-be process that cannot be supported by the existing systems. This requirements list will be the base for selecting the new software.

Select the Right E-Business Software

As a best practice, organizations need to identify the best-of-breed solution that is most suitable for the required functionality for their business, taking into consideration software-technology maturity and sustainability. Supporting leading industry standards for e-business technology like Java, XML, Linux, and Web services is crucial during the selection process.

Cost and Benefit Analysis

Cost and benefit analysis is the process that determines the potential benefits from implementing the combination of the best-practice process and the new application or software. It addresses questions like the following. What is the potential value of increasing the loyalty of our customers when new marketing-management software is implemented? What is the cost of implementing the new solution?

Adopt a Value-Driven Implementation Methodology

Adopting a value-driven approach to conduct the e-business transformation programs like Six Sigma is very critical. An effective transformation program typically takes 2 to 5 years, with several intermediate checkpoints (go-lives) to achieve the value needed to pay for the rest of the program.

Performance-Management System

A performance-management system consists of two phases. The first phase is to establish a consistent metrics-tracking and -publishing process, and this phase should finish before the implementation of the transformation program. The second phase is to continue measuring the benefits and ROI, which should start during and after the implementation.

Since the performance-management system depends mainly on monitoring the metrics (KPIs), it is critical to spend enough time on defining these metrics. The performance-management system should manage and coordinate the development of these metrics. Melnyk, Stewart, and Swink (2004) mentioned that metrics provide the following three basic functions.

- **Control:** Metrics enable managers to evaluate and control the performance of the resources.
- **Communication:** Metrics communicate performance to internal and external stakeholders.
- **Improvement:** Metrics identify gaps (between actual performance and expectation) that ideally point the way for intervention and improvement.

Sabri and Rehman (2004) provided guidelines for identifying and maintaining metrics based on best practice, recommended to capture all operational metrics because improvement in one area could be at the expense of another, and suggested to summarize benefits in six key areas: revenue increase, cost reduction, process lead-time reduction, asset reduction, customer benefits, and supplier benefits.

Change Management

Effective change management for e-business transformation programs should consider gaining and keeping executive sponsorship. Without executives' buy-in and support, a transformation program would be much closer to failure than success. It should also involve all SC partners in developing the new to-be process, and should establish a benefit-sharing and incentives mechanism.

Maintenance, Support, and Rollout

Although companies acknowledge the importance of ongoing operational management and support, few of them think ahead of time and allocate the right resources for it. Once the e-business application links are in place, companies find themselves with an urgent need to manage the ongoing maintenance and rollout. Ongoing monitoring and maintenance are necessary to ensure 100% uptime and compliance. The lack of a defined and clear plan for maintenance and rollout might impact the whole transformation program negatively. The ongoing maintenance and rollout process should include adding new SC organizations and removing existing ones as necessary. It includes training programs and process compliance by monitoring related metrics. It also includes the identification and description of all user groups, and the process of adding new users, making changes to user authorization levels, maintaining profiles, and deleting users. Finally, contingency plans should be reviewed periodically to make sure its readiness. Contingency plans represent predefined courses of actions to be followed in case of the occurrence of a drastic event like when the sources for inbound information go down.

Case Study

This case study is based on an article published in October 2004 by Reuben Slone in *Harvard Business Review* (HBR), which is about the supply-chain turnaround by Whirlpool in the last 4 years. Whirlpool makes a diverse line of products like washers, dryers, refrigerators, dishwashers, and ovens, with manufacturing facilities in 13 countries. This case study is a real-life example of a company that adopted many of the best-practice guidelines of implementing e-business applications that were highlighted previously in this section.

Strategy

Whirlpool needed a strategy that not only addresses the current needs, but also anticipates the challenges of the future. Whirlpool wanted a strategy that can optimize supply-chain performance at minimum cost, and include new e-business technology, processes, roles, and talents to achieve competitive advantage. Its strategy was to focus on customer requirements first and proceed backward. Therefore, Whirlpool and Sears as a customer studied consumers' desires with regard to appliance delivery. They found that consumers are asking for accurate promises as a first requirement: "Give a date, hit a date."

Benchmarking and Process Analysis

Whirlpool benchmarked its competitors and obtained cross-industry information and competitive intelligence from AMR, Gartner, and Forrester Research. Then it mapped out what is considered best-practice performance along 27 different SC-capability dimensions. This exercise helped identify areas of improvement.

Cost and Benefit Analysis

The program transformation team had to build a compelling business case to get the buy-in from upper management. They had to justify their program wholly on expense reductions and working capital improvements.

Effective Transformation Plans

Effective transformation plans include a value-driven implementation methodology (Six Sigma), performance-management system, change management, and rollout plans.

Whirlpool started with improving the S&OP process. Its current process was inadequate with Excel spreadsheet feeds. Now, Whirlpool is able to generate synchronized long and short plans that consider marketing, sales, finance, and manufacturing constraints or requirements. Then, it launched a CPFR pilot to share forecasts using a Web-based application and to collaborate on the exceptions, which enabled it to cut forecast accuracy error in half within 30 days of launch. In January 2002, Whirlpool implemented a suite of software products from i2 to reduce inventories while sustaining high service level. By May 2002, a blind Internet survey showed Whirlpool to be "most improved," "easiest to do business with," and "most progressive" in the eyes of their trade partners.

It segmented its products and followed a different strategy for each product group. For high-volume SKU like dishwashers, refrigerators, and washing machines, it used the build-to-stock replenishment technique with its customers. For smallest volume SKUs, they followed the pull replenishment technique with the more flexible build-to-order process. The inventory savings on the small-volume SKUs can balance out the costs of stocking up on the high-volume SKUs. Whirlpool also started to move away from having one service level across all products, recognizing that some products are more important or more profitable than others and should have higher service levels.

Recently, there has been a focus on system-to-system transactions, in which the Whirlpool system talks directly to a customer's system for purposes of transmitting purchase orders, exchanging sales data, and submitting invoices and payments. At the same time, customers can check availability and place orders via the Internet. Whirlpool is also looking to implement an event-management capability that provides a notification whenever an action in the process has taken place.

A couple things were absolutely critical to keep the transformation program schedule on track: a highly disciplined transformation program office and an effective management system. The key was to think big but focus relentlessly on near-term deadlines. Whirlpool organized the change effort into 30-day chunks, with three new capabilities or business releases rolling out monthly, some on the supply side and some on the demand side. The main job of the program transformation office, which adopted Six Sigma methodology, was to ensure the completion of projects on time, on budget, and on benefit.

The transformation program office contracted Michigan State University and the American Production and Inventory Control Society to develop a competency model that can outline the skills and roles required in a top-tier organization. Whirlpool also expanded the compensation system to allow employees to be rewarded for increasing their expertise even if they are not being promoted into supervisory roles. It also put a huge emphasis on developing employees' management skills and used a model developed by Project Management Institute (PMI) as a standard for evaluating and enhancing the organization's project-management capabilities. Finally, it assembled a supply-chain advisory board to provide guidance and assess the transformation program results and direction.

To summarize, Whirlpool followed the best practice in leveraging e-business technologies, and in return, it has much to show for its transformation efforts. Today, its product-availability service level is more than 95%. The inventory of finished goods has dropped from 32.8 to 26 days. In one year, it lowered its working capital by almost $100 million and supply-chain costs by $20 million with an ROI equal to 2.

Conclusion

E-business technologies present huge opportunities that are already being tapped by many companies and supply chains. Leveraging e-business technologies effectively is key to gaining competitive advantage, streamlining processes, slashing waste, and eventually achieving business agility, which is significantly needed in the new age of globalization and intensive competition.

More companies will start to realize that gaining competitive advantage is no longer feasible only by managing their own organizations. They need to get involved in the management of all upstream organizations that are responsible for the supply, as well as the downstream network that is responsible for delivery and the after-sales market. The challenge for companies for the rest of this decade is synchronization across supply-chain processes, from product design and procurement to marketing and customer-service management, in order to be more responsive to customer needs. The new trend of mergers and acquisitions will continue to rise, and big companies that are buying out smaller ones will grow even bigger in the complexity of their supply chains. This will increase the need for e-business technologies to streamline the process of collaboration between the different entities.

Therefore, in the next few years, we will see the explosion of e-business-applications use as companies utilize e-business to redefine supply-chain processes that span across suppliers and customers, which will result in a significant improvement in efficiency and will help companies achieve competitive advantage. Companies that do not come on board will realize that they are losing ground and customers soon.

The widespread use of e-business will lead to new options for improving business-to-business and business-to-consumer collaborations like multitier collaborations and root-cause analysis for exceptions in the supply-chain performance. In addition, it will open new ways of integration between supply-chain partners like system-to-system integration using Web services (e.g., integrating one firm's inventory-control system and another's logistics-scheduling environment), the use of wireless devices, and the tight integration of the Web site with the back-end systems of supply-chain partners. Eventually, e-business technologies will replace electronic data interchange, the benefits of which never materialized for midsized companies because of its high cost.

We also expect SMEs (small and medium-sized enterprises) to realize the importance of e-business and to follow one of the following arrangements in adopting e-business technologies depending on the business requirements and cost factors.

- **Microsoft arrangement:** Easy to implement due to wide familiarity with the product and its selling process through partners, cheap license, cheap maintenance, and tight integration with other Microsoft products like Excel

- **Public Web-enabled arrangement:** Prebuilt solution by a Web-enabled applications provider at a fixed monthly cost, no need for software to be present on the company's internal network, no maintenance fees, and lower risk due to almost zero-down investment

Intelligent performance-management systems that can capture negative performance trends and select the correct resolutions are expected to come into widespread use in the next few years.

To summarize, we will witness, for the rest of this decade, what is called a tightly integrated environment in which supply-chain interactions involve tightly integrated databases and applications; processes are significantly redesigned and streamlined to eliminate redundancies and non-value activities.

References

Anderson, D. M. (2003). *Build-to-order & mass customization.* Cambria, CA: CIM Press.

Ballou, R. H. (2004). *Business logistics/supply chain management* (5th ed.). Upper Saddle River, NJ: Prentice Hall.

Bowman, R. J. (2002). TaylorMade drives supply-chain efficiency with 24 hour club. *SupplyChainBrain.com.* Retrieved December 10, 2004, from *http://www.supplychainbrain.com/archives/10.02.TaylorMade.htm?adcode=5*

Curran, T. A., & Ladd, A. (2000). *SAP R3 business blueprint: Understanding enterprise supply chain management* (2nd ed.). Upper Saddle River, NJ: Prentice Hall.

Devaraj, S., & Kohli, R. (2002). *The IT payoff: Measuring the business value of information technology investment.* Upper Saddle River, NJ: Prentice Hall.

Handfield, R. B., & Nichols, E. R. (2002). *Supply chain redesign: Transforming supply chains into integrated value systems.* Upper Saddle River, NJ: Prentice Hall.

Harmon, P. (2003). *Business process chain: A manager's guide to improving, redesigning, and automating processes.* San Francisco: Morgan Kaufmann Publishers.

Lee, H., & Whang, S. (2001). E-business and supply chain integration. *Stanford Global Supply Chain Management Forum*, 1-20.

Melnyk, S. A., Stewart, D. M., & Swink, M. (2004). Metrics and performance measurement in operations management: Dealing with the metrics maze. *Journal of Operations Management, 22*, 209-217.

Rigby, D., Reichheld, F., & Schefter, P. (2002). Avoid the four perils of CRM. *Harvard Business Review*, 1-9.

Sabri, E. (2005). *Value chain management to achieve competitive advantage in retail industry.* Paper presented at the Middle East Retail Conference, United Arab Emirates.

Sabri, E., & Beamon, B. (2000). A multi-objective approach to simultaneous strategic and operational planning in supply chain design. *OMEGA: The International Journal of Management Science, 28*(5), 581-598.

Sabri, E., & Rehman, A. (2004). *ROI model for procurement order management process.* Paper presented at the Lean Management Solutions Conference, Los Angeles.

Slone, R. E. (2004). Leading supply chain turnaround. *Harvard Business Review*, 1-9.

Valencia, J. S., & Sabri, E. H. (2005). *E-business technologies impact on supply chain.* Paper presented at the 16th Annual Conference of POMS, Chicago.

Chapter XV

Concepts and Operations of Two Research Projects on Web Services and Context at Zayed University

Zakaria Maamar, Zayed University, UAE

Abstract

This chapter presents two research projects applying context in Web services. A Web service is an accessible application that other applications and humans can discover and invoke to satisfy multiple needs. While much of the work on Web services has up to now focused on low-level standards for publishing, discovering, and triggering Web services, several arguments back the importance of making Web services aware of their context. In the ConCWS project, the focus is on using context during Web-services composition, and in the ConPWS project, the focus is on using context during Web-services personalization. In both projects, various concepts are used such as software agents, conversations, and policies. For instance, software agents engage in conversations with their peers to agree on the Web services that participate in a composition. Agents' engagements are regulated using policies.

Introduction

With the latest development of information technologies, academia and industry communities are adopting Web services because of their integration capabilities (Papazoglou & Georgakopoulos, 2003). Indeed, Web services can connect business processes in a business-to-business fashion. This connection highlights the possibility of composing Web services into high-level business processes usually referred to as composite services. Composition primarily addresses a user's request that cannot be satisfied by any available Web service (called service in the rest of this document); in this situation, a composite service obtained by combining available Web services might be used.

A Web service presents the following properties (Benatallah, Sheng, & Dumas, 2003): They are independent as much as possible from specific platforms and computing paradigms, are primarily developed for interorganizational situations, and are easy to compose so that developing complex adapters for the needs of composition is not required. For composition purposes, a composite service is always associated with a specification, which describes among others the list of component Web services that take part in the composite service, the execution order of these component Web services, and the corrective strategies in case these component Web services raise exceptions. Different composition languages exist such as the business process execution language (Curbera, Khalaf, Mukhi, Tai, & Weerawarana, 2003) and Web services flow language (Leymann, 2001). The primary objective of these languages is to provide a high-level description of the composition process far away from any implementation concerns. The specification of composite services is also concerned with the semantics of information that the component Web services exchange (Sabou, Richards, & van Splunter, 2003). However, the semantic composition is outside this chapter's scope.

Despite the wide embracement of Web services, they still lack the capability that could propel them to the acceptance level that features traditional integration middleware such as common object request broker architecture (CORBA) and distributed component object model (DCOM). This lack of capability is primarily due to the trigger-response exchange pattern that is imposed on Web services and their interaction models with third parties. The compliance with this pattern means that a Web service has only to process the requests it receives, without, for example, considering its execution status or even questioning about the validity of these requests. However, there exist several situations that call for Web services' self-management so that the requirements of flexibility, autonomy, and stability are met. By flexibility, we mean the capacity of a Web service to adapt its behavior by selecting the appropriate operations that accommodate the ongoing situation in which it operates. By autonomy, we mean the capacity of a Web service to accept demands of participation in composite services, or to reject such demands in case of unappealing rewards. Last but not least, by stability, we mean the capacity of a Web service to resist change while maintaining function and to recover to normal levels of function after a disturbance. To meet these requirements, Web services have to assess their environment prior to engaging in any composition. In fact, Web services need to be context aware: "Context is not simply the state of a predefined environment with a fixed set of interaction resources. It is part of a process of interacting with an ever-changing environment composed of reconfigurable, migratory, distributed, and multiscale resources" (Coutaz, Crowley, Dobson, & Garlan, 2005). By developing context-aware Web services it would possible, for example, to consider the aspects of the environment in which the Web services are to be executed. These aspects are multiple and can be related to users (e.g., stationary, mobile), computing resources (e.g., fixed, handheld), time of day (e.g., in the afternoon, in the morning), and so forth.

In this chapter, we present two projects that are conducted in the college of information systems at Zayed University on context and Web services. The rest of this chapter is organized as follows. First the chapter provides some basic definitions about Web services and their composition. Then it presents the ConCWS project, which stands for Context for Composing Web Services. Next, it presents the ConPWS project, which stands for Context for Personalizing Web Services. The way context fits into Web services from the composition and personalization perspectives is discussed afterward. Finally, we conclude the chapter.

Web Services

Definitions

According to the World Wide Web Consortium, a Web service is a:

> ... *software application identified by a URI* [uniform resource identifier], *whose interfaces and binding are capable of being defined, described, and discovered by XML (extensible markup language) artifacts, that supports direct interactions with other software applications using XML-based messages via Internet-based applications.* (W3C, 2002)

Several standards are associated with Web services like electronic business extensible markup language (ebXML) registry services, Web service description language (WSDL), universal description, discovery, and integration (UDDI), simple object access protocol (SOAP), and Web services security (WSS).

Figure 1. Service chart diagram of a component Web service

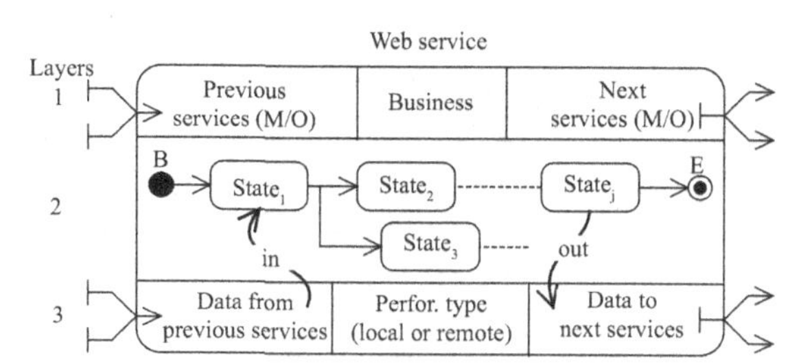

For our research on Web services (Maamar & Mansoor, 2003; Maamar, Sheng, & Benatallah, 2004; Maamar, Yahyaoui, & Mansoor, 2004), we developed service chart diagrams as a means for modeling and defining Web services (Maamar, Benatallah, & Mansoor, 2003). A service chart diagram enhances a state chart diagram, putting this time the emphasis on the context surrounding the execution of a service rather than only on the states that a service takes (Figure 1). To this end, the states of a service are wrapped into five perspectives, with each perspective having a set of parameters. The state perspective corresponds to the state chart diagram of the service. The flow perspective corresponds to the execution chronology of the composite service in which the service participates (previous-services and next-services parameters; M/O for mandatory or optional). The business perspective identifies the organizations (i.e., providers) that offer the service (business parameter). The information perspective identifies the data that are exchanged between the services of the composite service (data from previous services or data for next services). Because the services that participate in a composition can be either mandatory or optional, the information perspective is tightly coupled to the flow perspective with regard to mandatory and optional data. Finally, the performance perspective illustrates the ways by which the service is invoked for execution (performance-type parameter; more details on invocation types are given in Maamar, 2001).

Figure 2. Sample of the specification of a composite service

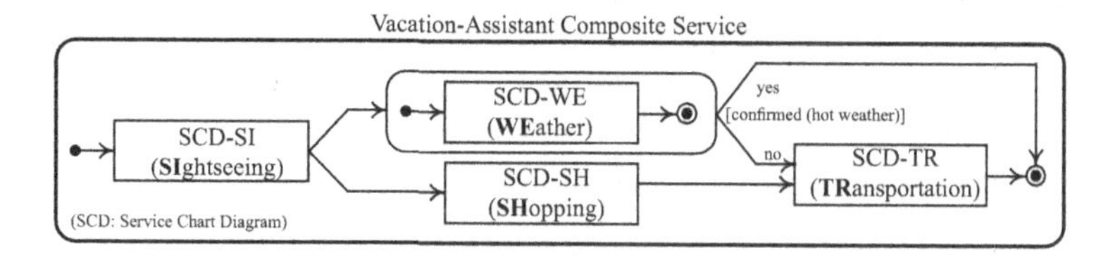

Composite Services

A composition approach connects Web services together in order to devise composite services. The connection of Web services implements a business logic, which depends on the application domain and control flow of the business case for which the composite service is being devised. Examples of business cases are various such as travel planning and chapter review. It is accepted that the efficiency and reliability of a composite service strongly depend on the commitments, performance, and delivery capabilities of each of the component services.

Because a composite service is made up of several component services, the process model underlying the composite service is specified as a state chart diagram (the value added from state charts to Web-services composition is discussed in Benatallah, Dumas, Sheng, & Ngu, 2002). In this diagram, states are associated with the service chart diagrams of the component Web services (Figure 1), and transitions are labeled with events, conditions, and variable assignment operations. For illustration purposes, Figure 2 presents the vacation-assistant composite service (VA-CS) as a state chart diagram. This diagram is about the orchestration of the following component Web services; each component is associated with a service chart diagram: sightseeing (SI), weather (WE), shopping search (SH), and transportation (TR).

ConCWS Project

An extensive description of the ConCWS project is given in Maamar, Kouadri Mostéfaoui, and Yahyaoui (2005). Besides context and Web services, additional concepts are used in this project, namely, software agent (Boudriga & Obaidat, 2004) and conversation (Ardissono, Goy, & Petrone, 2003).

Agents in ConCWS

In ConCWS, the Web-services instantiation principle is promoted. According to this principle, a Web service is a component that is instantiated each time it participates in a new composition. Prior to any instantiation, several elements of the Web service are checked. These elements constitute a part of the context, denoted by W-context, of the Web service, and are as follows: (a) the number of service instances currently running vs. the maximum number of Web-service instances that can be simultaneously run, (b) the execution status of each Web-service instance deployed, and (c) the request time of the Web-service instance vs. the availability time of the Web-service instance. The Web-services instantiation principle offers the possibility of organizing a Web service along three temporal categories (Figure

Figure 3. Organization of a Web service

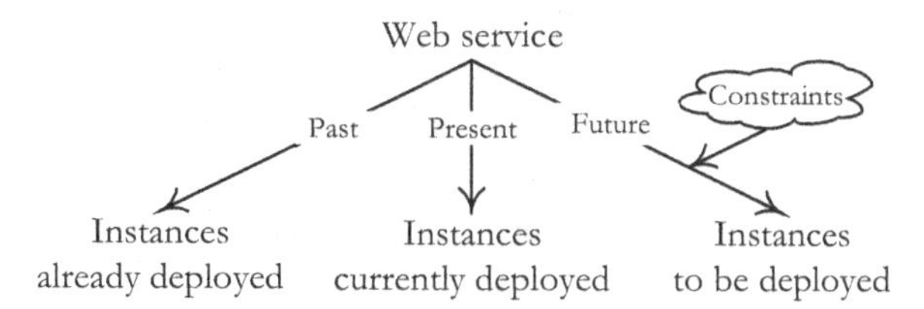

3): Web-service instances already deployed (past), Web-service instances currently deployed (present), and Web-service instances to be potentially deployed upon invitation acceptance (future). Invitation acceptance, which results in Web-service instantiation, is subject to the satisfaction of multiple constraints such as the maximum number of Web-service instances (Figure 3).

In ConCWS, Web-services composition results in identifying three types of software agents: composite service agent, master service agent, and service agent. The role of the master service agent is to track the multiple Web-service instances, which are obtained out of a Web service. Master service agents, Web services, and W-contexts are all stored in a pool (Figure 4). A master service agent processes the requests of instantiation that are submitted to a Web service. These requests originate from composite service agents that identify the composite services to set up. For instance, the master service agent makes decisions on whether a Web service is authorized

Figure 4. Agent deployment during the contextualizing of Web-services composition

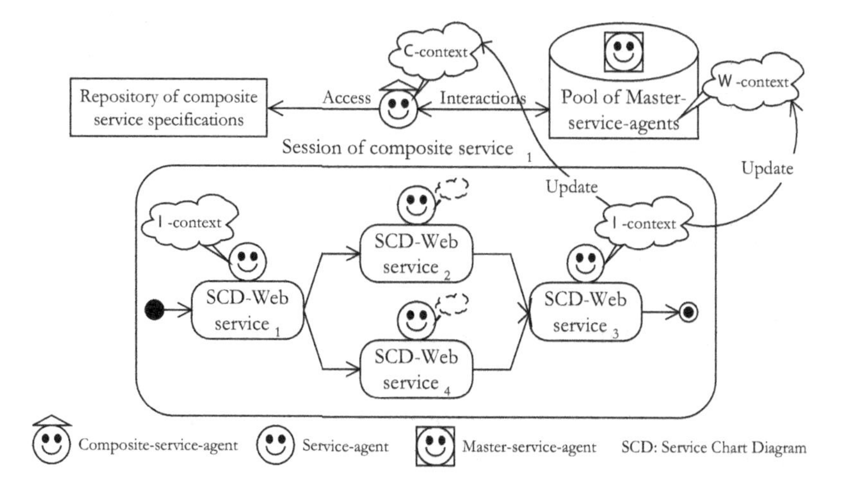

to join a composite service. Upon approval, a Web-service instance along with a context (denoted by I-context) is created. An authorization for joining a composite service can be eventually turned down because of multiple reasons: a period of nonavailability, the overloaded status, or an exception situation.

To be aware of the running instances of a Web service so its W-context is updated, the master service agent associates each service instance it creates with a service agent and I-context (Figure 4). The service agent manages the Web-service instance and its respective I-context. For example, the service agent knows the Web services that need to join the composite service after the execution of this Web-service instance is completed.

Master service agents and service agents are in constant interaction. The content of I-contexts feeds the content of W-contexts with various details: (a) what the execution status is (in progress, suspended, aborted, terminated) of a Web-service instance, (b) when the execution of a Web-service instance is supposed to resume in case it has been suspended, (c) when the execution of a Web-service instance is expected to complete, and (d) what the corrective actions are that need to be taken in case the execution of a Web-service instance fails.

With regard to composite service agents, their role is to trigger the specifications of the composite services (Figure 2) and monitor the deployment of these specifications. A composite service agent ensures that the appropriate component Web services are involved and collaborate according to a specific specification. When a composite service agent downloads the specification of a composite service from the repository of specifications, it establishes a context, denoted by C-context, for the composite service, and identifies the first Web services to be triggered. When the first Web service is identified, the composite service agent interacts with the master service agent of this Web service, asking for a Web-service instantiation. If the master service agent agrees on the instantiation upon checking the W-context, a service agent and I-context are both created. The service agent initiates the execution of the Web-service instance and notifies the master service agent about the execution status. Because of the regular notifications between service agents and master service agents, exceptional situations are immediately handled and corrective actions are carried out on time. In addition, while the Web-service instance is being performed, its service agent identifies the Web services that are due for execution after this Web-service instance completes its execution. In case there are Web services due for execution, the service agent requests from the composite service agent the ability to engage in conversations with their respective master service agents.

Modeling I-, W-, and C-Contexts in ConCWS

Besides the three types of agents that Figure 4 presents, three types of services are considered, namely, composite service, Web service, and Web-service instance.

Each service is attached to a specific context. I-context has the fine-grained content, whereas C-context has the coarse-grained content. The W-context is in between. Details on the I-context update the W-context, and details on the W-context update the C-context. The I-context of a Web-service instance consists of the following parameters.

- **Label:** corresponds to the identifier of the service instance
- **Service-agent label:** corresponds to the identifier of the service agent in charge of the service instance
- **Status:** informs about the current status of the service instance (in progress, suspended, aborted, terminated)
- **Previous service instances:** indicates whether there were service instances before the service instance (could be null)
- **Next service instances:** indicates whether there will be service instances after the service instance (could be null)
- **Regular actions:** illustrates the actions the service instance performs
- **Begin time:** informs when the execution of the service instance started
- **End time (expected and effective):** informs when the execution of the service instance is expected to terminate and has effectively terminated
- **Reasons of failure:** informs about the arguments that caused the failure of the execution of the service instance
- **Corrective actions:** illustrates the actions that the service instance performs because the execution has failed
- **Date:** identifies the time of updating the parameters above

The W-context of a Web service is built upon the I-contexts of its respective component Web-service instances and consists of the following parameters.

- **Label:** corresponds to the identifier of the Web service
- **Master-service-agent label:** corresponds to the identifier of the master service agent in charge of the Web service
- **Number of instances allowed:** corresponds to the maximum number of service instances that can be created from the Web service
- **Number of instances running:** corresponds to the number of service instances of the Web service that are currently running
- **Next service-instance availability:** corresponds to when a new service instance of the Web service will be made available

- **Status per service instance:** corresponds to the status of each service instance of the Web service that is deployed (based on the status parameter of the I-context)
- **Date:** identifies the time of updating the parameters above

The C-context of a composite service is built upon the W-contexts of its respective Web services and consists of the following parameters.

- **Label:** corresponds to the identifier of the composite service
- **Composite-service-agent label:** corresponds to the identifier of the composite service agent in charge of the composite service
- **Previous Web services:** indicates which Web services of the composite service have been executed with regard to the current Web services
- **Current Web services:** indicates which Web services of the composite service are currently under execution
- **Next Web services:** indicates which Web services of the composite service will be called for execution with regard to the current Web services
- **Beginning time:** informs when the execution of the composite service started
- **Status per Web service:** corresponds to the status of each Web-service instance of the composite service that is deployed (based on the status parameter of the I-context)
- **Date:** identifies the time of updating the parameters above

Conversations in ConCWS

In a reactive composition such as the one that features the approach of Figure 4, the selection of the component Web services of a composite service is performed on the fly. The selection operations are outsourced to composite service agents that engage in conversations with the respective master service agent of the appropriate Web services. In these conversations, master service agents decide if their Web services will join the composition process upon checking the W-contexts. In case of a positive decision, Web-service instances, service agents, and I-contexts are all deployed.

When a Web-service instance is under execution, its service agent checks if additional Web services have to be executed after this Web-service instance. If so, the service agent requests from the composite service agent permission to engage in conversations with the master service agents of these Web services. These conver-

Figure 5. Web services engaged in context-driven conversations

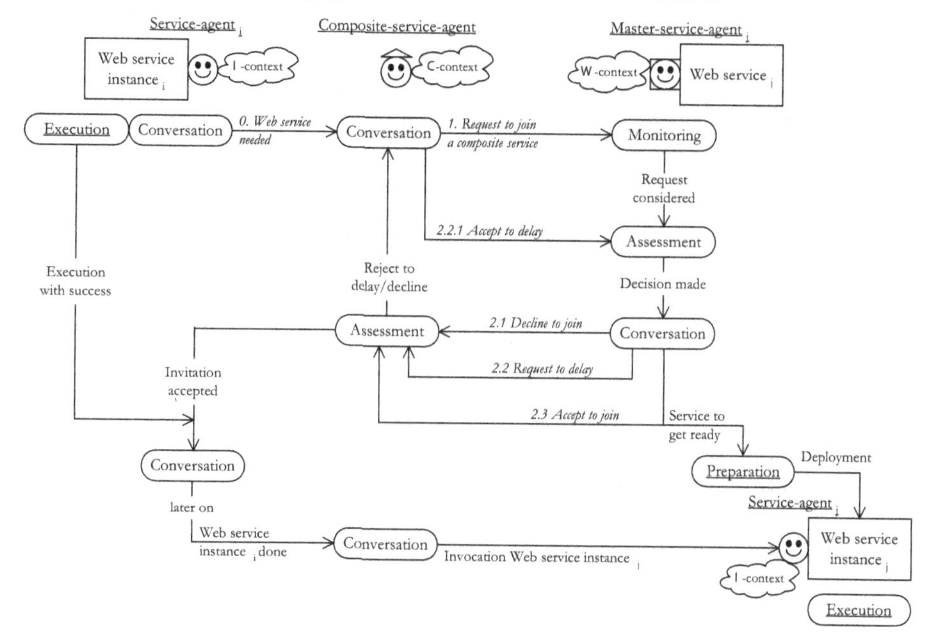

sations have two aims: to invite master service agents and, thus, their Web service to participate in the composition process, and to ensure that the Web services are ready for instantiation in case of invitation acceptance.

Figure 5 depicts a conversation diagram between a service agent, a composite service agent, and a master service agent. The composite service agent is in charge of a composite service of n component Web services$_{1...n}$. In this figure, rounded-corner rectangles are states (states with underlined labels belong to Web-service instances; other states belong to agents), italic sentences are conversations, and numbers are the chronology of conversations. Initially, Web-service instance$_i$ takes an execution state. Furthermore, service agent$_i$ and the composite service agent each take a conversation state. In these states, activities to request the participation of the next Web services (i.e., Web service$_j$) are conducted.

Upon receiving a request from service agent$_i$ about the need of involving Web service$_j$ (0), the composite service agent engages in conversations with master service agent$_j$ (1). This Web service is an element of the composite service and its preparation is in progress. A composite service is decomposed into three parts. The first part corresponds to the Web-service instances that have successfully completed their execution (Web services$_{1...(i-1)}$). The second part corresponds to the Web-service instance that is now being executed (Web service instance$_i$). Finally, the third part corresponds to the rest of the composite service that is due for execution and, hence, has to get ready for execution (Web services$_{j...n}$). Initially, master service agent$_j$ is in a monitoring mode in which it tracks the instances of Web service$_j$ that are already participating in different composite services. When it receives a request to create an additional instance, master service agent$_j$ enters the assessment state. Based on the W-context of Web service$_j$, master service agent$_j$ evaluates the request of the composite service agent and makes a decision on one of the following exclusive options: (a) decline the request, (b) delay making a decision, or (c) accept the request.

Option a. Master service agent$_j$ of Web service$_j$ declines the request of the composite service agent. A conversation message is sent back from master service agent$_j$ to the composite service agent for notification (2.1). Based on this response, the composite service agent enters again the conversation state, asking the master service agent for another Web service$_k$ to join the composite service (1).

Option b. Master service agent$_j$ of Web service$_j$ cannot make a decision before the deadline that the composite service agent has set. Thus, master service agent$_j$ requests from the composite service agent a deadline extension (2.2). The composite service agent has two alternatives taking into account the C-context of the composite service.

- Refuse to extend the deadline as per master service agent$_j$'s request. This means that the composite service agent has to reiterate the process described in Option a by conversing with master service agent$_k$.
- Accept to extend the deadline as per master service agent$_j$'s request. This means that master service agent$_j$ will be notified about the acceptance of the composite service agent (2.2.1). After receiving the acceptance, master service agent$_j$ of Web service$_j$ enters again the assessment state and checks the W-context in order to make a decision on whether to join the composite service (Option a). A master service agent may request a deadline extension for several reasons; for example, additional instances of Web service$_j$ cannot be committed before other instances finish their execution.

Option c. Master service agent$_j$ of Web service$_j$ accepts to join the composite service. Consequently, it informs its acceptance to the composite service agent (2.3). At the same time, master service agent$_j$ ensures that Web service$_j$ is ready for execution through the preparation state (i.e., deploy the I-context and service agent$_j$).

When the execution of Web-service instance$_i$ is completed, service agent$_i$ informs the composite service agent about that. According to the agreement reached in Option c, the composite service agent interacts with service agent$_j$ so that the newly created instance of Web service$_j$ is triggered. Therefore, Web-service instance$_j$ enters the execution state. At the same time, the composite service agent initiates conversations with the master service agents of the next Web services that follow Web service$_j$.

Summary

In this part of the chapter, we overviewed the ConCWS project for composing Web services using software agents, context, and conversations. Several types of software agents have been put forward, namely, composite service agents associated with composite services, master service agents associated with Web services, and finally service agents associated with Web-service instances. The different agents were aware of the context of their services using I-context, W-context, and C-context, respectively. Conversations between agents have also featured the composition of Web services. Before Web-service instances are created, agents engage in conversations to decide if Web-service instances can be created and annexed to a composite service. This decision is based on several factors such as the maximum number of Web-service instances that can be deployed at the same time and the availability of these Web-service instances for a certain period of time. As part of the implementation strategy in the ConCWS project, the Web service choreography interface (WSCI) could be used for implementing the conversations between Web services.

ConPWS Project

An extensive description of the ConPWS project is given in Maamar, Kouadri Mostéfaoui, and Mahmoud (2005). Besides context and Web services, additional concepts are considered in this project, namely, personalization (Bonett, 2001) and policies (Wright, Chadha, & Lapiotis, 2002).

Figure 6. Representation of the context-based personalization approach

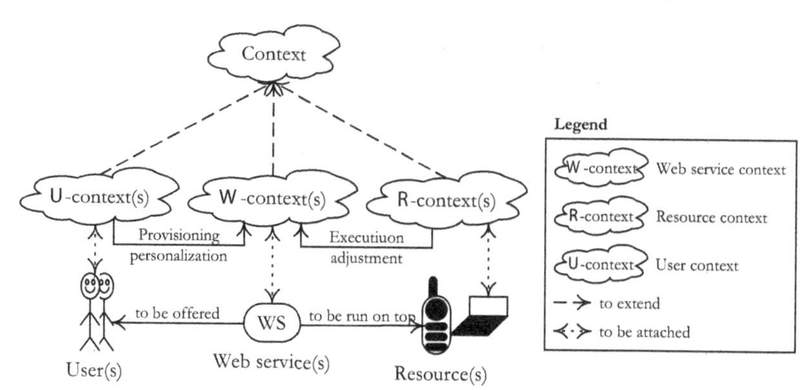

Foundations of ConPWS

Figure 6 illustrates the approach that ConPWS puts forward for personalizing Web services. The core concept is context, from which three subcontexts are derived: U-context, W-context, and R-context. The user context of a user is defined as an aggregation of his or her location, previous activities, and preferences (Muldoon, O'Hare, Phelan, Strahan, & Collier, 2003). We define the Web-service context of a Web service as an aggregation of its concurrent participations in composite services, locations of execution, times of execution, and constraints during execution. Finally, we define the resource context of a resource as an aggregation of its current status, periods of nonavailability, and capacities of meeting the execution requirements of Web services.

In Figure 6, the U-context, W-context, and R-context are interconnected. From the R-context to W-context, the execution-adjustment relationship identifies the execution constraints on a Web service (e.g., execution time, execution location, flow dependency) vs. the execution capabilities of a resource (e.g., next period of availability, scheduling policy) on which the Web service is supposed to operate. A resource has to check its status and assess its current commitments before it agrees on supporting the execution of an additional Web service. From the U-context to W-context, the provisioning-personalization relationship identifies the preferences of a user (e.g., when and where a Web service needs to be executed, and when and where the execution outcome needs to be returned) vs. the capabilities of a Web service to accommodate these preferences (e.g., can a Web service be executed at a certain time or in a certain user location?). A Web service needs to check its status before it agrees on satisfying user needs. The W-context is the common element between the U-context and R-context. A Web service reconciles elements between what a user wishes and what a resource permits.

A Web service is subject to personalization because of user preferences and adjustment because of resource availabilities. Users and resources are the triggers that make Web services change so that they can accommodate preferences and consider availabilities. In addition to both triggers, a third trigger exists that is, in fact, a personalized Web service. The personalization of a Web service can trigger the personalization of other peers that are in relationships with this personalized Web service. We call this type of relationship causal. For example, if a service is personalized so that it accommodates a certain time preference, it is important to ensure that the preceding services are all successfully executed before the execution time of this service. This means that the respective execution times of these services have to be checked and adjusted, if needed.

Figure 7. Interactions during Web-services personalization

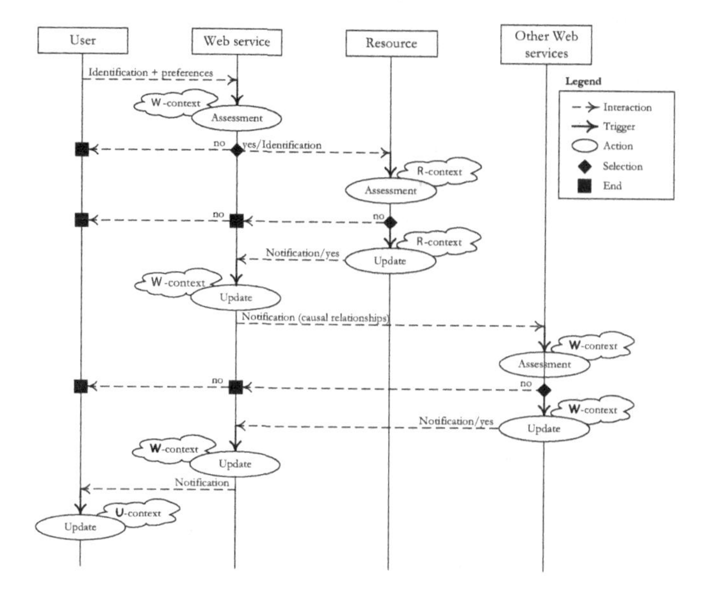

Operation of ConPWS

Figure 7 illustrates the interactions during the context-based personalization of Web services. When a user selects a Web service, he or she continues after with its personalization according to his or her time or location preferences. Time preference is organized along two parts: (a) when the execution of the service should start, and (b) when the outcome of this execution should be returned to the user. Location preference is user dependent (i.e., with respect to the user's current location) and is also organized along two parts: (a) where the execution of the service should occur, and (b) where the outcome of this execution should be returned to the user.

Once user preferences are submitted to the Web service, it is ensured that the dates and locations are valid and no conflicts might happen during deployment (e.g., the delivery time cannot occur before the execution time of a service, and the user cannot be committed to separate locations at the same time). Moreover, the user has to be continuously reminded that he or she has to explicitly identify his or her current location so that execution location and delivery location are both properly handled. While a manual feeding of the current location of users presents some limitations, this feeding allows a better handling of the privacy issue. Users only reveal to external systems the locations they wish to be known. The manual feeding is in line with privacy control as reported in Zuidweg, Pereira Filho, and van Sinderen (2003). Prior to identifying the resources on which it will be executed, the Web service checks its W-context with regard to (a) the number of Web services currently under execution vs. the maximum number of Web services under execution and (b) the next period of unavailability. After a positive check of the W-context, the identification of a resource is now launched. In ConPWS, we assume the existence of a mechanism supporting the identification of resources. A resource mainly needs to accommodate two things: (a) the beginning time of a Web-service execution, and (b) the time that the execution of a Web service lasts since the outcome of this execution depends on the delivery time as per user indication. To this purpose, a resource checks its R-context with regard to (a) the next periods of time that will feature the execution of Web services, and (b) the next period of maintenance. After a positive check, the resource notifies the Web service about its availability to support this service execution.

Before the personalized Web service notifies the user about the handling of his or her time and location preferences, an extra personalization process is triggered. This consists of adjusting the Web services that are linked, through the causal relationship, to the personalized Web service. The description given in the previous paragraphs also applies to the extra Web services, which assess their current status through their respective W-contexts and search for the resources on which they will operate. To keep Figure 7 clean, the interactions that the new personalized Web services undertake to search for the resources are not represented. Once all the Web services are personalized, a final notification is sent out to the user about the deployment of the Web service that he or she has requested.

Policies in ConPWS

Because of user preferences and resource availabilities, a Web service is adjusted so that it accommodates these preferences and deals with these availabilities. To ensure that the adjustment of a Web service is efficient, ConPWS integrates three types of policies (owners of Web services are normally responsible for developing the policies). The first type, called consistency, checks the status of a Web service after it has been personalized. The second type, called feasibility, guarantees that a personalized Web service finds a resource on which it can operate according to the constraints of time and location. Finally, the third type, called inspection, ensures that the deployment of a personalized Web service complies with the adjusted specification.

A consistency policy guarantees that a Web service still does what it is supposed to do after personalization. Personalization may alter the initial specification of a Web service when it comes, for instance, to the list of regular events that trigger the service. Indeed, time- and location-related parameters are new events that need to be added to the list of regular events. Moreover, because of the QoS (quality of service) -related parameters of a Web service (e.g., response time and throughput;

Menascé, 2002), it is important to verify that these QoS parameters did not change and are still satisfied despite the personalization. For illustration, because a user wishes to start the execution of a service at 2 p.m., which corresponds to the peak-time period of receiving requests, the response-time QoS cannot be satisfied.

A feasibility policy guarantees that an appropriate resource is always identified for the execution of a personalized Web service. Because Web services have different requirements (e.g., period of requests, period of deliveries) and resources have different constraints (e.g., period of availabilities, maximum capacity), an agreement has to be reached between what Web services need in terms of resources and what resources offer in terms of capabilities. Furthermore, the feasibility policy checks that the new operations of the personalized Web service are properly handled by the available resources. For example, if a new operation that is the result of a personalization requires a wireless connection, this connection has to be made available.

An inspection policy is a means by which various aspects are considered such as what to track (time, location, etc.), who asked to track (user, the service itself, or both), when to track (continuously, intermittently), when and how to update the arguments of contexts, and how to react if a discrepancy is noticed between what was requested and what has effectively happened. The inspection policy is mainly tightened to the W-context of a Web service. If there is a discrepancy between what was requested and what has effectively happened, the reasons have to be determined, assessed, and reported. One of the reasons could be the lack of appropriate resources on which the personalized service has to be executed.

Summary

In this part of the chapter, we reviewed ConPWS for personalizing Web services using preferences and policies. Personalization occurs when there is a need for accommodating preferences during the performance and outcome delivery of these Web services. Preferences are user related and are of different types varying from when the execution of a Web service should start to where the outcome of this execution should be delivered. Besides user preferences, ConPWS deals with the computing resources on which the Web services are carried out since resource availabilities impact their personalization. As part of the implementation strategy in the ConPWS project, the Web services policy language (WSPL) could be used for implementing the policies related to Web services.

How Context Fits into Web Services

Roman and Campbell (2002) observe that a user-centric context promotes applications that move with users, adapt to the changes in the available resources, and provide configuration mechanisms based on user preferences. Parallel to the user-centric context, ConCWS and ConPWS bind to a service-centric context in order to promote applications that permit service adaptability, track service execution, and support on-the-fly service composition. A user-centric context is associated with the U-context, whereas a service-centric context is associated with the W-context and C-context. Because Web services are the core components of a composition process, the W-context is organized in ConCWS and ConPWS along three perspectives (Figure 8): participation, execution, and location and time.

- The participation perspective is about overseeing the multiple composition scenarios in which a Web service concurrently takes part. This guarantees that the Web service is properly specified and is ready for execution in each composition scenario.
- The execution perspective is about looking for the computing resources on which a Web service operates, and monitoring the capabilities of these resources so that the Web service's requirements are constantly satisfied.
- The preference perspective is about ensuring that user preferences regarding execution time (e.g., at 2 p.m.) and execution location (e.g., user passing by meeting room) are integrated into the specification of a composite service.

Figure 8 also illustrates the connections between the participation, execution, and preference perspectives. First, deployment connects the participation and execution perspectives, and highlights the Web service that is executed once it agrees to par-

Figure 8. Perspective-based organization of W-context

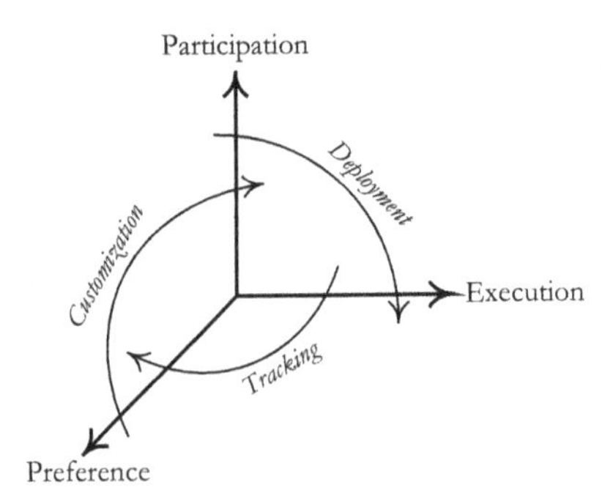

ticipate in a composition. Second, tracking connects the execution and preference perspectives, and highlights the significance of monitoring the execution of a Web service so that user preferences are properly handled. Finally, customization connects the preference and participation perspectives, and highlights the possibility of adjusting a Web service so that it can accommodate various user preferences. The integration of context into Web-services composition ensures that the requirements of and constraints on these Web services are taken into account. While current composition approaches rely on different selection criteria (e.g., execution cost and reliability), context supports Web services in their decision-making process when it comes to whether to accept or reject participating in a composition. Moreover, context is suitable for tracing the execution of Web services during exception handling. It would be possible to know at any time what happened and what is happening with a Web service. Predicting what will happen to a Web service would also be feasible in case the past contexts (i.e., what happened to a service) are stored. Web services can take advantage of the information that past contexts cater so that they can adapt their behavior for better actions and interactions with peers and users.

Conclusion

In this chapter, we reviewed two research projects denoted by ConCWS and ConPWS. Both are concerned with the integration of context into Web-services composition and personalization. We promoted the use of context because of the requirements of flexibility, autonomy, and stability that Web-services self-management situations have to satisfy. Additional requirements, namely, connectivity, nonfunctional quality-of-service properties, correctness, and scalability, also exist as reported in Milanovic and Malek (2004). The research field of context-aware Web services opens up the opportunity for further investigation since several obstacles still hinder the deployment of such Web services, including the fact that current standards for Web services do not enhance them with any context-awareness mechanisms, existing specification approaches for Web-services composition typically facilitate orchestration only while neglecting contexts and their impact on this orchestration, and guidelines supporting the operations of Web-services personalization and tracking are lacking.

Acknowledgments

The author acknowledges the contributions of S. K. Mostéfaoui, H. Yahyaoui, Q. Mahmoud, and W. J. van den Heuvel to the projects presented in this chapter.

References

Ardissono, L., Goy, A., & Petrone, G. (2003). Enabling conversations with Web services. In *Proceedings of the Second International Joint Conference on Autonomous Agents & Multi-Agent Systems (AAMAS 2003)*, Melbourne, Australia.

Benatallah, B., Dumas, M., Sheng, Q. Z., & Ngu, A. (2002). Declarative composition and peer-to-peer provisioning of dynamic Web services. In *Proceedings of the 18th International Conference on Data Engineering (ICDE 2002)*, San Jose, CA.

Benatallah, B., Sheng, Q. Z., & Dumas, M. (2003). The self-serve environment for Web services composition. *IEEE Internet Computing, 7*(1), 40-48.

Bonett, M. (2001). Personalization of Web services: Opportunities and challenges. *ARIADNE, 28*. Retrieved from http://www.ariadne.ac.uk/

Boudriga, N., & Obaidat, M. S. (2004). Intelligent agents on the Web: A review. *Computing in Science Engineering, 6*(4), 35-42.

Coutaz, J., Crowley, J. L., Dobson, S., & Garlan, D. (2005). Context is key. *Communications of the ACM, 48*(3), 49-53.

Curbera, F., Khalaf, R., Mukhi, N., Tai, S., & Weerawarana, S. (2003). The next step in Web services. *Communications of the ACM, 46*(10), 29-34.

Leymann, F. (2001). *Web services flow language (WSFL 1.0)* (Tech. Rep.). IBM Corporation.

Maamar, Z. (2001). Moving code (servlet strategy) vs. inviting code (applet strategy): Which strategy to suggest to software agents? In *Proceedings of the Third International Conference on Enterprise Information Systems (ICEIS 2001)*, Setúbal, Portugal.

Maamar, Z., Benatallah, B., & Mansoor, W. (2003). Service chart diagrams: Description & application. In *Proceedings of the Alternate Tracks of the 12th International World Wide Web Conference (WWW 2003),* Budapest, Hungary.

Maamar, Z., Kouadri Mostéfaoui, S., & Mahmoud, Q. (2005). On personalizing Web services using context. *International Journal of E-Business Research, 1*(3), 41-62.

Maamar, Z., Kouadri Mostéfaoui, S., & Yahyaoui, H. (2005). Towards an agent-based and context-oriented approach for Web services composition. *IEEE Transactions on Knowledge and Data Engineering, 17*(5), 686-697.

Maamar, Z., & Mansoor, W. (2003). Design and development of a software agent-based and mobile service-oriented environment. *e-Service Journal, 2*(3), 42-58.

Maamar, Z., Sheng, Q. Z., & Benatallah, B. (2004). On composite Web services provisioning in an environment of fixed and mobile computing resources. *Information Technology and Management Journal, 5*(3), 251-270.

Maamar, Z., Yahyaoui, H., & Mansoor, W. (2004). Design and development of an m-commerce environment: The E-CWE project. *Journal of Organizational Computing and Electronic Commerce, 14*(4), 285-303.

Menascé, D. A. (2002). QoS issues in Web services. *IEEE Internet Computing, 6*(6), 72-75.

Milanovic, N., & Malek, M. (2004). Current solutions for Web service composition. *IEEE Internet Computing, 8*(6), 51-59.

Muldoon, C., O'Hare, G., Phelan, D., Strahan, R., & Collier, R. (2003). ACCESS: An agent architecture for ubiquitous service delivery. In *Proceedings of the 7th International Workshop on Cooperative Information Agents (CIA 2003)*, Helsinki, Finland.

Papazoglou, M., & Georgakopoulos, D. (2003). Introduction to the special issue on service-oriented computing. *Communications of the ACM, 46*(10), 24-28.

Roman, M., & Campbell, R. H. (2002). *A user-centric, resource-aware, context-sensitive, multi-device application framework for ubiquitous computing environments* (Tech. Rep. No. UIUCDCS-R-2002-2282 UILU-ENG-2002-1728). Urbana, University of Illinois at Urbana-Champaign, Department of Computer Science.

Sabou, M., Richards, D., & van Splunter, D. (2003). An experience report on using DAML-S. In *Proceedings of the 12th International World Wide Web Conference (WWW 2003)*, Budapest, Hungary.

World Wide Web Consortium (W3C). (2002, October 28). *Web services description requirements* (W3C working draft). Retrieved August 23, 2006, from http://www.w3.org/TR/ws-desc-reqs/

Wright, S., Chadha, R., & Lapiotis, G. (2002). Special issue on policy-based networking. *IEEE Network, 16*(2), 8-9.

Zuidweg, M., Pereira Filho, J. G., & van Sinderen, M. (2003). Using P3P in a Web services-based context-aware application platform. In *Proceedings of the Ninth Open European Summer School and IFIP Workshop on Next Generation Networks (EUNICE 2003)*, Balatonfured, Hungary.

About the Authors

Robin G. Qiu is an assistant professor of information science at The Pennsylvania State University, USA, and is a university-endowed professor at Nanjing University of Aeronautics and Astronautics, China. His research interests include services operations and informatics, component business modeling and computing, business transformation and services innovations, automatic information retrieval (auto-IR), and the control and management of manufacturing systems. He has had about 90 publications including over 30 journal publications and 2 book chapters. He currently serves as the editor in chief of *International Journal of Services Operations and Informatics*, as an associate editor of *IEEE Transactions on Systems, Man and Cybernetics*, as an associate editor of *IEEE Transactions on Industrial Informatics*, and on the editorial board of *International Journal of Data Mining and Bioinformatics*. He was the founding and general chair of the 2005 IEEE International Conference on Service Operations and Logistics, and Informatics. He is a general co-chair of the 2006 IEEE International Conference on Service Operations and Logistics, and Informatics and the general chair of the 2007 International Conference on Flexible Automation and Intelligent Manufacturing. He also founded Services Science Global to promote the research on services science, management, and engineering worldwide. He holds a PhD in industrial engineering and a PhD (minor) in computer science and engineering from The Pennsylvania State University.

* * * * *

Lianjun An is a researcher at IBM's T.J. Watson Research Center and is currently studying the stability of supply chains through system-dynamics modeling and simulation, and designing a business-performance monitoring and management system. He received a PhD in applied mathematics from Duke University in 1991. He worked on the analysis of granular flow and plastic deformation, and the scientific simulation of oil reservoirs on parallel computers at McMaster University and the State University of New York at Stony Brook (1992-1997). He subsequently joined IBM and has worked on the Network Configuration Management System and Websphere Commerce Suite and Grid Computing Projects since 1998.

João Paulo Andrade Almeida is a PhD candidate in the Faculty of Electrical Engineering, Mathematics and Computer Science of the University of Twente. He currently works as a researcher at the Telematica Instituut, The Netherlands. His research interests are model-driven architecture, the design of distributed applications, and service-oriented architectures.

Jan vom Brocke is an assistant professor at the Department for Information Systems at the University of Muenster and a member of the European Research Center for Information Systems (ERCIS) in Germany. He graduated with a master's in information systems in 1998 and obtained his PhD at the Faculty of Business Administration and Economics of Muenster in 2003. He has research and teaching experience at the Universities of Muenster and Saarbrücken in Germany, the University of Bucharest in Romania, the University of Tartu in Estonia, and the University College Dublin in Ireland. At present, Jan vom Brocke is supervising two competence centers at ERCIS and running research projects funded by industry foundations, the German Federal Ministry of Education and Research, and the European Commission.

Barret R. Bryant is a professor and associate chair of computer and information sciences at the University of Alabama at Birmingham (UAB). He joined UAB after completing his PhD in computer science at Northwestern University. His primary research areas are the theory and implementation of programming languages, formal specification and modeling, and component-based software engineering. He has authored or coauthored over 100 technical papers in these areas. Bryant is a member of ACM, IEEE (senior member), and the Alabama Academy of Science. He is an ACM distinguished lecturer and chair of the ACM Special Interest Group on Applied Computing (SIGAPP).

Rajkumar Buyya is a senior lecturer, Storage Technology Coporation (StoreTek, USA) fellow of grid computing, and the director of the Grid Computing and Dis-

Fei Cao received his doctoral degree from the University of Alabama at Birmingham, USA. His research interests include model-driven software development, aspect-oriented programming, component-based software development, service-oriented computing, and generative programming. His work as a graduate assistant has been supported by the Naval Office of Research. He has been a research scientist in Avaya research labs working on a multimodal dialog system, and is now working in the Windows enterprise and server division at Microsoft.

Carl K. Chang is a professor and chair of the Department of Computer Science at Iowa State University, USA. Under his leadership, the department has grown in the past 3 years to almost 30 tenured and tenure-track faculty and over 100 PhD students. He received a PhD in computer science from Northwestern University. He worked for GTE Automatic Electric and Bell Laboratories before joining the University of Illinois at Chicago in 1984, where he directed the International Center for Software Engineering. He served as the inaugurating director for the Institute for Reconfigurable Smart Components (IRSC) at Auburn University from 2001to 2002 before moving to Iowa State University in July 2002 to become department chair in computer science. His research interests include requirements engineering, software architecture, and Net-centric computing, and he has published extensively in these areas. Having served as general chair and program chair for many international conferences, in 2005 he served as the general chair of the rapidly emerging IEEE International Conference on Web Services (ICWS) and IEEE Services Computing Conference (SCC). In 2006, he will lead the development of the first Congress on Software Technology and Engineering Practice (CoSTEP) for IEEE. He will also lead the IEEE International Computer Software and Applications Conference as chair of its standing committee to break a new page at its 30th anniversary in October 2006. Chang was the 2004 president of the IEEE computer society, which is the largest professional association in computing with 100,000 members worldwide from over 150 countries. Previously, he served as the editor in chief for *IEEE Software* (1991-1994). He received the Society's Meritorious Service Award, Outstanding Contribution Award, the Golden Core recognition, and the IEEE Third Millennium Medal. Chang is a fellow of IEEE and of AAAS.

S. C. Cheung was born in 1962. Before joining the Hong Kong University of Science and Technology, Hong Kong, he worked for the Distributed Software Engineering Group at the Imperial College in a major European ESPRIT II project on distributed reconfigurable systems. His effort led to the development of REX, which was adopted by various European firms like Siemens and Stollman to build in-house distributed software systems. More recently, he has been working on various research and industrial projects on object-oriented technologies and services computing. Dr. Cheung is an associate editor of *IEEE Transactions on Software Engineering*. He actively participates in the organization and program committees of many leading international conferences on software engineering and distributed computing, including ICSE, FSE, ASE, ISSTA, ICDCS, ER, and SCC. He is interested in technology transfer and has provided technical consultancy to various organizations, including banks, public organizations, and engineering companies on the use of object-oriented and component-based technologies.

Dickson K. W. Chiu is the founder of Dickson Computer Systems, Hong Kong. Besides being an experienced consultant, he also teaches part time at universities. He was born in Hong Kong and received the BSc (honors) degree in computer studies from the University of Hong Kong in 1987. He received the MSc (1994) and the PhD (2000) degrees in computer science from the Hong Kong University of Science and Technology, where he worked as a visiting assistant lecturer after graduation. He also started his own computer company while studying part time. From 2001 to 2003, he was an assistant professor at the Department of Computer Science at the Chinese University of Hong Kong. His research interests include information-systems engineering and service computers with a cross-disciplinary approach, involving Internet technologies, software engineering, agents, work flows, information-system management, security, and databases. His research results have been published in over 70 technical papers in international journals and conference proceedings, such as *IEEE Transactions*, *Information Systems*, and *Decision Support Systems*. He served in program committees of several international conferences, such as the IEEE International Conference on Web Services; IEEE International Conference on e-Technology, e-Commerce and e-Services; and International Conference on Web-Age Information Management. He received a best-paper award at the 37th Hawaii International Conference on System Sciences in 2004. Dr. Chiu is a senior member of the IEEE as well as a member of the ACM and the Hong Kong Computer Society.

Jen-Yao Chung received MS and PhD degrees in computer science from the University of Illinois at Urbana-Champaign. Currently, he is the senior manager for Engineering & Technology Services Innovation, where he was responsible for identifying and creating emergent solutions. He was chief technology officer for IBM Global Electronics Industry. Before that, he was senior manager of the electronic-commerce and supply-chain department, and program director for the IBM Institute for Advanced Commerce Technology office. Dr. Chung is the cofounder and cochair of the IEEE Technical Committee on e-Commerce (TCEC). He has served as general chair and program chair for many international conferences; most recently he served as the steering-committee chair for the IEEE International Conference on e-Commerce Technology (CEC05) and general chair for the IEEE International

Conference on e-Business Engineering (ICEBE05). He has authored or co-authored over 150 technical papers in published journals or conference proceedings. He is a senior member of the IEEE and a member of the ACM.

Mariagrazia Dotoli received the laureate degree in electronic engineering with honors in 1995 and a PhD in electrical engineering in 1999, both from Politecnico di Bari, Italy. She has been a visiting scholar at the Paris 6 University and at the Technical University of Denmark. In 1999, she joined Politecnico di Bari as an assistant professor in systems and control engineering. Her research interests include the modeling and control of discrete event systems, flexible production systems, and distributed manufacturing systems, as well as Petri nets and computational intelligence techniques. Dr. Dotoli has been cochairwoman of the Training and Education Committee of ERUDIT, the network of excellence for fuzzy logic and uncertainty modeling in information technology. She has also been key node representative of EUNITE, the European network of excellence on intelligent technologies. She has been a member of the IEEE since 1996. Since 2003, she has been an expert evaluator of the European Commission.

Jean-Jacques Dubray is a standards architect at SAP Laboratories, USA. He has been a pioneer of the business-process management field since 1997. He has been involved in developing composite application models since 2001. He has contributed to several standards (OAGIS, STAR-XML, ebXML BPSS and ebBP, WS-CDL, and WS-CAF) and is the editor of the OASIS ebBP specifications (v1.1 and v2.0). He enjoys spending time with his kids, Marie and Matthieu, and when he gets the chance, to travel to Corsica, to the village of Aquadilici where his family is from.

Maria Pia Fanti received the laureate degree in electronic engineering from the University of Pisa, Italy, in 1983. She was a visiting researcher at the Rensselaer Polytechnic Institute of Troy, New York, in 1999. Since 1983, she has been with the Department of Electrical and Electronic Engineering of the Polytechnic of Bari, Italy, where she was an assistant professor from 1990 until 1998, and where she is now an associate professor in automation. Her research interests include discrete event systems, Petri nets, the control and modeling of automated manufacturing systems and computer-integrated systems, and supply-chain modeling and management. Professor Fanti is an associate editor of *IEEE Transactions on Systems, Man, and Cybernetics: Part A* and of *IEEE Transactions on Automation Science and Engineering*. Since 2002, she has been a senior member of IEEE.

Xiang Gao is recognized as one of top specialists in Web information systems and Web services in the world. He stays abreast of information technology, initiated, planned, and completed several major cutting-edge IT projects, and published more than 10 highly acclaimed papers and technical reports on the latest information technology in academic- and industry-specific books, international journals, and the proceedings of IEEE international conferences. He worked as a researcher at York University in Canada and Tilburg University in the Netherlands.

Sushant Goel received his master's and PhD degrees in information technology from the School of Electrical and Computer Engineering, RMIT University, Australia, in 2001 and 2005, respectively. During the production of this book, he was a postdoctoral research fellow at the University of Melbourne, Australia. His current research interests are grid computing, grid databases, software engineering, and object-oriented systems.

Rainer Hauser is a research staff member in the services and software department at the IBM Zurich Research Laboratory, Switzerland. He received a diploma in mathematics and a PhD degree in computer science from the Swiss Federal Institute of Technology (ETH) in 1977 and 1984, respectively. He joined IBM at the Zurich Research Laboratory in 1980 initially as a PhD student, where he has worked on image processing, and joined after the completion of his PhD a team working on communication systems. He is currently working in the area of model-driven engineering and model transformations.

Jun-Jang ("J.-J.") Jeng is a researcher at IBM Research, USA. His research interests include business-performance management, policy-based management, model-driven development, agent technologies, and formal disciplines of system and software engineering. Before joining IBM, he served as a senior technical staff member at AT&T Labs. Jeng taught graduate-level courses at NJIT (New Jersey Institute of Technology), Rutgers University, and the George Washington University. Jeng obtained his doctoral degree in computer science from Michigan State University. He is a member of the ACM and IEEE.

Seong W. Kim, PhD, is currently a principle engineer and project manager of the Interaction Lab at Samsung, Korea. He received a PhD in computer science from the University of Illinois at Chicago in 2000. His research interests focus on component-based engineering.

Matthias Klein is a master's student in the Faculty of Computer Science at the University of New Brunswick, Canada. He was hired as a visiting worker by IIT (Institute of Information Technology) when his chapter was written. He is a specialist in ebXML standards and implementation. He holds interest in e-business applications and software engineering. He is the chief architect for several projects about content management and Web-site design within IIT.

Ho-fung Leung is currently an associate professor in the Department of Computer Science and Engineering at the Chinese University of Hong Kong, Hong Kong. He leads theoretical and applied research projects on intelligent agents, multiagent systems, game theory, artificial intelligence, and agent-mediated electronic-commerce technologies, and offers postgraduate courses in game theory and multiagent systems. He is a participating member of the Center for the Advancement of E-Commerce Technologies (AECT) of the engineering faculty. Professor Leung reviews for many major journals and has served on the program committee of many conferences. Currently, he is serving on the program committees of CEC06 and EEE06, EDOC 2006, ISA2006, and PRIMA 2006. Professor Leung was the chairperson of the ACM (Hong Kong chapter) in 1998. He serves as the university's nominee in the HKCE Computer and Information Technology Subject Committee of the Hong Kong Examinations and Assessment Authority. Professor Leung is a professional member of the ACM, a senior member of the IEEE, a chartered member of the BCS, and a chartered IT professional. He is a chartered engineer registered by the ECUK and was awarded the designation of chartered scientist by the Science Council of the United Kingdom. Professor Leung received his BSc and MPhil degrees in computer science from the Chinese University of Hong Kong, and his PhD degree and DIC (diploma of Imperial College) in computing from the Imperial College of Science, Technology and Medicine, University of London.

Zakaria Maamar received his PhD in computer science from Laval University, Quebec, Canada, in 1998. Currently, he is an associate professor at the College of Information Systems at Zayed University, Dubai, UAE. His research interests lie in the areas of mobile computing, Web services, and software agents.

Carlo Meloni graduated with a degree in electronic engineering at Università degli Studi di Roma La Sapienza, Italy, in 1997. He received a PhD in operations research from the same university in 2000. He joined Università degli Studi Roma Tre in 2000 as a postdoctoral fellow in operations research. In July 2002, he became an assistant professor of systems engineering and optimization at Politecnico di Bari, Italy. His major research interests concern combinatorial optimization, graph theory and algorithms, planning and scheduling, and decision-support systems.

Luís Ferreira Pires is an associate professor of the Faculty of Electrical Engineering, Mathematics and Computer Science of the University of Twente, The Netherlands.

His research interests include design methods and architectures for distributed (context-aware and mobile) applications.

Dick Quartel is an assistant professor of the Faculty of Electrical Engineering, Mathematics and Computer Science of the University of Twente, The Netherlands. His research interests include architectural modeling and implementation techniques for distributed applications and service-oriented architectures.

Ehap H. Sabri has 15 years of professional experience in software development, process mapping, project management, system analysis, advanced planning and scheduling, and e-business solutions in a variety of industries, with a proven record of success in adding value and achieving cost reductions. Currently, he holds the position of senior solution architect at i2 Technologies in Dallas, Texas, which is considered a well-known leader in supply-chain and e-business solutions. He teaches advanced supply-chain management and logistics courses at the University of Dallas (UD), USA, and the University of Texas at Dallas for MBA and PhD programs, and has published several conference and journal papers. Ehap is also the cofounder of JAICO, a staff-development and consulting company.

Raghvinder S. Sangwan holds a PhD in computer and information sciences from Temple University in Philadelphia. He is a member of the ACM, IEEE, and SPIE. His research interests include the analysis, design, and development of large-scale object-oriented distributed systems; their communication, connectivity, portability, security, and interoperability; and approaches to the assessment and improvement of their quality. Prior to joining Penn State University, USA, he worked as a lead architect for Siemens on geographically distributed development projects building information systems for large integrated health networks. He still serves as a consulting technical staff member for Siemens Corporate Research in Princeton, NJ, investigating approaches to managing global software-development projects.

Tao Tao received his PhD from the University of Alabama at Birmingham in 2001. He had worked with United Airlines as a software engineer for four years since his graduation, developing applications for airport operations. Since January 2005, he has been working on the New Application Development Team of U.S. Cellular as a senior system analyst. Dr. Tao's interest areas are distributed computing, object-oriented systems, service-oriented architectures, business-process management, and agent-oriented mobile computing.

Marten J. van Sinderen is an associate professor of the Faculty of Electrical Engineering, Mathematics and Computer Science of the University of Twente, The

Netherlands. His research interests include the design and architectural modeling of open distributed systems, and context-aware, proactive, and mobile applications and service-support infrastructures.

Jan vom Brocke is an assistant professor at the Department for Information Systems at the University of Muenster and a member of the European Research Center for Information Systems (ERCIS) in Germany. He graduated with a master's in information systems in 1998 and obtained his PhD at the Faculty of Business Administration and Economics of Muenster in 2003. He has research and teaching experience at the Universities of Muenster and Saarbrücken in Germany, the University of Bucharest in Romania, the University of Tartu in Estonia, and the University College Dublin in Ireland. At present, Jan vom Brocke is supervising two competence centers at ERCIS and running research projects funded by industry foundations, the German Federal Ministry of Education and Research, and the European Commission.

Yuhong Yan is a research officer at the Institute of Information Technology, a government-funded research institute in the National Research Council system, Canada. She has been an adjunct professor in the Faculty of Computer Science at the University of New Brunswick since 2004. She got her PhD from Tsinghua University, China, in 1999. She worked as a postdoc at the University of Toronto and University of Paris 13 for 2 years, and worked as a software consultant in the United States for 1 year before she joined IIT. Her research interests include Web services, service-oriented architectures, and model-based reasoning for Web services. She holds general interests in distributed systems and artificial intelligence. She published widely on these topics.

Jia Zhang, PhD, is an assistant professor in the Department of Computer Science at Northern Illinois University, USA. She is now with BEA Systems Inc. and is also a guest scientist of the National Institute of Standards and Technology (NIST). Her current research interests center on software trustworthiness in the domain of Web services, with a focus on reliability, integrity, security, and interoperability. Zhang has published over 60 technical papers in journals, book chapters, and conference proceedings. She also has 7 years of industrial experience as a software technical lead in Web-application development. Zhang is an associate editor of the *International Journal of Web Services Research* (JWSR). Zhang received a PhD in computer science from the University of Illinois at Chicago in 2000. She is a member of the IEEE and ACM.

Zhijun Zhang received his bachelor's degree in computer science from Peking University, Beijing, China, in 1990. He developed computer software for 3 years before starting his graduate studies at the University of Maryland, where he conducted research in software engineering and human-computer interaction. He received his PhD degree in computer science from the University of Maryland in 1999. Besides teaching at the University of Phoenix, USA, Dr. Zhang was a technology researcher at a large financial-service company, focusing on emerging mobile technologies and their impact on human-computer interaction. He is now an enterprise architect for the same company, working on service-oriented architectures and other strategic architecture initiatives.

Wei Zhao is a doctoral candidate in computer and information sciences at the University of Alabama at Birmingham, USA. Her primary research areas include model-driven development, business-process modeling and management, component-based software engineering, and the theory and implementation of programming languages. Her research has been supported by the Naval Office of Research. She completed 8 months of internship at the IBM T.J. Watson Research Center in 2004. She is a student member of the IEEE.

Mengchu Zhou received his BS from Nanjing University of Science and Technology, his MS from the Beijing Institute of Technology, and his PhD from Rensselaer Polytechnic Institute. He joined the New Jersey Institute of Technology, USA, in 1990, and is currently a professor of electrical and computer engineering and director of the Discrete-Event Systems Laboratory. His interests are in computer-integrated systems, Petri nets, networks, and manufacturing. He has over 200 publications including 6 books and over 80 journal papers. He is managing editor of *IEEE Transactions on Systems, Man and Cybernetics*, associate editor of *IEEE Transactions on Automation Science and Engineering*, and editor in chief of *International Journal of Intelligent Control and Systems*. He served as general and program chair of many international conferences. Dr. Zhou has led or participated in 28 research and education projects with a total budget of over $10 million, funded by the NSF (National Science Foundation), DoD, and industry. He was the recipient of the CIM University-LEAD Award by the Society of Manufacturing Engineers, the Perlis Research Award by NJIT, and the Humboldt Research Award from U.S. Senior Scientists; he is also a distinguished lecturer of the IEEE SMC Society. He is a life member of the Chinese Association for Science and Technology-USA and served as its president in 1999. He is a fellow of the IEEE.

Index

C

D

E

F

G

H

I

J

K

L

M

N

O

S

T

U

V

W

X

Y

Z